Computer Methods in Power System Analysis

McGraw-Hill Series in Electronic Systems

John G. Truxal and Ronald A. Rohrer *Consulting Editors*

Schultz and Melsa ***State Functions and Linear Control Systems***
Stagg and El-Abiad ***Computer Methods in Power System Analysis***
Timothy and Bona ***State Space Analysis: An Introduction***

Computer Methods in Power System Analysis

Glenn W. Stagg

President, Stagg Systems Inc.

Ahmed H. El-Abiad

Professor of Electrical Engineering
Purdue University

McGraw-Hill Book Company

New York St. Louis
San Francisco
Toronto London
Sydney

Computer Methods in Power System Analysis

Library of Congress Catalog Card Number 67-12963

07-060658-7

567890 KPKP 79876543

Preface

This book presents techniques that have been applied successfully in solving power system problems with a digital computer. It can thus serve as a text for advanced power system courses to inform prospective power engineers of methods currently employed in the electric utility industry. Because of the increasing use of the computer as an indispensable tool in power system engineering, this book will also serve as a basic reference for power system engineers responsible for the development of computer applications.

The material contained in the text has been developed from notes for special two-week courses offered since 1964 at Purdue University, The University of Wisconsin and the University of Santa Clara. These courses were attended by representatives of universities, electric utilities, and equipment manufacturers.

Solution techniques are presented for the three problems encountered most frequently in power system analysis, namely, short circuit, load flow, and power system stability. In addition to an engineering description of these problems, the mathematical techniques that are required for a computer solution are described. Thus, relevant material is included from matrix algebra and numerical analysis. It is assumed, however, that the reader has a general understanding of elementary power system analysis.

Chapter 1 presents, as a brief introduction, the impact of computers on power system engineering, the orientation of engineering problems to computers, and the advantages of digital computation. Chapter 2 covers the basic principles of matrix algebra and provides sufficient background in matrix theory for the remainder of the book. For readers familiar with matrix techniques, this chapter serves as a review and establishes the notation used throughout the text. Incidence and network matrices are introduced in Chapter 3, which presents the techniques for describing the geometric structure of a network and outlines the transformations required to derive network matrices. The formation of these matrices is the first step in the analysis of power system problems. Chapter 4 presents algorithms which can be used in an alternative method for the formation of certain network matrices. These algorithms have proved to be effective for use in computer calculation. The methods described in Chapters 3 and 4 are developed for single-phase representation of power systems. Chapter 5 extends these methods for three-phase representation. The application of network matrices to short circuit calculations is presented in Chapter 6. Several methods are included and a typical computer program is described to illustrate a practical application of the techniques.

Chapter 7 contains a brief introduction to the solution of linear and nonlinear simultaneous algebraic equations. This material is presented in a manner that affords direct application to the solution of the load flow problem. The formulation and solution of the load flow problem is presented in Chapter 8. This chapter also describes the procedures for handling voltage-controlled buses, transformers, and tie line control. The different methods are compared from several points of view and a description is given of an actual program used for load flow calculations. In a manner similar to that in Chapter 7, Chapter 9 introduces methods for the numerical solution of the differential equations that are required for transient stability studies. Chapter 10 covers the formulation and solution techniques employed in transient studies and presents procedures for the detailed representation of synchronous and induction machines, exciter and governor systems, and the distance relays. An actual transient stability computer program is described.

The first efforts in the development of this material were made in the early 1950s at the American Electric Power Service Corporation as a result of the interest in the application of computers to the planning and operation of electric power systems. In 1959, the authors had an opportunity to work together as members of the staff of the American Electric Power Service Corporation and continued to work together on a part-time basis for several years. This made possible the further development of basic computer methods established in previous years.

This research work was endorsed enthusiastically by the management of the American Electric Power Company. The authors wish to express their appreciation for this support.

It is a pleasure also to acknowledge the contributions of those who have helped in the preparation of this book. The authors would like particularly to thank Jorge F. Dopazo, who studied the text in detail and made many suggestions; Marjorie Watson, for her contribution related to the mathematical techniques and for editing the manuscript; and G. Robert Bailey, Dennis W. Johnston, Kasi Nagappan, Janice F. Hohenstein, and other members of the Engineering Analysis and Computer Division. The authors would like also to thank Professors Arun G. Phadke and Daniel K. Reitan of The University of Wisconsin for their helpful comments in reviewing the text. Last, but certainly not least, sincere thanks to Constance Aquila for her excellent work in the typing and general preparation of the manuscript.

Glenn W. Stagg
Ahmed H. El-Abiad

Contents

chapter 1
Introduction

1.1 Historical note

The great technical advances in the design and production of commercial and scientific general-purpose digital computers since the early 1950s have placed a powerful tool at the disposal of the engineering profession. This advancement has made economically feasible the utilization of digital computers for routine calculations encountered in everyday engineering work. In addition, it has provided the capability for performing more advanced engineering and scientific computations that were previously impossible because of their complex or time-consuming nature. All these trends have increased immensely the interest in digital computers and have necessitated a better understanding of the engineering and mathematical bases for problem solving.

The planning, design, and operation of power systems require continuous and comprehensive analysis to evaluate current system performance and to ascertain the effectiveness of alternative plans for system expansion. These studies play an important role in providing a high standard of power system reliability and ensuring the maximum utilization of capital investment.

The computational task of determining power flows and voltage levels resulting from a single operating condition for even a small network is all but insurmountable if performed by manual methods. The need for computational aids in power system engineering led in 1929 to the design of a special-purpose analog computer called an ac network analyzer. This device made possible the study of a greater variety of system operating conditions for both present and future system designs. It provided the ability to determine power flows and system voltages during normal and emergency conditions and to study the transient behavior of the system resulting from fault conditions and switching operations. By the middle 1950s 50 network analyzers were in operation in the United States and Canada and were indispensable tools to planning, relaying, and operating engineers.

The earliest application of digital computers to power system problems dates back to the late 1940s. However, most of the early applications were limited in scope because of the small capacity of the punched card calculators generally in use at that time. The availability of large-scale digital computers in the middle 1950s provided equipment of sufficient capacity and speed to meet the requirements of major power system problems. In 1957 the American Electric Power Service Corporation completed a large-scale load flow program for the IBM 704 which calculated the voltages and power flows for a specified power system network.

The initial application of the load flow program to transmission planning studies proved so successful that all subsequent studies employed the digital computer instead of the network analyzer. The success of this program led to the development of programs for short circuit and transient stability calculations. Today the computer is an indispensable tool in all phases of power system planning, design, and operation.

1.2 Impact of computers

The development of computer technology has provided the following advantages to power system engineering:

1. More efficient and economic means of performing routine engineering calculations required in the planning, design, and operation of a power system
2. A better utilization of engineering talent by relieving the engineer from tedious hand calculations and permitting him to spend more time on technical work
3. The ability to perform more effective engineering studies by applying calculating procedures to obtain a number of alternate solutions for a particular problem to provide a broad base for engineering decisions
4. The capability of performing studies which heretofore were not possible because of the volume of calculations involved

Two major factors which have contributed to the realization of these benefits are the declining cost of computing equipment and the development of efficient computational techniques. Now that a substantial reduction in computing cost has been effected, principal effort must be directed toward the orientation of engineering problems to computer solutions.

1.3 Orientation of engineering problems to computers

The process of applying a computer to the solution of engineering problems involves a number of distinct steps. These steps are:

1. Problem definition Initially, the problem must be defined precisely and the objectives determined. This may be the most difficult step in the entire process. Consideration must be given to the pertinent data available for input, the scope of the problem and its limitations, the desired results, and their relative importance in making an engineering decision. This phase requires the judgement of experienced and capable engineers.

2. Mathematical formulation After the problem has been defined, it is necessary to develop a mathematical model to represent the physical system. This requires specifying the characteristics of individual system components as well as the relations which govern the interconnection of the elements. Different mathematical models may be used to represent the same system and, for many problems, complementary (dual) formulations may be obtained. One formulation may result in a different number of equations than another as, for example, in the case of network problems which can be solved using either loop equations or node equations. The mathematical formulation of the problem, therefore, includes the design of a number of models and the selection of the best model to describe the physical system.

3. Selection of a solution technique The formulation of most engineering problems involves mathematical expressions, such as sets of nonlinear equations, differential equations, and trigonometric functions, which cannot be evaluated directly by a digital computer. A computer is capable of performing only the four basic arithmetic operations of addition, subtraction, multiplication, and division. A solution for any problem, therefore, must be obtained by numerical techniques which employ the four basic arithmetic operations. It is important in this phase to select a method which is practical for machine computation and, in particular, will produce the desired results in a reasonable amount of computer time. Since numerical approaches involve a number of assumptions, careful consideration must be given to the degree of accuracy required.

4. Program design The sequence of logical steps by which a particular problem is to be solved, the allocation of memory, the access of data, and the assignment of input and output units are important aspects of computer program design. The objectives are primarily to develop a procedure which eliminates unnecessary repetitive calculations and remains within the capability of the computer. The program design is usually prepared in the form of a diagram called a flow chart.

5. Programming A digital computer has a series of instructions consisting of operation codes and addresses which it is able to interpret and execute. In addition to the arithmetic and input/output instructions, logical instructions are available which are used to direct the sequence of calculations. The translation of the precise detailed steps to be performed in the solution of a problem into an organized list of computer instructions is the process of programming. A program can be developed by using computer instructions in actual or symbolic form, or it can be written in a generalized programming language, such as FORTRAN.

6. Program verification There are many opportunities to introduce errors in the development of a complete computer program. Therefore, a systematic series of checks must be performed to ensure the correctness of problem formulation, method of solution, and operation of the program.

7. Application Engineering programs, in general, can be classified into two groups. The first consists of special-purpose programs, which are developed in a relatively short period for the solution of simple engineering problems. Such a problem is usually well defined, and often the program completely serves its purpose after the first series of calculations has been completed. Some small programs are used on a continuing basis but are restricted in their use because of their special-purpose nature.

The second group consists of general-purpose programs that are designed for the analysis of large engineering problems. These programs are applied extensively in the regular studies of one or more engineering departments. Their use may have an effect on the approach to an engineering problem and the organization of a study. Thus, it is important that consideration be given to the manner in which a program is to be employed in an engineering activity. Some aspects which must be considered are means of collecting and preparing data, processing time, and presentation of results. Programs of this type are becoming an integral part of power system engineering.

The relative importance of each of these steps varies from problem to problem. Moreover, all steps are closely related and play an important role in the decisions that must be made. Of primary importance is the interrelation of the mathematical formulation of a problem and the selection of a solution technique. Frequently, it is difficult to evaluate the influence of these two steps on each other without developing a complete program and performing actual calculations to compare the alternatives.

The material covered in this book pertains to the first three steps, with particular emphasis on the interrelations of steps 2 and 3. Simplified flow charts are used to illustrate the methods presented.

Bibliography

Gross, Eric T. B.: Network Analyzer Installations in Canada and the United States, *Proc. Am. Power Conf.*, vol. 21, pp. 665–669, 1959.

St. Clair, H. P., and G. W. Stagg: Digital Computer Solution of Power System Problems, *Proc. Am. Power Conf.*, vol. 20, pp. 614–629, 1958.

St. Clair, H. P., G. W. Stagg, and Maria Tscherne: Digital Computer Takes Over Load Flow Calculations, *Electrical World*, Sept. 30, 1957.

Stagg, G. W., A. F. Gabrielle, and J. F. Hohenstein: Digital Computer Calculates Transient Stability Problems, *Electrical World*, Sept. 1, 1958.

Stagg, G. W., and E. L. Wizemann: Computer Program for Load Flow Study Handles Ten-system Interconnection, *Electrical World*, Aug. 1, 1960.

Zuckernick, S. J., and G. W. Stagg: Computer Solves Relay Problems, *Electrical World*, July 20, 1959.

chapter 2
Matrix algebra

2.1 Introduction

In recent years, the use of matrix algebra for the formulation and solution of complex engineering problems has become increasingly important with the advent of digital computers to perform the required calculations. The application of matrix notation provides a concise and simplified means of expressing many problems. The use of matrix operations presents a logical and ordered process which is readily adaptable for a computer solution of a large system of simultaneous equations.

2.2 Basic concepts and definitions

Matrix notation

Matrix notation is a shorthand means of writing systems of simultaneous equations in a concise form. A matrix is defined as a rectangular array of numbers, called *elements*, arranged in a systematic manner with m rows and n columns. These elements can be real or complex numbers. A double-subscript notation a_{ij} is used to designate a matrix element. The first subscript i designates the row in which the element lies, and the second subscript j designates the column.

In the following system of equations,

$$\begin{aligned} a_{11}x_1 + a_{12}x_2 + a_{13}x_3 &= y_1 \\ a_{21}x_1 + a_{22}x_2 + a_{23}x_3 &= y_2 \\ a_{31}x_1 + a_{32}x_2 + a_{33}x_3 &= y_3 \end{aligned} \tag{2.2.1}$$

x_1, x_2, and x_3 are unknown variables; a_{11}, a_{12}, a_{13}, . . . , a_{33} are the coefficients of these variables; y_1, y_2, and y_3 are known parameters. The coefficients form an array

$$\begin{bmatrix} a_{11} & a_{12} & a_{13} \\ a_{21} & a_{22} & a_{23} \\ a_{31} & a_{32} & a_{33} \end{bmatrix} \tag{2.2.2}$$

which is the *coefficient matrix* of the system of equations (2.2.1).

Similarly, the variables and parameters can be written in matrix form:

$$\begin{bmatrix} x_1 \\ x_2 \\ x_3 \end{bmatrix} \quad \text{and} \quad \begin{bmatrix} y_1 \\ y_2 \\ y_3 \end{bmatrix} \tag{2.2.3}$$

The matrix (2.2.2) is designated by a capital letter A and the matrices (2.2.3) by X and Y, respectively. In matrix notation the equations (2.2.1) are written

$$AX = Y$$

A matrix with m rows and n columns is said to be of dimension m by n, or $m \times n$. A matrix with a single row and more than one column ($m = 1$ and $n > 1$) is called a *row matrix* or *row vector*. A matrix with a single column and more than one row is called a *column matrix* or *column vector*.

Types of matrices

Some matrices with special characteristics are significant in matrix operations. These are:

Square matrix When the number of rows equals the number of columns, that is, $m = n$, the matrix is called a *square matrix* and its order is equal to the number of rows (or columns). The elements in a square matrix a_{ij} for which $i = j$ are called *diagonal elements*. Those for which $i \neq j$ are called *off-diagonal elements*. For elements a_{ij} to the right of the diagonal i is less than j, and for those to the left of the diagonal i is greater than j.

Upper triangular matrix If the elements a_{ij} of a square matrix are zero for $i > j$, then the matrix is an *upper triangular matrix*. For example:

$$A = \begin{bmatrix} a_{11} & a_{12} & a_{13} \\ 0 & a_{22} & a_{23} \\ 0 & 0 & a_{33} \end{bmatrix}$$

Lower triangular matrix If the elements a_{ij} of a square matrix are zero for $i < j$, then the matrix is a *lower triangular matrix*. For example:

$$A = \begin{bmatrix} a_{11} & 0 & 0 \\ a_{21} & a_{22} & 0 \\ a_{31} & a_{32} & a_{33} \end{bmatrix}$$

Diagonal matrix If all off-diagonal elements of a square matrix are zero ($a_{ij} = 0$ for all $i \neq j$), then the matrix is a *diagonal matrix*. For example:

$$A = \begin{bmatrix} a_{11} & 0 & 0 \\ 0 & a_{22} & 0 \\ 0 & 0 & a_{33} \end{bmatrix}$$

Unit or identity matrix If all diagonal elements of a square matrix equal one and all other elements are zero ($a_{ij} = 1$ for $i = j$ and $a_{ij} = 0$ for $i \neq j$), the matrix is the *unit* or *identity matrix*, designated by the letter U. For example:

$$U = \begin{bmatrix} 1 & 0 & 0 \\ 0 & 1 & 0 \\ 0 & 0 & 1 \end{bmatrix}$$

Null matrix If all elements of a matrix are zero, it is a *null matrix*.

Transpose of a matrix If the rows and columns of an $m \times n$ matrix are interchanged, the resultant $n \times m$ matrix is the *transpose* and is designated by A^t. For the matrix

$$A = \begin{bmatrix} a_{11} & a_{12} \\ a_{21} & a_{22} \\ a_{31} & a_{32} \end{bmatrix}$$

the transpose is

$$A^t = \begin{bmatrix} a_{11} & a_{21} & a_{31} \\ a_{12} & a_{22} & a_{32} \end{bmatrix}$$

Symmetric matrix If the corresponding off-diagonal elements of a square matrix are equal ($a_{ij} = a_{ji}$), the matrix is a *symmetric matrix*. For example:

$$A = \begin{bmatrix} 1 & 5 & 3 \\ 5 & 2 & 6 \\ 3 & 6 & 4 \end{bmatrix}$$

The transpose of a symmetric matrix is identical to the matrix itself, that is, $A^t = A$.

Skew-symmetric matrix If $A = -A^t$ for a square matrix, A is a *skew-symmetric matrix*. The corresponding off-diagonal elements are equal but of opposite sign ($a_{ij} = -a_{ji}$) and the diagonal elements are zero. For example:

$$A = \begin{bmatrix} 0 & -5 & 3 \\ 5 & 0 & 6 \\ -3 & -6 & 0 \end{bmatrix}$$

Orthogonal matrix If $A^tA = U = AA^t$ for a square matrix with real elements, then A is an *orthogonal matrix*.

Conjugate of a matrix If all the elements of a matrix are replaced by their conjugates (replace the element $a + jb$ by $a - jb$), the resultant matrix is the *conjugate* and is designated by A^*. For a matrix

$$A = \begin{bmatrix} j3 & 5 \\ 4 + j2 & 1 + j1 \end{bmatrix}$$

the conjugate is

$$A^* = \begin{bmatrix} -j3 & 5 \\ 4 - j2 & 1 - j1 \end{bmatrix}$$

If all the elements of A are real, then $A = A^*$. If all elements are pure imaginary, then $A = -A^*$.

Hermitian matrix If $A = (A^*)^t$ for a square complex matrix, A is a *Hermitian matrix* in which all diagonal elements are real. For example:

$$A = \begin{bmatrix} 4 & 2 - j3 \\ 2 + j3 & 5 \end{bmatrix}$$

Skew-Hermitian matrix If $A = -(A^*)^t$ for a square complex matrix, A is a *skew-Hermitian matrix* in which all diagonal elements are either zero or pure imaginary. For example:

$$A = \begin{bmatrix} 0 & 2 - j3 \\ -2 - j3 & 0 \end{bmatrix}$$

Unitary matrix If $(A^*)^tA = U = A(A^*)^t$ for a square complex matrix, A is a *unitary matrix*. A unitary matrix with real elements is an orthogonal matrix.

Table 2.1 summarizes some types of special matrices.

Table 2.1 Types of special matrices

Condition	*Type of matrix*
$A = -A$	Null
$A = A^t$	Symmetric
$A = -A^t$	Skew-symmetric
$A = A^*$	Real
$A = -A^*$	Pure imaginary
$A = (A^*)^t$	Hermitian
$A = -(A^*)^t$	Skew-Hermitian
$A^tA = U$	Orthogonal
$(A^*)^tA = U$	Unitary

2.3 Determinants

Definition and properties of determinants

The solution of two simultaneous equations

$$\begin{aligned} a_{11}x_1 + a_{12}x_2 &= k_1 \\ a_{21}x_1 + a_{22}x_2 &= k_2 \end{aligned} \tag{2.3.1}$$

can be obtained by eliminating the variables one at a time. Solving for x_2 in terms of x_1 from the second equation and substituting this expression for x_2 in the first equation, the following is obtained:

$$a_{11}x_1 + a_{12}\left(\frac{k_2}{a_{22}} - \frac{a_{21}}{a_{22}}x_1\right) = k_1$$

$$a_{11}a_{22}x_1 + a_{12}k_2 - a_{12}a_{21}x_1 = a_{22}k_1$$

$$(a_{11}a_{22} - a_{12}a_{21})x_1 = a_{22}k_1 - a_{12}k_2$$

$$x_1 = \frac{a_{22}k_1 - a_{12}k_2}{a_{11}a_{22} - a_{12}a_{21}}$$

Then, substituting x_1 in either of the equations (2.3.1), x_2 is obtained:

$$x_2 = \frac{a_{11}k_2 - a_{21}k_1}{a_{11}a_{22} - a_{12}a_{21}}$$

The expression $(a_{11}a_{22} - a_{12}a_{21})$ is the value of the determinant of the coefficient matrix A, where $|A|$ denotes the determinant

$$|A| = \begin{vmatrix} a_{11} & a_{12} \\ a_{21} & a_{22} \end{vmatrix}$$

The solution of the equations (2.3.1) by means of determinants is

$$x_1 = \frac{\begin{vmatrix} k_1 & a_{12} \\ k_2 & a_{22} \end{vmatrix}}{\begin{vmatrix} a_{11} & a_{12} \\ a_{21} & a_{22} \end{vmatrix}} = \frac{a_{22}k_1 - a_{12}k_2}{a_{11}a_{22} - a_{12}a_{21}}$$

and

$$x_2 = \frac{\begin{vmatrix} a_{11} & k_1 \\ a_{21} & k_2 \end{vmatrix}}{\begin{vmatrix} a_{11} & a_{12} \\ a_{21} & a_{22} \end{vmatrix}} = \frac{a_{11}k_2 - a_{21}k_1}{a_{11}a_{22} - a_{12}a_{21}}$$

A determinant is defined only for a square matrix and has a single value. A method for evaluating the determinant of an $n \times n$ matrix is given in Chap. 7.

Determinants have the following properties:

1. The value of a determinant is zero if
   a. All elements of a row or column are zero
   b. The corresponding elements of two rows (or columns) are equal
   c. A row (or column) is a linear combination of one or more rows (or columns)
2. If two rows (or columns) of a determinant are interchanged, the value of the determinant is changed in sign only
3. The value of a determinant is not changed if
   a. All corresponding rows and columns are interchanged, i.e.,

   $$|A| = |A^t|$$

   b. k times the elements of any row (or column) are added to the corresponding elements of another row (or column)
4. If all elements of a row (or column) are multiplied by a factor k, the value of the determinant is multiplied by k
5. The determinant of the product of matrices is equal to the product of the determinants of the matrices, i.e.,

   $$|A\ B\ C| = |A|\ |B|\ |C|$$

6. The determinant of the sum (or difference) of matrices is *not* equal to the sum (or difference) of the individual determinants, i.e.,

   $$|A + B - C| \neq |A| + |B| - |C|$$

The application of these properties can reduce the work in evaluating determinants.

Minors and cofactors

The determinant obtained by striking out the ith row and jth column is called the *minor* of the element a_{ij}. Thus, for

$$|A| = \begin{vmatrix} a_{11} & a_{12} & a_{13} \\ a_{21} & a_{22} & a_{23} \\ a_{31} & a_{32} & a_{33} \end{vmatrix}$$

the minor of a_{21} is

$$\begin{vmatrix} a_{12} & a_{13} \\ a_{32} & a_{33} \end{vmatrix}$$

The order of this minor is one less than that of the original determinant. By striking out any two rows and columns a minor of order two less than the original determinant is obtained, etc.

The *cofactor* of an element is

$$(-1)^{i+j}(\text{minor of } a_{ij})$$

where the order of the minor of a_{ij} is $n - 1$. The cofactor of a_{21}, designated by A_{21}, is

$$A_{21} = (-1)^{2+1}\begin{vmatrix} a_{12} & a_{13} \\ a_{32} & a_{33} \end{vmatrix} = -\begin{vmatrix} a_{12} & a_{13} \\ a_{32} & a_{33} \end{vmatrix}$$

The following relationships between a determinant and cofactors exist:

1. The sum of the products of the elements in any row (or column) and their cofactors is equal to the determinant:

$$|A| = a_{21}A_{21} + a_{22}A_{22} + a_{23}A_{23} \tag{2.3.2}$$

2. The sum of the products of the elements in any row (or column) and the cofactors of the corresponding elements in another row (or column) is equal to zero:

$$a_{21}A_{31} + a_{22}A_{32} + a_{23}A_{33} = 0 \tag{2.3.3}$$

Adjoint

If each element of a square matrix is replaced by its cofactor and then the matrix is transposed, the resulting matrix is an *adjoint* which is designated by A^+

$$A^+ = \begin{bmatrix} A_{11} & A_{21} & A_{31} \\ A_{12} & A_{22} & A_{32} \\ A_{13} & A_{23} & A_{33} \end{bmatrix}$$

2.4 Matrix operations

Equality of matrices

If A and B are matrices with the same dimension and each element a_{ij} of A is equal to the corresponding element b_{ij} of B, the matrices are equal, i.e.,

$$A = B$$

Addition and subtraction of matrices

Matrices of the same dimension are conformable for addition and subtraction. The sum or difference of two $m \times n$ matrices, A and B, is a matrix C of the same dimension, i.e.,

$$A \pm B = C$$

where each element of C is

$$c_{ij} = a_{ij} \pm b_{ij}$$

For n conformable matrices the sum or difference is

$$A \pm B \pm C \pm D \pm \cdots \pm N = R$$

where the elements of the resultant matrix R are

$$r_{ij} = a_{ij} \pm b_{ij} \pm c_{ij} \pm d_{ij} \pm \cdots \pm n_{ij}$$

The commutative and associative laws apply to addition of matrices as follows:

$$A + B = B + A \qquad \text{commutative law}$$

i.e., the sum of the matrices is independent of the order of the addition.

$$A + B + C = A + (B + C) = (A + B) + C \qquad \text{associative law}$$

i.e., the sum of the matrices is independent of the order in which the matrices are associated for addition.

Multiplication of a matrix by a scalar

When a matrix is multiplied by a scalar, the elements of the resultant matrix are equal to the product of the original elements and the scalar. For example:

$$kA = B$$

where $b_{ij} = ka_{ij}$ for all i and j.

The multiplication of a matrix by a scalar obeys the commutative law and the distributive law as follows:

$$kA = Ak \qquad \text{commutative law}$$
$$k(A + B) = kA + kB = (A + B)k \qquad \text{distributive law}$$

Multiplication of matrices

Multiplication of two matrices

$$AB = C$$

is defined only if the number of columns of the first matrix A equals the number of rows of B. Thus, for the product of matrix A of dimension $m \times q$ and matrix B of dimension $q \times n$, the matrix C is of dimension $m \times n$. Any element c_{ij} of C is the sum of the products of the corresponding elements of the ith row of A and the jth column of B, that is,

$$c_{ij} = a_{i1}b_{1j} + a_{i2}b_{2j} + \cdots + a_{iq}b_{qj}$$

or

$$c_{ij} = \sum_{k=1}^{q} a_{ik}b_{kj} \qquad i = 1, 2, \ldots, m; j = 1, 2, \ldots, n$$

For example:

$$AB = \begin{bmatrix} a_{11} & a_{12} \\ a_{21} & a_{22} \\ a_{31} & a_{32} \end{bmatrix} \begin{bmatrix} b_{11} & b_{12} \\ b_{21} & b_{22} \end{bmatrix} = \begin{bmatrix} a_{11}b_{11} + a_{12}b_{21} & a_{11}b_{12} + a_{12}b_{22} \\ a_{21}b_{11} + a_{22}b_{21} & a_{21}b_{12} + a_{22}b_{22} \\ a_{31}b_{11} + a_{32}b_{21} & a_{31}b_{12} + a_{32}b_{22} \end{bmatrix}$$

In the product AB, A premultiplies B or B postmultiplies A. The product BA is not defined since the number of columns of B is not equal to the number of rows of A. When the products AB and BA are defined for a square matrix, it can be shown that, in general,

$$AB \neq BA$$

Therefore, the commutative law does not hold for matrix multiplication. If the matrices A, B, and C satisfy the dimension requirements for multiplication and addition, the following properties hold:

$$A(B + C) = AB + AC \qquad \text{distributive law}$$
$$A(BC) = (AB)C = ABC \qquad \text{associative law}$$

However,

$AB = 0$ does not necessarily imply that $A = 0$ or $B = 0$
$CA = CB$ does not necessarily imply that $A = B$

If $C = AB$, then the transpose of C is equal to the product of the transposed matrices in reverse order, i.e.,

$$C^t = B^t A^t$$

This is the *reversal rule.*

Inverse of a matrix

Division does not exist in matrix algebra except in the case of the division of a matrix by a scalar. This operation is performed by dividing each element of a matrix by the scalar. However, for a given set of equations,

$$\begin{aligned} a_{11}x_1 + a_{12}x_2 + a_{13}x_3 &= y_1 \\ a_{21}x_1 + a_{22}x_2 + a_{23}x_3 &= y_2 \\ a_{31}x_1 + a_{32}x_2 + a_{33}x_3 &= y_3 \end{aligned} \tag{2.4.1}$$

or, in matrix form,

$$AX = Y \tag{2.4.2}$$

it is desirable to express x_1, x_2, and x_3 as functions of y_1, y_2, and y_3, that is,

$$X = BY$$

If there is a unique solution for the equations (2.4.1), then matrix B exists and is the *inverse* of A.

If the determinant of A is not zero, the equations can be solved for the x_i's as follows:

$$\begin{aligned} x_1 &= \frac{A_{11}}{|A|} y_1 + \frac{A_{21}}{|A|} y_2 + \frac{A_{31}}{|A|} y_3 \\ x_2 &= \frac{A_{12}}{|A|} y_1 + \frac{A_{22}}{|A|} y_2 + \frac{A_{32}}{|A|} y_3 \\ x_3 &= \frac{A_{13}}{|A|} y_1 + \frac{A_{23}}{|A|} y_2 + \frac{A_{33}}{|A|} y_3 \end{aligned}$$

where $A_{11}, A_{12}, \ldots, A_{33}$ are the cofactors of $a_{11}, a_{12}, \ldots, a_{33}$ and $|A|$ is the determinant of A. Thus

$$B = \begin{bmatrix} \frac{A_{11}}{|A|} & \frac{A_{21}}{|A|} & \frac{A_{31}}{|A|} \\ \frac{A_{12}}{|A|} & \frac{A_{22}}{|A|} & \frac{A_{32}}{|A|} \\ \frac{A_{13}}{|A|} & \frac{A_{23}}{|A|} & \frac{A_{33}}{|A|} \end{bmatrix} = \frac{A^+}{|A|}$$

where A^+ is the adjoint of A. It should be noted that the elements of the adjoint A^+ are the cofactors of the elements of A, but are placed in transposed position. The matrix B is the inverse of A and is written A^{-1}.

Multiplying A by its inverse,

$$\begin{bmatrix} a_{11} & a_{12} & a_{13} \\ a_{21} & a_{22} & a_{23} \\ a_{31} & a_{32} & a_{33} \end{bmatrix} \begin{bmatrix} \frac{A_{11}}{|A|} & \frac{A_{21}}{|A|} & \frac{A_{31}}{|A|} \\ \frac{A_{12}}{|A|} & \frac{A_{22}}{|A|} & \frac{A_{32}}{|A|} \\ \frac{A_{13}}{|A|} & \frac{A_{23}}{|A|} & \frac{A_{33}}{|A|} \end{bmatrix} = \begin{bmatrix} 1 & 0 & 0 \\ 0 & 1 & 0 \\ 0 & 0 & 1 \end{bmatrix} = U$$

results in the unit matrix. This follows from the relationships (2.3.2) and (2.3.3). A diagonal term of U, such as u_{11}, equals 1 since

$$a_{11}\frac{A_{11}}{|A|} + a_{12}\frac{A_{12}}{|A|} + a_{13}\frac{A_{13}}{|A|} = \frac{|A|}{|A|} = 1$$

and an off-diagonal term, such as u_{12}, equals zero since

$$a_{11}\frac{A_{21}}{|A|} + a_{12}\frac{A_{22}}{|A|} + a_{13}\frac{A_{23}}{|A|} = \frac{0}{|A|} = 0$$

Thus

$$AA^{-1} = A^{-1}A = U$$

To solve for X from the matrix equation (2.4.2) both sides of the equation are premultiplied by A^{-1}.

$$\begin{aligned} AX &= Y \\ A^{-1}AX &= A^{-1}Y \\ UX &= A^{-1}Y \\ X &= A^{-1}Y \end{aligned}$$

The order of the matrices in the product must be maintained.

If the determinant of a matrix is zero, the inverse does not exist. Such a matrix is called a *singular matrix*. If the determinant of a matrix is not zero, the matrix is a *nonsingular matrix* and has a unique inverse.

The inverse of the product of matrices can be obtained by the reversal rule, i.e.,

$$(AB)^{-1} = B^{-1}A^{-1}$$

The transpose and inverse operations on a matrix can be interchanged, i.e.,

$$(A^t)^{-1} = (A^{-1})^t$$

Partitioning of matrices

A large matrix can be subdivided into several submatrices of smaller dimensions:

$$A = \left[\begin{array}{c|c} A_1 & A_2 \\ \hline A_3 & A_4 \end{array}\right]$$

If the diagonal submatrices A_1 and A_4 are square, the subdivision is called *principal partitioning.*

Partitioning can be used to show the specific structure of A and to simplify matrix computation. Each submatrix is considered as an element in the partitioned matrix. Addition or subtraction is performed as follows:

$$\left[\begin{array}{c|c} A_1 & A_2 \\ \hline A_3 & A_4 \end{array}\right] \pm \left[\begin{array}{c|c} B_1 & B_2 \\ \hline B_3 & B_4 \end{array}\right] = \left[\begin{array}{c|c} A_1 \pm B_1 & A_2 \pm B_2 \\ \hline A_3 \pm B_3 & A_4 \pm B_4 \end{array}\right]$$

where the dimensions of corresponding submatrices must be conformable.

Multiplication is performed as follows:

$$\left[\begin{array}{c|c} A_1 & A_2 \\ \hline A_3 & A_4 \end{array}\right] \left[\begin{array}{c|c} B_1 & B_2 \\ \hline B_3 & B_4 \end{array}\right] = \left[\begin{array}{c|c} C_1 & C_2 \\ \hline C_3 & C_4 \end{array}\right]$$

where

$$C_1 = A_1B_1 + A_2B_3$$
$$C_2 = A_1B_2 + A_2B_4$$
$$C_3 = A_3B_1 + A_4B_3$$
$$C_4 = A_3B_2 + A_4B_4$$

The rule for partitioning two matrices whose product is to be found is: the n columns of the premultiplier are grouped into k and $n - k$ columns from left to right, and the n rows of the postmultiplier are grouped into k and $n - k$ rows from top to bottom in order that the submatrices are conformable for multiplication.

The transpose of a partitioned matrix is shown below.

$$A = \left[\begin{array}{c|c} A_1 & A_2 \\ \hline A_3 & A_4 \end{array}\right]$$

$$A^t = \left[\begin{array}{c|c} A_1^t & A_3^t \\ \hline A_2^t & A_4^t \end{array}\right]$$

The inverse of a partitioned matrix is obtained as follows:

$$A = \left[\begin{array}{c|c} A_1 & A_2 \\ \hline A_3 & A_4 \end{array}\right]$$

$$A^{-1} = \left[\begin{array}{c|c} B_1 & B_2 \\ \hline B_3 & B_4 \end{array}\right]$$

where

$$\begin{aligned} B_1 &= (A_1 - A_2A_4^{-1}A_3)^{-1} \\ B_2 &= -B_1A_2A_4^{-1} \\ B_3 &= -A_4^{-1}A_3B_1 \\ B_4 &= A_4^{-1} - A_4^{-1}A_3B_2 \end{aligned} \qquad (2.4.3.)$$

and A_1 and A_4 must be square matrices.

2.5 Linear dependence and rank of a matrix

Linear dependence

The columns of an $m \times n$ matrix A can be written as n column vectors.

$\{c_1\}\{c_2\} \cdots \{c_n\}$

Also, the rows of matrix A can be written as m row vectors.

$\{r_1\}\{r_2\} \cdots \{r_m\}$

The column vectors are *linearly independent* if the equation

$$p_1\{c_1\} + p_2\{c_2\} + \cdots + p_n\{c_n\} = 0 \tag{2.5.1}$$

is satisfied only for all $p_k = 0$ ($k = 1, 2, \ldots, n$). Similarly, the row vectors are linearly independent if only zero values for the scalars q_r ($r = 1, 2, \ldots, m$) satisfy the equation

$$q_1\{r_1\} + q_2\{r_2\} + \cdots + q_m\{r_m\} = 0 \tag{2.5.2}$$

It is not possible to express one or more linearly independent column vectors (or row vectors) as a linear combination of others.

If some $p_k \neq 0$ satisfies (2.5.1), the column vectors are *linearly dependent.* If some $q_r \neq 0$ satisfies (2.5.2), the row vectors are linearly dependent. That is, it is possible to express one or more column vectors (or row vectors) as a linear combination of others. If the column vectors (or row vectors) of a matrix A are linearly dependent, then the determinant of A is zero.

Rank of a matrix

The *rank* of an $m \times n$ matrix A is equal to the maximum number of linearly independent columns of A or the maximum number of linearly independent rows of A. The former is called the *column rank* and the latter the *row rank.* The column rank is equal to the row rank. The rank of a matrix is equal to the order of the largest nonvanishing determinant in A. For example, consider the matrix

$$A = \begin{bmatrix} 1 & 2 & 4 \\ 2 & 4 & 8 \\ 3 & 8 & 10 \end{bmatrix}$$

The rows are linearly dependent since the equation

$$q_1\{1\ 2\ 4\} + q_2\{2\ 4\ 8\} + q_3\{3\ 8\ 10\} = 0$$

is satisfied for

$$q_1 = 2$$
$$q_2 = -1$$
$$q_3 = 0$$

Similarly, the columns are linearly dependent since the equation

$$p_1 \begin{Bmatrix} 1 \\ 2 \\ 3 \end{Bmatrix} + p_2 \begin{Bmatrix} 2 \\ 4 \\ 8 \end{Bmatrix} + p_3 \begin{Bmatrix} 4 \\ 8 \\ 10 \end{Bmatrix} = 0$$

is satisfied for

$$p_1 = 6$$
$$p_2 = -1$$
$$p_3 = -1$$

However, no two columns are linearly dependent and, therefore, the rank of the matrix is 2.

2.6 Linear equations

A linear system of m equations in n unknowns is written

$$\begin{aligned} a_{11}x_1 + a_{12}x_2 + \cdots + a_{1n}x_n &= y_1 \\ a_{21}x_1 + a_{22}x_2 + \cdots + a_{2n}x_n &= y_2 \\ \cdots\cdots\cdots\cdots\cdots\cdots\cdots\cdots \\ a_{m1}x_1 + a_{m2}x_2 + \cdots + a_{mn}x_n &= y_m \end{aligned} \tag{2.6.1}$$

where a_{ij} = known coefficients or parameters of the system
x_j = unknown variables of the system
y_i = known constants of the system

The system of equations (2.6.1) in matrix form is

$$AX = Y$$

The *augmented matrix* of A, designated by $\hat{A}$, is formed by adjoining the column vector Y as the $(n + 1)$st column to A.

$$\hat{A} = \begin{bmatrix} a_{11} & a_{12} & \cdots & a_{1n} & y_1 \\ a_{21} & a_{22} & \cdots & a_{2n} & y_2 \\ \cdots & \cdots & \cdots & \cdots & \cdots \\ a_{m1} & a_{m2} & \cdots & a_{mn} & y_m \end{bmatrix}$$

If $y_1, y_2, \ldots, y_m$ are all zero in (2.6.1), the linear equations are *homogeneous* and

$$AX = 0$$

If one or more y_i are nonzero, the linear equations are *nonhomogeneous*.

The necessary and sufficient condition for a system of linear equations to have a solution is that the rank of the coefficient matrix A be equal to the rank of the augmented matrix $\hat{A}$. A unique solution exists when A is a square matrix and the rank of A is equal to the number of columns (variables). The unique solution is nontrivial for nonhomogeneous equations and trivial (i.e., zero) for homogeneous equations. If the rank of A is less than the number of equations, some of the equations are redundant and do not place any further constraint on the variables. If the rank of A is less than the number of variables of the system, there are an infinite number of nontrivial solutions.

Problems

2.1 Given:

$$A = \begin{bmatrix} 1 & 1 & 5 \\ 3 & 2 & 4 \\ -7 & 2 & 3 \end{bmatrix} \quad \text{and} \quad B = \begin{bmatrix} 2 & 1 & 5 \\ -3 & 4 & 4 \\ 7 & -2 & 2 \end{bmatrix}$$

Determine:

a. $C = A + B$
b. What type of matrix C is
c. $D = A - B$
d. What type of matrix D is

2.2 Given:

$$A = \begin{bmatrix} -3 & 1 & 4 \\ -2 & -5 & 3 \end{bmatrix} \quad \text{and} \quad B = \begin{bmatrix} 1 & 2 \\ 3 & 4 \\ 5 & 6 \end{bmatrix}$$

Determine:

a. $C = AB$
b. $D = BA$
c. $E = A^tB^t$
d. What the relationship of matrix E to matrix D is

2.3 Given:

$$A = \begin{bmatrix} 11 & 8 & 5 \\ 8 & 1 & 5 \\ 5 & 5 & 2 \end{bmatrix} \quad \text{and} \quad B = \begin{bmatrix} -2 & 1 & 1 \\ 1 & -1 & 0 \\ 3 & -1 & -1 \end{bmatrix}$$

Determine:

a. $C = B^tAB$
b. What type of matrix A is
c. What type of matrix C is
d. $D = C^{-1}$
e. $E = CC^{-1}$
f. What type of matrix E is

2.4 Given:

$$A = \begin{bmatrix} 1 & 0 & 1 \\ 1 & -1 & 1 \\ 2 & 1 & 1 \end{bmatrix} \quad \text{and} \quad Y = \begin{bmatrix} 1 \\ 2 \\ 3 \end{bmatrix}$$

Determine:

a. $B = A^{-1}$

b. X from $AX = Y$

2.5 Given:

$$A = \begin{bmatrix} 0 & 5 & 7 & 1 \\ -5 & 0 & -1 & 7 \\ -7 & 1 & 0 & -5 \\ -1 & -7 & 5 & 0 \end{bmatrix}$$

Determine:

a. $B = 0.1A + 0.5U$

b. What type of matrix A is

c. That B is orthogonal

2.6 Given:

$$A = \frac{1}{\sqrt{3}} \begin{bmatrix} 1 & 1 & 1 \\ 1 & -\frac{1}{2} - j\frac{\sqrt{3}}{2} & -\frac{1}{2} + j\frac{\sqrt{3}}{2} \\ 1 & -\frac{1}{2} + j\frac{\sqrt{3}}{2} & -\frac{1}{2} - j\frac{\sqrt{3}}{2} \end{bmatrix}$$

Determine:

a. $B = (A^*)^t A$

b. What type of matrix A is

2.7 Given:

$$A = \begin{bmatrix} 2 & -j2 & 1 + j3 \\ j2 & 1 & 2 - j \\ 1 - j3 & 2 + j & 3 \end{bmatrix}$$

Determine:

a. $B = (A^*)^t$

b. What type of matrix A is

2.8 Given:

$$A = \begin{bmatrix} 1 & -1 \\ 0 & -3 \end{bmatrix} \qquad jB = \begin{bmatrix} j2 & j \\ j4 & 0 \end{bmatrix} \qquad \text{and} \qquad Y = \begin{bmatrix} 1 \\ j \end{bmatrix}$$

Determine:

a. X from $(A + jB)X^* = Y$

b. What type of matrix A is

c. What type of matrix B is

2.9 Given the partitioned matrix:

$$A = \left[\begin{array}{cc|c} 1 & 1 & 2 \\ -1 & -2 & 1 \\ \hline -6 & 4 & 2 \end{array}\right]$$

Determine $B = A^{-1}$ using the formulas (2.4.3) for the inverse of a partitioned matrix.

2.10 Given:

$$\left[\begin{array}{c} A_1 \\ \hline A_2 \end{array}\right] = \left[\begin{array}{cc|ccc} 1 & 2 & 0 & 3 & 4 \\ 2 & 6 & 0 & 0 & 1 \\ \hline 0 & 0 & 14 & 5 & 1 \\ 3 & 0 & 5 & 10 & 2 \\ 4 & 1 & 1 & 2 & 12 \end{array}\right] \left[\begin{array}{c} 1 \\ 1 \\ \hline 1 \\ 1 \\ 1 \end{array}\right]$$

Determine A_2.

2.11 Given the partitioned matrices:

$$A = \left[\begin{array}{cc|ccc|c} 1 & 2 & 0 & 0 & 0 & 0 \\ 1 & 1 & 0 & 0 & 0 & 0 \\ \hline 0 & 0 & 1 & 3 & 2 & 0 \\ 0 & 0 & 2 & 3 & 4 & 0 \\ 0 & 0 & 1 & 5 & 2 & 0 \\ \hline 0 & 0 & 0 & 0 & 0 & 5 \end{array}\right]$$

$$B = \left[\begin{array}{cc|ccc|c} 1 & 3 & 2 & 5 & 6 & 1 \\ 4 & 7 & 1 & 3 & 2 & 4 \\ \hline 3 & 4 & 2 & 6 & 5 & 1 \\ 4 & 6 & 3 & 1 & 2 & 5 \\ 2 & 6 & 7 & 3 & 8 & 1 \\ \hline 7 & 2 & 3 & 1 & 4 & 5 \end{array}\right]$$

Determine $C = AB$.

2.12 Given:

$$C = \left[\begin{array}{c|c} A & N \\ \hline N & B \end{array}\right]$$

where

$$A = \begin{bmatrix} 2 & 1 \\ 5 & 3 \end{bmatrix} \qquad B = \begin{bmatrix} 2 & -1 \\ 6 & -4 \end{bmatrix}$$

and N is a null matrix.
Show that the inverse of C is

$$C^{-1} = \left[\begin{array}{c|c} A^{-1} & N \\ \hline N & B^{-1} \end{array}\right]$$

2.13 Given:

$$A = \begin{bmatrix} 2 & 1 & -1 \\ 5 & 6 & 3 \\ 3 & 5 & 4 \end{bmatrix}$$

Show that A is a singular matrix and determine its rank.

2.14 Given:

$$A = \left[\begin{array}{c|c} A_1 & A_2 \\ \hline A_3 & A_4 \end{array}\right] = \left[\begin{array}{ccc|c} 6 & 0 & 3 & 1 \\ 2 & 4 & 2 & 2 \\ 1 & 7 & 5 & 3 \\ \hline 6 & 4 & 2 & 2 \end{array}\right]$$

Determine $B = A_1 - A_2A_4^{-1}A_3$.

2.15 Given:

$$A = \begin{bmatrix} 3 & -2 & 1 \\ -2 & 5 & -8 \\ 1 & -8 & 12 \end{bmatrix} \qquad B = \begin{bmatrix} 2 & -1 \\ -1 & 5 \end{bmatrix} \qquad \text{and}$$

$$C = \begin{bmatrix} 1 & 0 & -1 \\ 0 & -1 & 1 \end{bmatrix}$$

Determine $D = A - C^tBC$.

Bibliography

Aitken, A. C.: "Determinants and Matrices," Interscience Publishers, Inc., New York, 1959.

Ayres, Frank, Jr.: "Theory and Problems of Matrices," Schaum Publishing Company, New York, 1962.

Braae, R.: "Matrix Algebra for Electrical Engineers," Addison-Wesley Publishing Company, Inc., Reading, Mass., 1963.

Stigant, S. Austen: "The Elements of Determinants, Matrices, and Tensors for Engineers," MacDonald and Co., Ltd., London, 1959.

Tropper, A. M.: "Matrix Theory for Electrical Engineers," Addison-Wesley Publishing Company, Inc., Reading, Mass., 1962.

chapter 3
Incidence and network matrices

3.1 Introduction

The formulation of a suitable mathematical model is the first step in the analysis of an electrical network. The model must describe the characteristics of individual network components as well as the relations that govern the interconnection of these elements. A network matrix equation provides a convenient mathematical model for a digital computer solution.

The elements of a network matrix depend on the selection of the independent variables, which can be either currents or voltages. Correspondingly, the elements of the network matrix will be impedances or admittances.

The electrical characteristics of the individual network components can be presented conveniently in the form of a primitive network matrix. This matrix, while adequately describing the characteristics of each component, does not provide any information pertaining to the network connections. It is necessary, therefore, to transform the primitive network matrix into a network matrix that describes the performance of the interconnected network.

The form of the network matrix used in the performance equation depends on the frame of reference, namely, bus or loop. In the bus frame of reference the variables are the nodal voltages and nodal currents. In the loop frame of reference the variables are loop voltages and loop currents.

The formation of the appropriate network matrix is an integral part of a digital computer program for the solution of power system problems.

3.2 Graphs

In order to describe the geometrical structure of a network it is sufficient to replace the network components by single line segments irrespective of the characteristics of the components. These line segments are called *elements* and their terminals are called *nodes.* A node and an element are *incident* if the node is a terminal of the element. Nodes can be incident to one or more elements.

A *graph* shows the geometrical interconnection of the elements of a network. A *subgraph* is any subset of elements of the graph. A *path* is a subgraph of connected elements with no more than two elements connected to any one node. A graph is *connected* if and only if there is a path between every pair of nodes. If each element of the connected graph is assigned a direction it is then *oriented.* A representation of a power system and the corresponding oriented graph are shown in Fig. 3.1.

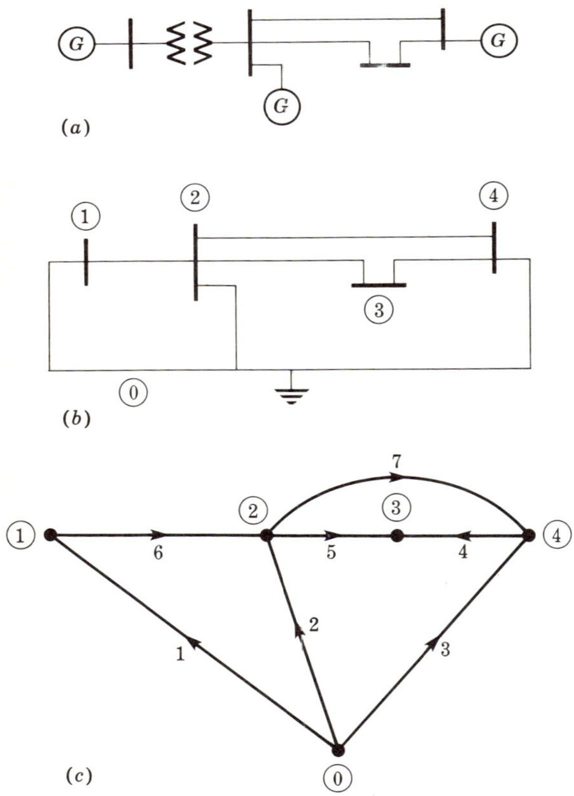

Fig. 3.1 Power system representations. (a) Single line diagram; (b) positive sequence network diagram; (c) oriented connected graph.

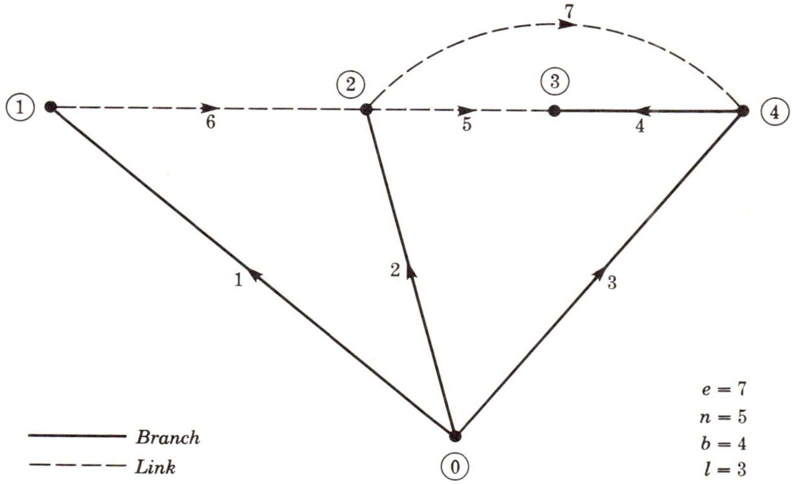

Fig. 3.2 ***Tree and cotree of the oriented connected graph.***

A connected subgraph containing all nodes of a graph but no closed path is called a *tree*. The elements of a tree are called *branches* and form a subset of the elements of the connected graph. The number of branches b required to form a tree is

$$b = n - 1 \tag{3.2.1}$$

where n is the number of nodes in the graph.

Those elements of the connected graph that are not included in the tree are called *links* and form a subgraph, not necessarily connected, called the *cotree*. The cotree is the complement of the tree. The number of links l of a connected graph with e elements is

$$l = e - b$$

From equation (3.2.1) it follows that

$$l = e - n + 1 \tag{3.2.2}$$

A tree and the corresponding cotree of the graph given in Fig. 3.1*c* are shown in Fig. 3.2.

If a link is added to the tree, the resulting graph contains one closed path, called a *loop*. The addition of each subsequent link forms one or more additional loops. Loops which contain only one link are independent and are called *basic loops*. Consequently, the number of basic

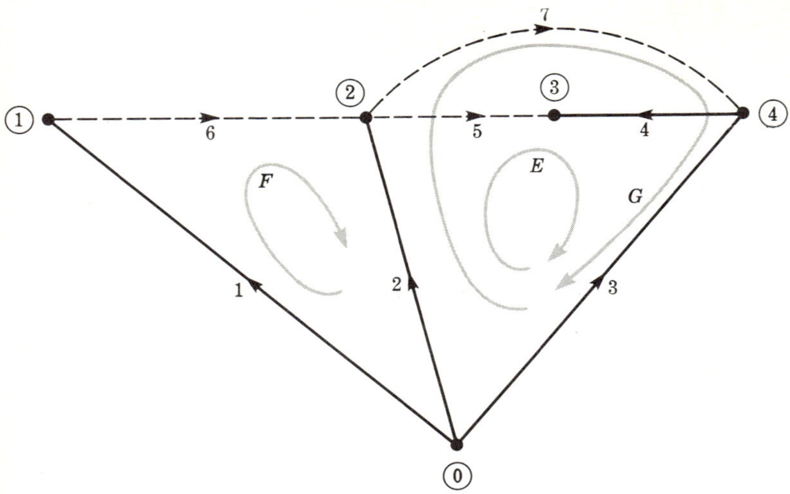

Fig. 3.3 Basic loops of the oriented connected graph.

loops is equal to the number of links given by equation (3.2.2). Orientation of a basic loop is chosen to be the same as that of its link. The basic loops of the graph given in Fig. 3.2 are shown in Fig. 3.3.

A *cut-set* is a set of elements that, if removed, divides a connected graph into two connected subgraphs. A unique independent group of

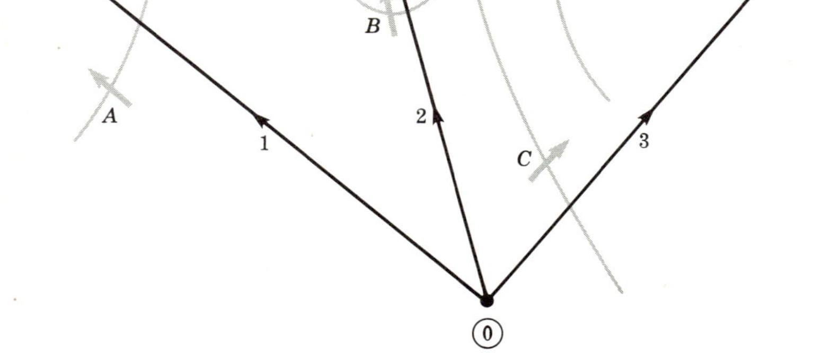

Fig. 3.4 Basic cut-sets of the oriented connected graph.

cut-sets may be chosen if each cut-set contains only one branch. Independent cut-sets are called *basic cut-sets.* The number of basic cut-sets is equal to the number of branches. Orientation of a basic cut-set is chosen to be the same as that of its branch. The basic cut-sets of the graph given in Fig. 3.2 are shown in Fig. 3.4.

3.3 Incidence matrices

Element-node incidence matrix $\hat{A}$

The incidence of elements to nodes in a connected graph is shown by the element-node incidence matrix. The elements of the matrix are as follows:

$a_{ij} = 1$ if the ith element is incident to and oriented away from the jth node

$a_{ij} = -1$ if the ith element is incident to and oriented toward the jth node

$a_{ij} = 0$ if the ith element is not incident to the jth node

The dimension of the matrix is $e \times n$, where e is the number of elements and n is the number of nodes in the graph. The element-node incidence matrix for the graph shown in Fig. 3.2 is

$\hat{A} =$

e \ n	⓪	①	②	③	④
1	1	−1			
2	1		−1		
3	1				−1
4				−1	1
5			1	−1	
6		1	−1		
7			1		−1

Since

$$\sum_{j=0}^{4} a_{ij} = 0 \qquad i = 1, 2, \ldots, e$$

the columns of $\hat{A}$ are linearly dependent. Hence, the rank of $\hat{A} < n$.

Bus incidence matrix A

Any node of a connected graph can be selected as the reference node. Then, the variables of the other nodes, referred to as buses, can be measured with respect to the assigned reference. The matrix obtained from $\hat{A}$ by deleting the column corresponding to the reference node is the element-bus incidence matrix A, which will be called the bus incidence matrix. The dimension of this matrix is $e \times (n - 1)$ and the rank is $n - 1 = b$, where b is the number of branches in the graph. Selecting node 0 as reference for the graph shown in Fig. 3.2,

$A =$

e \ bus	①	②	③	④
1	−1			
2		−1		
3				−1
4			−1	1
5		1	−1	
6	1	−1		
7		1		−1

This matrix is rectangular and therefore singular.

If the rows of A are arranged according to a particular tree, the matrix can be partitioned into submatrices A_b of dimension $b \times (n - 1)$ and A_l of dimension $l \times (n - 1)$, where the rows of A_b correspond to branches and the rows of A_l to links. The partitioned matrix for the graph shown in Fig. 3.2 is

$A =$

e \ bus	①	②	③	④
1	−1			
2		−1		
3				−1
4			−1	1
5		1	−1	
6	1	−1		
7		1		−1

$=$

e \ bus	Buses
Branches	A_b
Links	A_l

A_b is a nonsingular square matrix with rank $(n - 1)$.

Branch-path incidence matrix K

The incidence of branches to paths in a tree is shown by the branch-path incidence matrix, where a path is oriented from a bus to the reference node. The elements of this matrix are:

$k_{ij} = 1$ if the ith branch is in the path from the jth bus to reference and is oriented in the same direction

$k_{ij} = -1$ if the ith branch is in the path from the jth bus to reference but is oriented in the opposite direction

$k_{ij} = 0$ if the ith branch is not in the path from the jth bus to reference

With node 0 as reference the branch-path incidence matrix associated with the tree shown in Fig. 3.2 is

$K =$

b \ path	①	②	③	④
1	−1			
2		−1		
3			−1	−1
4			−1	

This is a nonsingular square matrix with rank $(n - 1)$.

The branch-path incidence matrix and the submatrix A_b relate the branches to paths and branches to buses, respectively. Since there is a one-to-one correspondence between paths and buses,

$$A_bK^t = U \tag{3.3.1}$$

Therefore,

$$K^t = A_b^{-1} \tag{3.3.2}$$

Basic cut-set incidence matrix B

The incidence of elements to basic cut-sets of a connected graph is shown by the basic cut-set incidence matrix B. The elements of this matrix are:

$b_{ij} = 1$ if the ith element is incident to and oriented in the same direction as the jth basic cut-set
$b_{ij} = -1$ if the ith element is incident to and oriented in the opposite direction as the jth basic cut-set
$b_{ij} = 0$ if the ith element is not incident to the jth basic cut-set

The basic cut-set incidence matrix, of dimension $e \times b$, for the graph shown in Fig. 3.4 is

$$
B = \begin{array}{c|c|c|c|c|}
{}_{e}\diagdown^{b} & \multicolumn{4}{c}{\begin{array}{c}\text{Basic cut-sets}\\ A \quad\quad B \quad\quad C \quad\quad D\end{array}} \\
\hline
1 & 1 & & & \\
\hline
2 & & 1 & & \\
\hline
3 & & & 1 & \\
\hline
4 & & & & 1 \\
\hline
5 & & -1 & 1 & 1 \\
\hline
6 & -1 & 1 & & \\
\hline
7 & & -1 & 1 & \\
\hline
\end{array}
$$

The matrix B can be partitioned into submatrices U_b and B_l where the rows of U_b correspond to branches and the rows of B_l to links. The partitioned matrix is

$$B = \begin{array}{c|cccc} e \backslash b & A & B & C & D \\ \hline 1 & 1 & & & \\ 2 & & 1 & & \\ 3 & & & 1 & \\ 4 & & & & 1 \\ \hline 5 & & -1 & 1 & 1 \\ 6 & -1 & 1 & & \\ 7 & & -1 & 1 & \end{array} = \begin{array}{c|c} e \backslash b & \text{Basic cut-sets} \\ \hline \text{Branches} & U_b \\ \hline \text{Links} & B_l \end{array}$$

The identity matrix U_b shows the one-to-one correspondence of the branches and basic cut-sets.

The submatrix B_l can be obtained from the bus incidence matrix A. The incidence of links to buses is shown by the submatrix A_l and the incidence of branches to buses is shown by the submatrix A_b. Since there is a one-to-one correspondence of the branches and basic cut-sets, B_lA_b shows the incidence of links to buses, that is,

$$B_lA_b = A_l$$

Therefore,

$$B_l = A_lA_b^{-1}$$

In addition, as shown in equation (3.3.2),

$$A_b^{-1} = K^t$$

Therefore,

$$B_l = A_lK^t \tag{3.3.3}$$

Augmented cut-set incidence matrix $\hat{B}$

Fictitious cut-sets, called *tie cut-sets*, can be introduced in order that the number of cut-sets equals the number of elements. Each tie cut-set contains only one link of the connected graph. The tie cut-sets for the graph given in Fig. 3.4 are shown in Fig. 3.5. An augmented cut-set incidence matrix is formed by adjoining to the basic cut-set incidence matrix additional columns corresponding to these tie cut-sets. A tie

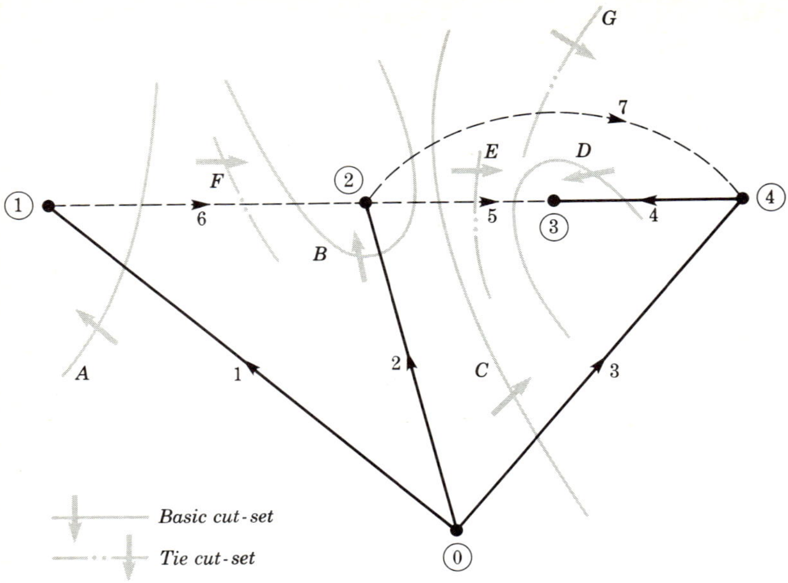

Fig. 3.5 Basic and tie cut-sets of the oriented connected graph.

cut-set is oriented in the same direction as the associated link. The augmented cut-set incidence matrix for the graph shown in Fig. 3.5 is

$\hat{B} =$

e \ e	Basic cut-sets A	B	C	D	Tie cut-sets E	F	G
1	1						
2		1					
3			1				
4				1			
5		−1	1	1	1		
6	−1	1				1	
7		−1	1				1

This is a square matrix of dimension $e \times e$ and is nonsingular.

The matrix $\hat{B}$ can be partitioned as follows:

	e \ e	Basic cut-sets				Tie cut-sets		
		A	*B*	*C*	*D*	*E*	*F*	*G*
	1	1						
	2		1					
	3			1				
$\hat{B} =$	4				1			
	5		−1	1	1	1		
	6	−1	1				1	
	7		−1	1				1

	e \ e	Basic cut-sets	Tie cut-sets
=	Branches	U_b	0
	Links	B_l	U_l

Basic loop incidence matrix C

The incidence of elements to basic loops of a connected graph is shown by the basic loop incidence matrix C. The elements of this matrix are:

$c_{ij} = 1$ if the ith element is incident to and oriented in the same direction as the jth basic loop

$c_{ij} = -1$ if the ith element is incident to and oriented in the opposite direction as the jth basic loop

$c_{ij} = 0$ if the ith element is not incident to the jth basic loop

The basic loop incidence matrix, of dimension $e \times l$, for the graph shown in Fig. 3.3 is

$C =$

e \ l	Basic loops E	F	G
1		1	
2	1	−1	1
3	−1		−1
4	−1		
5	1		
6		1	
7			1

The matrix C can be partitioned into submatrices C_b and U_l where the rows of C_b correspond to branches and the rows of U_l to links. The partitioned matrix is

$C =$

e \ l	Basic loops E	F	G
1		1	
2	1	−1	1
3	−1		−1
4	−1		
5	1		
6		1	
7			1

$=$

e \ l	Basic loops
Branches	C_b
Links	U_l

The identity matrix U_l shows the one-to-one correspondence of links to basic loops.

Augmented loop incidence matrix $\hat{C}$

The number of basic loops in a connected graph is equal to the number of links. In order to have a total number of loops equal to the number of

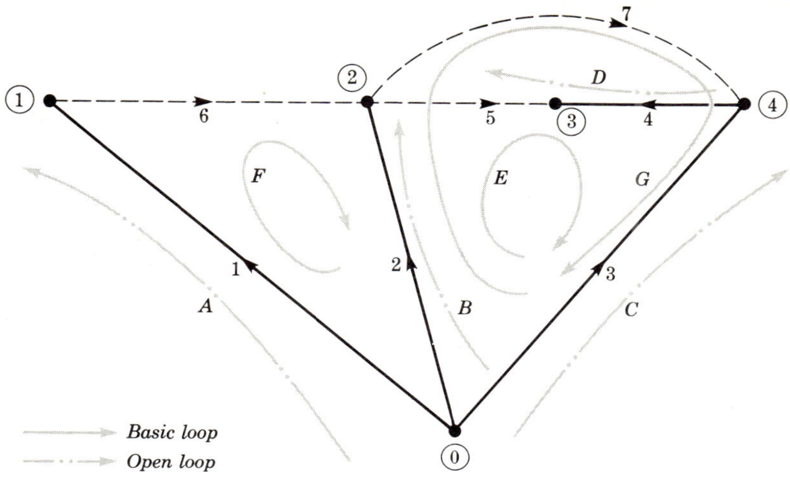

Fig. 3.6 Basic and open loops of the oriented connected graph.

elements, let $(e - l)$ loops, corresponding to the b branches, be designated as *open loops*. An open loop, then, is defined as a path between adjacent nodes connected by a branch. The open loops for the graph given in Fig. 3.3 are shown in Fig. 3.6. The orientation of an open loop is the same as that for the associated branch.

The augmented loop incidence matrix is formed by adjoining to the basic loop incidence matrix the columns showing the incidence of elements to open loops. This matrix, for the graph shown in Fig. 3.6, is

$\hat{C} =$

e \ e	Open loops A	B	C	D	Basic loops E	F	G
1	1					1	
2		1			1	−1	1
3			1		−1		−1
4				1	−1		
5					1		
6						1	
7							1

This is a square matrix, of dimension $e \times e$, and is nonsingular.

The matrix $\hat{C}$ can be partitioned as follows:

$\hat{C} =$

e \ e	Open loops A	B	C	D	Basic loops E	F	G
1	1					1	
2		1			1	−1	1
3			1		−1		−1
4				1	−1		
5					1		
6						1	
7							1

$=$

e \ e	Open loops	Basic loops
Branches	U_b	C_b
Links	0	U_l

3.4 Primitive network

Network components represented both in impedance form and in admittance form are shown in Fig. 3.7. The performance of the components can be expressed using either form. The variables and parameters are:

v_{pq} is the voltage across the element p–q
e_{pq} is the source voltage in series with element p–q
i_{pq} is the current through element p–q

j_{pq} is the source current in parallel with element p–q
z_{pq} is the self-impedance of element p–q
y_{pq} is the self-admittance of element p–q

Each element has two variables, v_{pq} and i_{pq}. In steady state these variables and the parameters of the elements z_{pq} and y_{pq} are real numbers for direct current circuits and complex numbers for alternating current circuits.

The performance equation of an element in impedance form is

$$v_{pq} + e_{pq} = z_{pq}i_{pq} \tag{3.4.1}$$

or in admittance form is

$$i_{pq} + j_{pq} = y_{pq}v_{pq} \tag{3.4.2}$$

The parallel source current in admittance form is related to the series source voltage in impedance form by

$$j_{pq} = -y_{pq}e_{pq}$$

A set of unconnected elements is defined as a primitive network. The performance equations of a primitive network can be derived from (3.4.1) or (3.4.2) by expressing the variables as vectors and the parame-

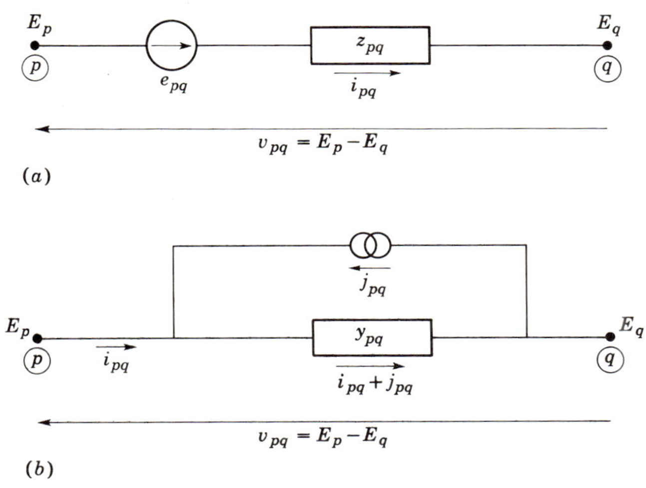

***Fig. 3.7** Representations of a network component. (a) Impedance form; (b) admittance form.*

ters as matrices. The performance equation in impedance form is

$$\bar{v} + \bar{e} = [z]\bar{\imath}$$

or in admittance form is

$$\bar{\imath} + \bar{\jmath} = [y]\bar{v}$$

A diagonal element of the matrix $[z]$ or $[y]$ of the primitive network is the self-impedance $z_{pq,pq}$ or self-admittance $y_{pq,pq}$. An off-diagonal element is the mutual impedance $z_{pq,rs}$ or the mutual admittance $y_{pq,rs}$ between the elements p–q and r–s. The primitive admittance matrix $[y]$ can be obtained by inverting the primitive impedance matrix $[z]$. The matrices $[z]$ and $[y]$ are diagonal matrices if there is no mutual coupling between elements. In this case the self-impedances are equal to the reciprocals of the corresponding self-admittances.

3.5 Formation of network matrices by singular transformations

Network performance equations

A network is made up of an interconnected set of elements. In the bus frame of reference, the performance of an interconnected network is described by $n - 1$ independent nodal equations, where n is the number of nodes. In matrix notation, the performance equation in impedance form is

$$\bar{E}_{BUS} = Z_{BUS}\bar{I}_{BUS}$$

or in admittance form is

$$\bar{I}_{BUS} = Y_{BUS}\bar{E}_{BUS}$$

where $\bar{E}_{BUS}$ = vector of bus voltages measured with respect to the reference bus

$\bar{I}_{BUS}$ = vector of impressed bus currents

Z_{BUS} = bus impedance matrix whose elements are open circuit driving point and transfer impedances

Y_{BUS} = bus admittance matrix whose elements are short circuit driving point and transfer admittances

In the branch frame of reference the performance of the interconnected network is described by b independent branch equations where b is the number of branches. In matrix notation, the performance equation in impedance form is

$$\bar{E}_{BR} = Z_{BR}\bar{I}_{BR}$$

or in admittance form is

$$\bar{I}_{BR} = Y_{BR}\bar{E}_{BR}$$

where $\bar{E}_{BR}$ = vector of voltages across the branches
$\bar{I}_{BR}$ = vector of currents through the branches
Z_{BR} = branch impedance matrix whose elements are open circuit driving point and transfer impedances of the branches of the network
Y_{BR} = branch admittance matrix whose elements are short circuit driving point and transfer admittances of the branches of the network

In the loop frame of reference, the performance of an interconnected network is described by l independent loop equations where l is the number of links or basic loops. The performance equation in impedance form is

$$\bar{E}_{LOOP} = Z_{LOOP}\bar{I}_{LOOP}$$

or in admittance form is

$$\bar{I}_{LOOP} = Y_{LOOP}\bar{E}_{LOOP}$$

where $\bar{E}_{LOOP}$ = vector of basic loop voltages
$\bar{I}_{LOOP}$ = vector of basic loop currents
Z_{LOOP} = loop impedance matrix
Y_{LOOP} = loop admittance matrix

Bus admittance and bus impedance matrices

The bus admittance matrix Y_{BUS} can be obtained by using the bus incidence matrix A to relate the variables and parameters of the primitive network to bus quantities of the interconnected network. The performance equation of the primitive network

$$\bar{\imath} + \bar{\jmath} = [y]\bar{v}$$

is premultiplied by A^t, the transpose of the bus incidence matrix, to obtain

$$A^t\bar{\imath} + A^t\bar{\jmath} = A^t[y]\bar{v} \qquad (3.5.1)$$

Since the matrix A shows the incidence of elements to buses, $A^t\bar{\imath}$ is a vector in which each element is the algebraic sum of the currents through the network elements terminating at a bus. In accordance with Kirchhoff's current law, the algebraic sum of the currents at a bus is zero. Then

$$A^t\bar{\imath} = 0 \qquad (3.5.2)$$

Similarly, $A^t\bar{\jmath}$ gives the algebraic sum of the source currents at each bus and equals the vector of impressed bus currents. Therefore

$$\bar{I}_{BUS} = A^t\bar{\jmath} \tag{3.5.3}$$

Substituting from equations (3.5.2) and (3.5.3) into (3.5.1) yields

$$\bar{I}_{BUS} = A^t[y]\bar{v} \tag{3.5.4}$$

Power into the network is $(\bar{I}^*_{BUS})^t\bar{E}_{BUS}$ and the sum of the powers in the primitive network is $(\bar{\jmath}^*)^t\bar{v}$. The power in the primitive and interconnected networks must be equal, that is, the transformation of variables must be power-invariant. Hence

$$(\bar{I}^*_{BUS})^t\bar{E}_{BUS} = (\bar{\jmath}^*)^t\bar{v} \tag{3.5.5}$$

Taking the conjugate transpose of equation (3.5.3)

$$(\bar{I}^*_{BUS})^t = (\bar{\jmath}^*)^t A^*$$

Since A is a real matrix

$$A^* = A$$

and

$$(\bar{I}^*_{BUS})^t = (\bar{\jmath}^*)^t A \tag{3.5.6}$$

Substituting from equation (3.5.6) into (3.5.5)

$$(\bar{\jmath}^*)^t A\bar{E}_{BUS} = (\bar{\jmath}^*)^t\bar{v}$$

Since this equation is valid for all values of $\bar{\jmath}$, it follows that

$$A\bar{E}_{BUS} = \bar{v} \tag{3.5.7}$$

Substituting from equation (3.5.7) into (3.5.4),

$$\bar{I}_{BUS} = A^t[y]A\bar{E}_{BUS} \tag{3.5.8}$$

Since the performance equation of the network is

$$\bar{I}_{BUS} = Y_{BUS}\bar{E}_{BUS} \tag{3.5.9}$$

it follows from equations (3.5.8) and (3.5.9) that

$$Y_{BUS} = A^t[y]A$$

The bus incidence matrix A is singular and therefore $A^t[y]A$ is a singular transformation of $[y]$.

The bus impedance matrix can be obtained from

$$Z_{BUS} = Y^{-1}_{BUS} = (A^t[y]A)^{-1}$$

Branch admittance and branch impedance matrices

The branch admittance matrix Y_{BR} can be obtained by using the basic cut-set incidence matrix B to relate the variables and parameters of the primitive network to branch quantities of the interconnected network. The performance equation of the primitive network in admittance form is premultiplied by B^t to obtain

$$B^t\bar{\imath} + B^t\bar{\jmath} = B^t[y]\bar{v} \tag{3.5.10}$$

Since the matrix B shows the incidence of elements to basic cut-sets, $B^t\bar{\imath}$ is a vector in which each element is the algebraic sum of the currents through the elements incident to a basic cut-set.

The elements of a basic cut-set if removed divide the network into two connected subnetworks. Therefore, an element of the vector $B^t\bar{\imath}$ is the algebraic sum of the current entering a subnetwork and by Kirchhoff's current law is zero. Therefore

$$B^t\bar{\imath} = 0 \tag{3.5.11}$$

Similarly, $B^t\bar{\jmath}$ is a vector in which each element is the algebraic sum of the source currents of the elements incident to the basic cut-set and is the total source current in parallel with a branch. Therefore

$$\bar{I}_{BR} = B^t\bar{\jmath} \tag{3.5.12}$$

Substituting from equations (3.5.11) and (3.5.12) into (3.5.10) yields

$$\bar{I}_{BR} = B^t[y]\bar{v} \tag{3.5.13}$$

Power into the network is $(\bar{I}^*_{BR})^t(\bar{E}_{BR})$ and since power is invariant

$$(\bar{I}^*_{BR})^t\bar{E}_{BR} = (\bar{\jmath}^*)^t\bar{v}$$

Obtaining $(\bar{I}^*_{BR})^t$ from equation (3.5.12), then

$$(\bar{\jmath}^*)^tB^*\bar{E}_{BR} = (\bar{\jmath}^*)^t\bar{v}$$

Since B is a real matrix

$$B^* = B \qquad \text{and} \qquad (\bar{\jmath}^*)^tB\bar{E}_{BR} = (\bar{\jmath}^*)^t\bar{v}$$

Since this equation is valid for all values of $\bar{\jmath}$, it follows that

$$\bar{v} = B\bar{E}_{BR} \tag{3.5.14}$$

Substituting from equation (3.5.14) into (3.5.13) yields

$$\bar{I}_{BR} = B^t[y]B\bar{E}_{BR} \tag{3.5.15}$$

The relation between the branch currents and the branch voltages is

$$\bar{I}_{BR} = Y_{BR}\bar{E}_{BR} \tag{3.5.16}$$

It follows from equations (3.5.15) and (3.5.16) that

$$Y_{BR} = B^t[y]B$$

The basic cut-set matrix B is a singular matrix and therefore $B^t[y]B$ is a singular transformation of $[y]$.

The branch impedance matrix can be obtained from

$$Z_{BR} = Y_{BR}^{-1} = (B^t[y]B)^{-1}$$

Loop impedance and loop admittance matrices

The loop impedance matrix Z_{LOOP} can be obtained by using the basic loop incidence matrix C to relate the variables and parameters of the primitive network to loop quantities of the interconnected network. The performance equation of the primitive network

$$\bar{v} + \bar{e} = [z]\bar{\imath}$$

is premultiplied by C^t to obtain

$$C^t\bar{v} + C^t\bar{e} = C^t[z]\bar{\imath} \tag{3.5.17}$$

Since the matrix C shows the incidence of elements to basic loops, $C^t\bar{v}$ gives the algebraic sum of the voltages around each basic loop. In accordance with Kirchhoff's voltage law, the algebraic sum of the voltages around a loop is zero. Hence

$$C^t\bar{v} = 0 \tag{3.5.18}$$

Similarly $C^t\bar{e}$ gives the algebraic sum of the source voltages around each basic loop. Therefore

$$\bar{E}_{LOOP} = C^t\bar{e} \tag{3.5.19}$$

Since power is invariant

$$(\bar{I}^*_{LOOP})^t\bar{E}_{LOOP} = (\bar{\imath}^*)^t\bar{e}$$

Substituting for $\bar{E}_{LOOP}$ from equation (3.5.19), then

$$(\bar{I}^*_{LOOP})^tC^t\bar{e} = (\bar{\imath}^*)^t\bar{e}$$

Since this equation is valid for all values of $\bar{e}$, it follows that

$$(\bar{\imath}^*)^t = (\bar{I}^*_{LOOP})^tC^t$$

Hence,

$$\bar{\imath} = C^*\bar{I}_{LOOP}$$

Since C is a real matrix,

$$C^* = C$$

and

$$\bar{\imath} = C\bar{I}_{LOOP} \tag{3.5.20}$$

Substituting from equations (3.5.18), (3.5.19), and (3.5.20) into (3.5.17) yields

$$\bar{E}_{LOOP} = C^t[z]C\bar{I}_{LOOP} \tag{3.5.21}$$

The performance equation of the network in the loop frame of reference is

$$\bar{E}_{LOOP} = Z_{LOOP}\bar{I}_{LOOP} \tag{3.5.22}$$

and it follows from equations (3.5.21) and (3.5.22) that

$$Z_{LOOP} = C^t[z]C$$

Since C is a singular matrix, $C^t[z]C$ is a singular transformation of $[z]$.

Table 3.1 Formation of network matrices by singular transformations

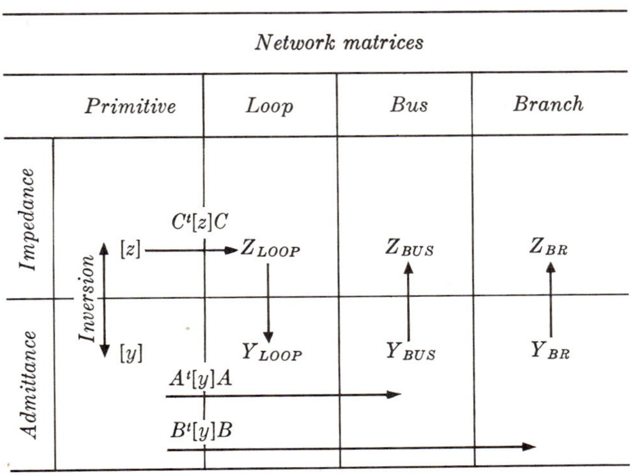

Table 3.2 Current and voltage relations between primitive and interconnected networks

	Frame of reference		
	Loop	Bus	Branch
Current	$\bar{i} = C\bar{I}_{LOOP}$	$\bar{I}_{BUS} = A^t\bar{j}$	$\bar{I}_{BR} = B^t\bar{j}$
Voltage	$\bar{E}_{LOOP} = C^t\bar{e}$	$\bar{v} = A\bar{E}_{BUS}$	$\bar{v} = B\bar{E}_{BR}$

The loop admittance matrix can be obtained from

$$Y_{LOOP} = Z_{LOOP}^{-1} = (C^t[z]C)^{-1}$$

The singular transformations for obtaining network matrices are summarized in Table 3.1. The current and voltage relations between the primitive and interconnected networks are summarized in Table 3.2.

3.6 Formation of network matrices by nonsingular transformations

Branch admittance and branch impedance matrices

The branch admittance matrix Y_{BR} can be obtained also by using the augmented cut-set incidence matrix $\hat{B}$ to relate the variables and parameters of the primitive network to those of an augmented interconnected network. The augmented network is obtained by connecting a fictitious branch in series with each link of the original network. In order to preserve the performance of the interconnected network the admittance of each fictitious branch is set to zero and its current source is set equal to the current through the associated link, as shown in Fig. 3.8*a*. The voltage across a fictitious branch is zero. Then a tie cut-set can be interpreted as a cut-set containing a link and a fictitious branch, as shown in Fig. 3.8*b*.

The performance equation of the augmented network in the branch frame of reference is

$$\hat{I}_{BR} = \hat{Y}_{BR}\hat{E}_{BR}$$

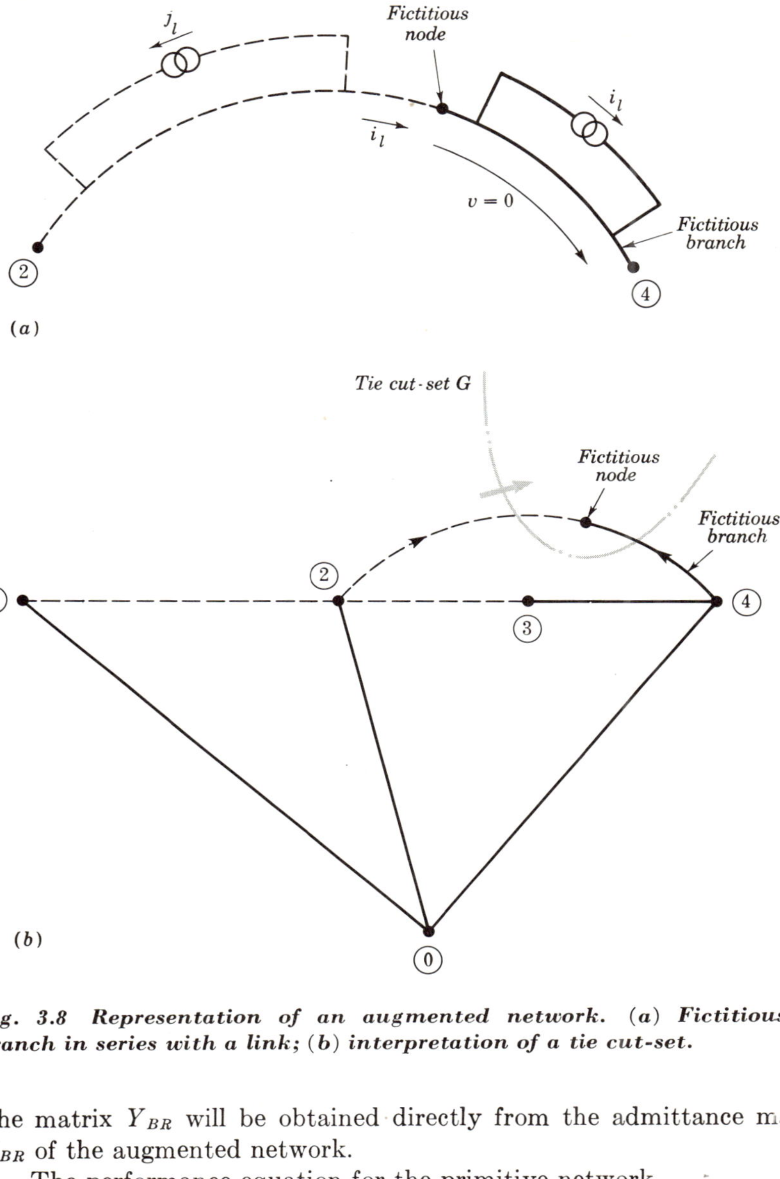

***Fig. 3.8** Representation of an augmented network. (a) Fictitious branch in series with a link; (b) interpretation of a tie cut-set.*

The matrix Y_{BR} will be obtained directly from the admittance matrix $\hat{Y}_{BR}$ of the augmented network.

The performance equation for the primitive network

$$\bar{\imath} + \bar{\jmath} = [y]\bar{v}$$

is premultiplied by $\hat{B}^t$ to obtain

$$\hat{B}^t\bar{\imath} + \hat{B}^t\bar{\jmath} = \hat{B}^t[y]\bar{v} \tag{3.6.1}$$

Equation (3.6.1) can be written in the partitioned matrix form:

$$\left[\begin{array}{c|c} U_b & B_l^t \\ \hline 0 & U_l \end{array}\right]\left[\begin{array}{c} i_b \\ \hline i_l \end{array}\right] + \left[\begin{array}{c|c} U_b & B_l^t \\ \hline 0 & U_l \end{array}\right]\left[\begin{array}{c} j_b \\ \hline j_l \end{array}\right]$$

$$= \left[\begin{array}{c|c} U_b & B_l^t \\ \hline 0 & U_l \end{array}\right]\left[\begin{array}{c} y \end{array}\right]\left[\begin{array}{c} v \end{array}\right] \qquad (3.6.2)$$

where the primitive current vectors $\bar{\imath}$ and $\bar{\jmath}$ are partitioned into the current vectors $\bar{\imath}_b$ and $\bar{\jmath}_b$, which are associated with branches of the network, and the current vectors $\bar{\imath}_l$ and $\bar{\jmath}_l$, which are associated with links. The left side of equation (3.6.2) is

$$\left[\begin{array}{c} i_b + B_l^t i_l \\ \hline i_l \end{array}\right] + \left[\begin{array}{c} j_b + B_l^t j_l \\ \hline j_l \end{array}\right]$$

where

$$\bar{\imath}_b + B_l^t \bar{\imath}_l = B^t \bar{\imath} \qquad \text{and} \qquad \bar{\jmath}_b + B_l^t \bar{\jmath}_l = B^t \bar{\jmath}$$

However

$$B^t \bar{\imath} = 0 \qquad \text{and} \qquad B^t \bar{\jmath} = \bar{I}_{BR}$$

Then the left side of equation (3.6.2) is

$$\left[\begin{array}{c} 0 \\ \hline i_l \end{array}\right] + \left[\begin{array}{c} I_{BR} \\ \hline j_l \end{array}\right] = \left[\begin{array}{c} I_{BR} \\ \hline i_l + j_l \end{array}\right]$$

Since each element of $\bar{\imath}_l$ is equal to a current source of a fictitious branch, $\bar{\imath}_l + \bar{\jmath}_l$ is a vector in which each element is equal to the algebraic sum of the source currents of a fictitious branch and its associated link. Therefore,

$$\hat{I}_{BR} = \left[\begin{array}{c} I_{BR} \\ \hline i_l + j_l \end{array}\right]$$

and equation (3.6.1) becomes

$$\hat{I}_{BR} = \hat{B}^t[y]\bar{v} \tag{3.6.3}$$

Since the voltages across the fictitious branches are zero, the voltage vector of the augmented network is

$$\hat{E}_{BR} = \begin{bmatrix} E_{BR} \\ \hline 0 \end{bmatrix}$$

The voltages across the elements of the original network from equation (3.5.14) are

$$\bar{v} = B\bar{E}_{BR}$$

However

$$B\bar{E}_{BR} = \hat{B}\hat{E}_{BR}$$

then

$$\bar{v} = \hat{B}\hat{E}_{BR} \tag{3.6.4}$$

Substituting from equation (3.6.4) into equation (3.6.3)

$$\hat{I}_{BR} = \hat{B}^t[y]\hat{B}\hat{E}_{BR} \tag{3.6.5}$$

Since the performance equation of the augmented network is

$$\hat{I}_{BR} = \hat{Y}_{BR}\hat{E}_{BR} \tag{3.6.6}$$

it follows from equations (3.6.5) and (3.6.6) that the admittance matrix of the augmented network is

$$\hat{Y}_{BR} = \hat{B}^t[y]\hat{B} \tag{3.6.7}$$

Equation (3.6.7) can be written in the partitioned form

$$\left[\begin{array}{c|c} Y_1 & Y_2 \\ \hline Y_3 & Y_4 \end{array}\right] = \left[\begin{array}{c|c} U_b & B_l^t \\ \hline 0 & U_l \end{array}\right] \left[\begin{array}{c|c} y_{bb} & y_{bl} \\ \hline y_{lb} & y_{ll} \end{array}\right] \left[\begin{array}{c|c} U_b & 0 \\ \hline B_l & U_l \end{array}\right] \tag{3.6.8}$$

where $[y_{bb}]$ = primitive admittance matrix of branches

$[y_{bl}] = [y_{lb}]^t$ = primitive admittance matrix whose elements are the mutual admittances between branches and links

$[y_{ll}]$ = primitive admittance matrix of links

It follows from equation (3.6.8) that

$$Y_1 = [y_{bb}] + B_l^t[y_{lb}] + [y_{bl}]B_l + B_l^t[y_{ll}]B_l \tag{3.6.9}$$

Since

$$Y_{BR} = B^t[y]B$$

or

$$Y_{BR} = \begin{bmatrix} U_b & B_l^t \end{bmatrix} \begin{bmatrix} y_{bb} & y_{bl} \\ y_{lb} & y_{ll} \end{bmatrix} \begin{bmatrix} U_b \\ B_l \end{bmatrix}$$

then

$$Y_{BR} = [y_{bb}] + B_l^t[y_{lb}] + [y_{bl}]B_l + B_l^t[y_{ll}]B_l \qquad (3.6.10)$$

From equations (3.6.9) and (3.6.10), therefore,

$$Y_{BR} = Y_1$$

The branch impedance matrix can be obtained from

$$Z_{BR} = Y_1^{-1}$$

Loop impedance and loop admittance matrices

The loop impedance matrix Z_{LOOP} can be obtained also by using the augmented loop incidence matrix $\hat{C}$ to relate the variables and parameters of the primitive network to those of an augmented interconnected network. The augmented network is obtained by connecting a fictitious link in parallel with each branch of the original network. In order to preserve the performance of the interconnected network the impedance of each fictitious link is set to zero and its voltage source is set equal and opposite to the voltage across the associated branch, as shown in Fig. 3.9*a*. The current through a fictitious link is zero. Then an open loop can be interpreted as a loop containing a branch and a fictitious link as shown in Fig. 3.9*b*.

The performance equation of the augmented network in the loop frame of reference is

$$\hat{E}_{LOOP} = \hat{Z}_{LOOP}\hat{I}_{LOOP}$$

The matrix Z_{LOOP} will be obtained directly from the impedance matrix $\hat{Z}_{LOOP}$ of the augmented network.

The performance equation for the primitive network

$$\bar{v} + \bar{e} = [z]\bar{\imath}$$

is premultiplied by $\hat{C}^t$ to obtain

$$\hat{C}^t\bar{v} + \hat{C}^t\bar{e} = \hat{C}^t[z]\bar{\imath} \qquad (3.6.11)$$

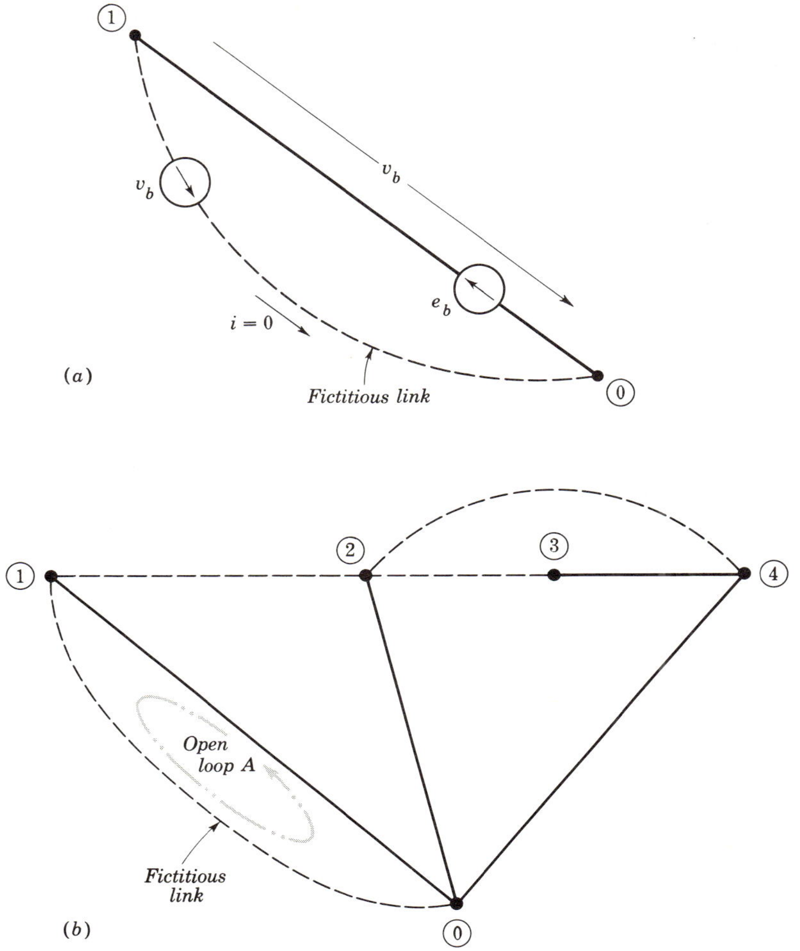

Fig. 3.9 Representation of an augmented network. **(*a*) *Fictitious link in parallel with a branch;* (*b*) *interpretation of an open loop.***

Equation (3.6.11) can be written in the partitioned form

$$\left[\begin{array}{c|c} U_b & 0 \\ \hline C_b{}^t & U_l \end{array}\right] \left[\begin{array}{c} v_b \\ \hline v_l \end{array}\right] + \left[\begin{array}{c|c} U_b & 0 \\ \hline C_b{}^t & U_l \end{array}\right] \left[\begin{array}{c} e_b \\ \hline e_l \end{array}\right] = \left[\begin{array}{c|c} U_b & 0 \\ \hline C_b{}^t & U_l \end{array}\right] \left[\begin{array}{c} z \end{array}\right] \left[\begin{array}{c} i \end{array}\right] \tag{3.6.12}$$

where the primitive voltage vectors $\bar{v}$ and $\bar{e}$ are partitioned into the voltage vectors $\bar{v}_b$ and $\bar{e}_b$, which are associated with the branches of the network, and the voltage vectors $\bar{v}_l$ and $\bar{e}_l$, which are associated with the links. The left side of equation (3.6.12) is

$$\left[\begin{array}{c} v_b \\ \hline C_b{}^t v_b + v_l \end{array}\right] + \left[\begin{array}{c} e_b \\ \hline C_b{}^t e_b + e_l \end{array}\right]$$

where

$$C_b{}^t\bar{v}_b + \bar{v}_l = C^t\bar{v} \qquad \text{and} \qquad C_b{}^t\bar{e}_b + \bar{e}_l = C^t\bar{e}$$

However

$$C^t\bar{v} = 0 \qquad \text{and} \qquad C^t\bar{e} = \bar{E}_{LOOP}$$

The left side of equation (3.6.12) is then

$$\left[\begin{array}{c} v_b \\ \hline 0 \end{array}\right] + \left[\begin{array}{c} e_b \\ \hline E_{LOOP} \end{array}\right] = \left[\begin{array}{c} v_b + e_b \\ \hline E_{LOOP} \end{array}\right]$$

Since each element of $\bar{v}_b$ is equal to a voltage source of a fictitious link, $\bar{v}_b + \bar{e}_b$ is a vector in which each element is equal to the algebraic sum of the source voltages in an open loop. Therefore,

$$\hat{E}_{LOOP} = \left[\begin{array}{c} v_b + e_b \\ \hline E_{LOOP} \end{array}\right] \tag{3.6.13}$$

and from equations (3.6.11) and (3.6.13)

$$\hat{E}_{LOOP} = \hat{C}^t[z]\bar{\imath} \tag{3.6.14}$$

Since the currents in the open loops are zero, the current vector of the augmented network is

$$\hat{I}_{LOOP} = \left[\begin{array}{c} 0 \\ \hline I_{LOOP} \end{array}\right]$$

The currents through the elements of the original network from equation (3.5.20) are

$$\bar{\imath} = C\bar{I}_{LOOP}$$

However,

$$C\bar{I}_{LOOP} = \hat{C}\hat{I}_{LOOP}$$

then

$$\bar{i} = \hat{C}\hat{I}_{LOOP} \qquad (3.6.15)$$

Substituting from equation (3.6.15) into equation (3.6.14),

$$\hat{E}_{LOOP} = \hat{C}^t[z]\hat{C}\hat{I}_{LOOP} \qquad (3.6.16)$$

Since the performance equation of the augmented network is

$$\hat{E}_{LOOP} = \hat{Z}_{LOOP}\hat{I}_{LOOP} \qquad (3.6.17)$$

it follows from equations (3.6.16) and (3.6.17) that the impedance matrix of the augmented network is

$$\hat{Z}_{LOOP} = \hat{C}^t[z]\hat{C} \qquad (3.6.18)$$

Equation (3.6.18) can be written in the partitioned form

$$\begin{bmatrix} Z_1 & Z_2 \\ Z_3 & Z_4 \end{bmatrix} = \begin{bmatrix} U_b & 0 \\ C_b^t & U_l \end{bmatrix} \begin{bmatrix} z_{bb} & z_{bl} \\ z_{lb} & z_{ll} \end{bmatrix} \begin{bmatrix} U_b & C_b \\ 0 & U_l \end{bmatrix} \qquad (3.6.19)$$

where $[z_{bb}]$ = primitive impedance matrix of branches
$[z_{bl}] = [z_{lb}]^t$ = primitive impedance matrix whose elements are the mutual impedances between branches and links
$[z_{ll}]$ = primitive impedance matrix of links

It follows from equation (3.6.19) that

$$Z_4 = C_b^t[z_{bb}]C_b + [z_{lb}]C_b + C_b^t[z_{bl}] + [z_{ll}] \qquad (3.6.20)$$

Since

$$Z_{LOOP} = C^t[z]C$$

or

$$Z_{LOOP} = \begin{bmatrix} C_b^t & U_l \end{bmatrix} \begin{bmatrix} z_{bb} & z_{bl} \\ z_{lb} & z_{ll} \end{bmatrix} \begin{bmatrix} C_b \\ U_l \end{bmatrix}$$

then

$$Z_{LOOP} = C_b^t[z_{bb}]C_b + [z_{lb}]C_b + C_b^t[z_{bl}] + [z_{ll}] \qquad (3.6.21)$$

From equations (3.6.20) and (3.6.21), therefore,

$$Z_{LOOP} = Z_4$$

The loop admittance matrix can be obtained from

$$Y_{LOOP} = Z_4^{-1}$$

Derivation of loop admittance matrix from augmented network admittance matrix

The loop admittance matrix Y_{LOOP} can be obtained from the augmented admittance matrix $\hat{Y}_{BR}$. From equations (3.6.7) and (3.6.18),

$$\hat{Z}_{LOOP}\hat{Y}_{BR} = \hat{C}^t[z]\hat{C}\hat{B}^t[y]\hat{B} \tag{3.6.22}$$

In partitioned form,

$$\hat{C}\hat{B}^t = \left[\begin{array}{c|c} U_b & C_b \\ \hline 0 & U_l \end{array}\right]\left[\begin{array}{c|c} U_b & B_l^t \\ \hline 0 & U_l \end{array}\right] = \left[\begin{array}{c|c} U_b & B_l^t + C_b \\ \hline 0 & U_l \end{array}\right] \tag{3.6.23}$$

The currents through the elements of the primitive network from equation (3.5.20) are

$$\bar{\imath} = C\bar{I}_{LOOP}$$

Premultiplying by B^t,

$$B^t\bar{\imath} = B^tC\bar{I}_{LOOP} \tag{3.6.24}$$

However, from equation (3.5.11) the left side of equation (3.6.24) is zero. Therefore, equation (3.6.24) can be written

$$(C_b + B_l^t)\bar{I}_{LOOP} = 0$$

It follows that

$$C_b = -B_l^t \tag{3.6.25}$$

Substituting from equation (3.6.25) into equation (3.6.23),

$$\hat{C}\hat{B}^t = U \tag{3.6.26}$$

In a similar manner it can be shown that

$$\hat{C}^t\hat{B} = U \tag{3.6.27}$$

Substituting from equation (3.6.26) into (3.6.22),

$$\hat{Z}_{LOOP}\hat{Y}_{BR} = \hat{C}^t[z][y]\hat{B}$$

Since

$$[z][y] = U$$

then

$$\hat{Z}_{LOOP}\hat{Y}_{BR} = \hat{C}^t\hat{B}$$

Therefore, from equation (3.6.27),

$$\hat{Z}_{LOOP}\hat{Y}_{BR} = U \tag{3.6.28}$$

Equation (3.6.28) in partitioned form is

$$\left[\begin{array}{c|c} Z_1 & Z_2 \\ \hline Z_3 & Z_4 \end{array}\right] \left[\begin{array}{c|c} Y_1 & Y_2 \\ \hline Y_3 & Y_4 \end{array}\right] = \left[\begin{array}{c|c} U_b & 0 \\ \hline 0 & U_l \end{array}\right]$$

It follows that

$$Z_1Y_1 + Z_2Y_3 = U_b \tag{3.6.29}$$
$$Z_1Y_2 + Z_2Y_4 = 0$$
$$Z_3Y_1 + Z_4Y_3 = 0 \tag{3.6.30}$$
$$Z_3Y_2 + Z_4Y_4 = U_l \tag{3.6.31}$$

Solving for Z_3 from equation (3.6.30),

$$Z_3 = -Z_4Y_3Y_1^{-1}$$

and substituting into equation (3.6.31),

$$-Z_4Y_3Y_1^{-1}Y_2 + Z_4Y_4 = U_l$$

or

$$Z_4(Y_4 - Y_3Y_1^{-1}Y_2) = U_l$$

Since

$$Z_4Y_{LOOP} = U_l$$

it follows that

$$Y_{LOOP} = Y_4 - Y_3Y_1^{-1}Y_2$$

Derivation of branch impedance matrix from augmented impedance matrix

The branch impedance matrix Z_{BR} can be obtained from the augmented impedance matrix $\hat{Z}_{LOOP}$. Combining equations (3.6.29) and (3.6.30) yields

$$(Z_1 - Z_2Z_4^{-1}Z_3)Y_1 = U_b$$

Since

$$Z_{BR}Y_1 = U_b$$

it follows that

$$Z_{BR} = Z_1 - Z_2Z_4^{-1}Z_3$$

Derivation of branch admittance and impedance matrices from bus admittance and impedance matrices

Using the branch-path incidence matrix K the branch admittance matrix Y_{BR} can be obtained from Y_{BUS}. From equation (3.3.1),

$$A_bK^t = U_b$$

and from equation (3.3.3),

$$B_l = A_lK^t$$

Postmultiplying A by K^t,

$$AK^t = \begin{bmatrix} A_b \\ \hline A_l \end{bmatrix} K^t = \begin{bmatrix} A_bK^t \\ \hline A_lK^t \end{bmatrix} \tag{3.6.32}$$

Substituting from equations (3.3.1) and (3.3.3) into (3.6.32),

$$AK^t = \begin{bmatrix} U_b \\ \hline B_l \end{bmatrix} = B$$

Transposing,

$$KA^t = B^t$$

Postmultiplying by $[y]AK^t$ yields

$$KA^t[y]AK^t = B^t[y]AK^t$$

or

$$K(A^t[y]A)K^t = B^t[y]B \tag{3.6.33}$$

From the singular transformations,

$$Y_{BUS} = A^t[y]A \qquad \text{and} \qquad Y_{BR} = B^t[y]B$$

Hence equation (3.6.33) becomes

$$Y_{BR} = KY_{BUS}K^t \tag{3.6.34}$$

The branch impedance matrix is

$$Z_{BR} = Y_{BR}^{-1} = (K^t)^{-1}Y_{BUS}^{-1}K^{-1} \tag{3.6.35}$$

From equation (3.3.2),

$$K^t = A_b^{-1} \tag{3.6.36}$$

Substituting from equation (3.6.36) into equation (3.6.35),

$$Z_{BR} = A_b Z_{BUS} A_b^t$$

Derivation of bus admittance and impedance matrices from branch admittance and impedance matrices

Equation (3.6.34) is premultiplied by K^{-1} and postmultiplied by $(K^t)^{-1}$ to obtain

$$K^{-1}Y_{BR}(K^t)^{-1} = Y_{BUS} \tag{3.6.37}$$

Substituting from equation (3.6.36) into equation (3.6.37),

$$Y_{BUS} = A_b^t Y_{BR} A_b$$

Since

$$Z_{BUS} = Y_{BUS}^{-1}$$

then

$$Z_{BUS} = (A_b^t Y_{BR} A_b)^{-1} \qquad \text{or} \qquad Z_{BUS} = K^t Z_{BR} K$$

The nonsingular transformations for obtaining network matrices are summarized in Table 3.3.

3.7 Example of formation of incidence and network matrices

The method of forming the incidence and network matrices will be illustrated for the network shown in Fig. 3.10. The incidence matrices for a given network are not unique and depend on the orientation of the graph and the selection of branches, basic cut-sets, and basic loops. However, the network matrices are unique.

Table 3.3 Formation of network matrices by nonsingular transformations

		Network matrices				
		Primitive	Augmented	Loop	Bus	Branch
Impedance		$[z] \xrightarrow{\hat{C}^t[z]\hat{C}}$	$\hat{Z}_{LOOP} = \begin{bmatrix} Z_1 & Z_2 \\ Z_3 & Z_4 \end{bmatrix}$; $Z_1 - Z_2Z_4^{-1}Z_3 \rightarrow$ Branch; $Z_4 = Z_{LOOP} \rightarrow$ Loop	Z_{LOOP}	Z_{BUS}; $A_bZ_{BUS}A_b^t \rightarrow$ Branch; $\leftarrow K^tZ_{BR}K$	Z_{BR}
	Inversion $\updownarrow$		$\updownarrow$	$\updownarrow$	$\updownarrow$	$\updownarrow$
Admittance		$[y] \xrightarrow{\hat{B}^t[y]\hat{B}}$	$\hat{Y}_{BR} = \begin{bmatrix} Y_1 & Y_2 \\ Y_3 & Y_4 \end{bmatrix}$; $Y_4 - Y_3Y_1^{-1}Y_2 \rightarrow$ Loop; $Y_1 = Y_{BR} \rightarrow$ Branch	Y_{LOOP}	Y_{BUS}; $KY_{BUS}K^t \rightarrow$ Branch; $\leftarrow A_b^tY_{BR}A_b$	Y_{BR}

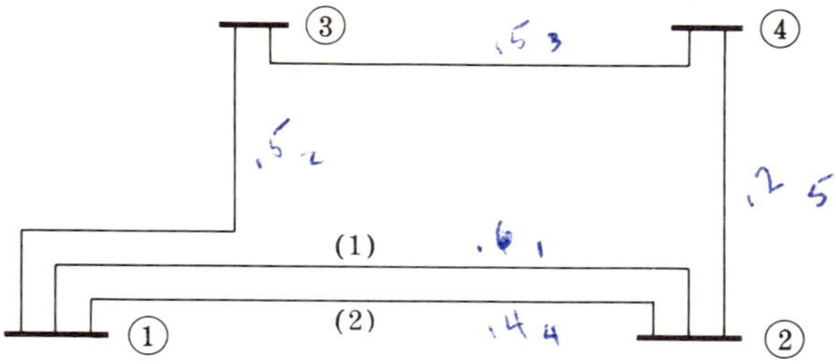

Fig. 3.10 Sample network.

Problem

a. Form the incidence matrices $\hat{A}$, A, K, B, $\hat{B}$, C, and $\hat{C}$ for the network shown in Fig. 3.10.
b. Form the network matrices Y_{BUS}, Y_{BR}, and Z_{LOOP} by singular transformations.
c. Form the network matrices Z_{LOOP}, Z_{BR}, and Z_{BUS} by nonsingular transformations.

Solution

The impedance data for the sample network is given in Table 3.4.

Table 3.4 Impedances for sample network

	Self		Mutual	
Element number	Bus code p–q	Impedance $z_{pq,pq}$	Bus code r–s	Impedance $z_{pq,rs}$
1	1–2(1)	0.6		
2	1–3	0.5	1–2(1)	0.1
3	3–4	0.5		
4	1–2(2)	0.4	1–2(1)	0.2
5	2–4	0.2		

The network contains four nodes and five elements, that is, $n = 4$ and $e = 5$. The .umber of branches is

$$b = n - 1 = 3$$

and the number of basic loops is

$$l = e - n + 1 = 2$$

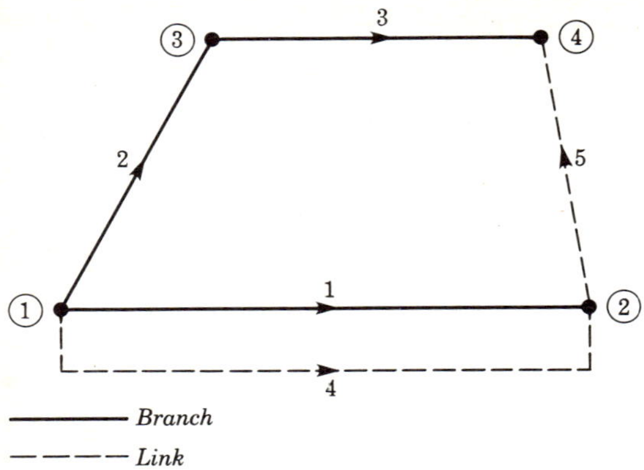

***Fig. 3.11** Tree and cotree of the oriented connected graph of sample network.*

a. The branches and links of the oriented connected graph of the network are shown in Fig. 3.11. The element-node incidence matrix is

$\hat{A} =$

e \ n	①	②	③	④
1	1	−1		
2	1		−1	
3			1	−1
4	1	−1		
5		1		−1

Selecting node 1 as the reference, the bus incidence matrix is

$A =$

e \ bus	②	③	④
1	−1		
2		−1	
3		1	−1
4	−1		
5	1		−1

The branch-path incidence matrix is

$K =$

b \ path	②	③	④
1	−1		
2		−1	−1
3			−1

The basic and tie cut-sets of the oriented connected graph of the network are shown in Fig. 3.12. The basic cut-set incidence matrix is

$B =$

e \ b	A	B	C
1	1		
2		1	
3			1
4	1		
5	−1	1	1

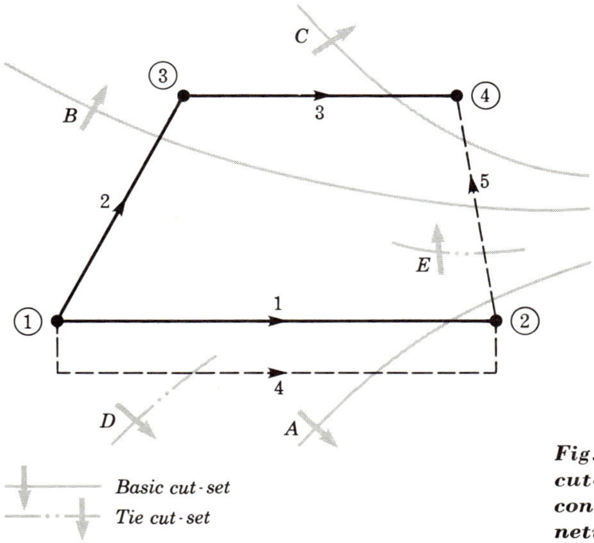

Fig. 3.12 Basic and tie cut-sets of the oriented connected graph of sample network.

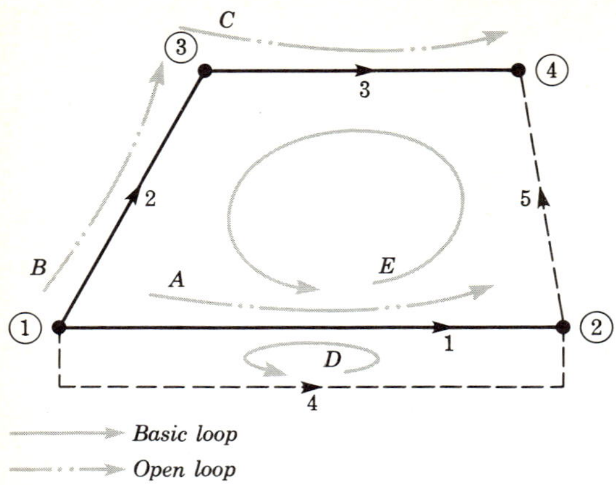

Fig. 3.13 Basic and open loops of the oriented connected graph of sample network.

The augmented cut-set incidence matrix is

$\hat{B} =$

e \ e	A	B	C	D	E
1	1				
2		1			
3			1		
4	1			1	
5	−1	1	1		1

The basic and open loops of the oriented connected graph are shown in Fig. 3.13. The basic loop incidence matrix is

$C =$

e \ l	D	E
1	−1	1
2		−1
3		−1
4	1	
5		1

The augmented loop incidence matrix is

$\hat{C} =$

e \ e	A	B	C	D	E
1	1			−1	1
2		1			−1
3			1		−1
4				1	
5					1

b. The primitive impedance matrix of the sample network from Table 3.4 is

$[z] =$

e \ e	1	2	3	4	5
1	0.6	0.1		0.2	
2	0.1	0.5			
3			0.5		
4	0.2			0.4	
5					0.2

By inversion, the primitive admittance matrix is

$[y] =$

e \ e	1	2	3	4	5
1	2.083	−0.417		−1.042	
2	−0.417	2.083		0.208	
3			2.000		
4	−1.042	0.208		3.021	
5					5.000

The bus admittance matrix obtained by a singular transformation is

$Y_{BUS} = A^t[y]A$

	1	2	3	4	5
②	−1			−1	1
③		−1	1		
④			−1		−1

	1	2	3	4	5
1	2.083	−0.417		−1.042	
2	−0.417	2.083		0.208	
3			2.000		
4	−1.042	0.208		3.021	
5					5.000

	②	③	④
1	−1		
2		−1	
3		1	−1
4	−1		
5	1		−1

=

	1	2	3	4	5
②	−1.041	0.209		−1.979	5.000
③	0.417	−2.083	2.000	−0.208	
④			−2.000		−5.000

	②	③	④
1	−1		
2		−1	
3		1	−1
4	−1		
5	1		−1

=

	②	③	④
②	8.020	−0.209	−5.000
③	−0.209	4.083	−2.000
④	−5.000	−2.000	7.000

The loop impedance matrix obtained by a singular transformation is

$Z_{LOOP} = C^t[z]C$

=

	1	2	3	4	5
D	−1			1	
E	1	−1	−1		1

	1	2	3	4	5
1	0.6	0.1		0.2	
2	0.1	0.5			
3			0.5		
4	0.2			0.4	
5					0.2

	D	E
1	−1	1
2		−1
3		−1
4	1	
5		1

=

	1	2	3	4	5
D	−0.4	−0.1		0.2	
E	0.5	−0.4	−0.5	0.2	0.2

	D	E
1	−1	1
2		−1
3		−1
4	1	
5		1

=

	D	E
D	0.6	−0.3
E	−0.3	1.6

The branch admittance matrix obtained by a singular transformation is

$Y_{BR} = B^t[y]B$

	1	2	3	4	5
A	1			1	−1
B		1			1
C			1		1

	1	2	3	4	5
1	2.083	−0.417		−1.042	
2	−0.417	2.083		0.208	
3			2.000		
4	−1.042	0.208		3.021	
5					5.000

	A	B	C
1	1		
2		1	
3			1
4	1		
5	−1	1	1

$=$

	1	2	3	4	5
A	1.041	−0.209		1.979	−5.000
B	−0.417	2.083		0.208	5.000
C			2.000		5.000

	A	B	C
1	1		
2		1	
3			1
4	1		
5	−1	1	1

$=$

	A	B	C
A	8.020	−5.209	−5.000
B	−5.209	7.083	5.000
C	−5.000	5.000	7.000

c. The augmented network impedance matrix obtained by nonsingular transformation is

$$\hat{Z}_{LOOP} = \hat{C}^t[z]\hat{C} = \begin{array}{|c|c|} \hline Z_1 & Z_2 \\ \hline Z_3 & Z_4 \\ \hline \end{array}$$

=

	1	2	3	4	5
A	1				
B		1			
C			1		
D	−1			1	
E	1	−1	−1		1

	1	2	3	4	5
1	.6	.1		.2	
2	.1	.5			
3			.5		
4	.2			.4	
5					.2

	A	*B*	*C*	*D*	*E*
1	1			−1	1
2		1			−1
3			1		−1
4				1	
5					1

	1	2	3	4	5
A	.6	.1		.2	
B	.1	.5			
= C			.5		
D	−.4	−.1		.2	
E	.5	−.4	−.5	.2	.2

	A	B	C	D	E
1	1			−1	1
2		1			−1
3			1		−1
4				1	
5					1

	A	B	C	D	E
A	.6	.1		−.4	.5
B	.1	.5		−.1	−.4
= C			.5		−.5
D	−.4	−.1		.6	−.3
E	.5	−.4	−.5	−.3	1.6

Then the loop impedance matrix is

$Z_{LOOP} = Z_4 =$

	D	E
D	0.6	−0.3
E	−0.3	1.6

The branch impedance matrix is

$$Z_{BR} = Z_1 - Z_2 Z_4^{-1} Z_3$$

$$= \begin{bmatrix} 0.6 & 0.1 & \\ 0.1 & 0.5 & \\ & & 0.5 \end{bmatrix} - \begin{bmatrix} -0.4 & 0.5 \\ -0.1 & -0.4 \\ & -0.5 \end{bmatrix} \begin{bmatrix} 1.839 & 0.345 \\ 0.345 & 0.690 \end{bmatrix} \begin{bmatrix} -0.4 & -0.1 & \\ 0.5 & -0.4 & -0.5 \end{bmatrix}$$

$$= \begin{bmatrix} 0.6 & 0.1 & \\ 0.1 & 0.5 & \\ & & 0.5 \end{bmatrix} - \begin{bmatrix} 0.329 & -0.026 & -0.104 \\ -0.026 & 0.156 & 0.155 \\ -0.104 & 0.155 & 0.172 \end{bmatrix} = \begin{array}{c} \\ 1 \\ 2 \\ 3 \end{array} \begin{array}{c} \begin{array}{ccc} 1 & 2 & 3 \end{array} \\ \begin{bmatrix} 0.271 & 0.126 & 0.104 \\ 0.126 & 0.344 & -0.155 \\ 0.104 & -0.155 & 0.328 \end{bmatrix} \end{array}$$

The bus impedance matrix obtained by a nonsingular transformation is

$Z_{BUS} = K^t Z_{BR} K$

	1	2	3
②	−1		
③		−1	
④		−1	−1

	1	2	3
1	0.271	0.126	0.104
2	0.126	0.344	−0.155
3	0.104	−0.155	0.328

	②	③	④
1	−1		
2		−1	−1
3			−1

=

	A	B	C
②	−0.271	−0.126	−0.104
③	−0.126	−0.344	0.155
④	−0.230	−0.189	−0.173

	②	③	④
1	−1		
2		−1	−1
3			−1

=

	②	③	④
②	0.271	0.126	0.230
③	0.126	0.344	0.189
④	0.230	0.189	0.362

Problems

3.1 Select for the sample network shown in Fig. 3.10 a different tree than that used in the example. Retain node 1 as the reference and form:

a. The incidence matrices $\hat{A}$, A, K, B, $\hat{B}$, C, and $\hat{C}$ and verify the following relations:

i. $A_b K^t = U$

ii. $B_l = A_l K^t$

iii. $C_b = -B_l^t$

iv. $\hat{C}\hat{B}^t = U$

b. The network matrices Y_{BUS}, Y_{BR}, and Z_{LOOP} by singular transformations

c. The network matrices Z_{LOOP}, Z_{BR}, and Z_{BUS} by nonsingular transformations

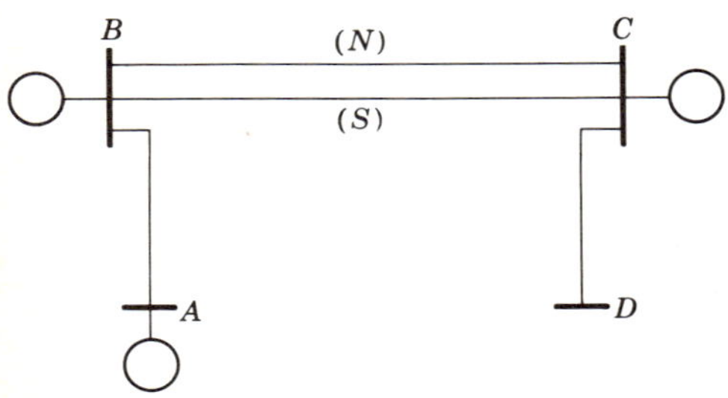

***Fig. 3.14** Sample power system for Prob. 3.2.*

3.2 The positive and zero sequence impedance data for the sample power system shown in Fig. 3.14 is given in Table 3.5. For this system:

a. Draw the positive sequence diagram and an oriented connected graph.

***Table 3.5** Positive and zero sequence impedance data of sample power system for Prob. 3.2*

Element	*Positive sequence impedance*	*Zero sequence impedance*	*Element*	*Mutual impedance*
Generator A	$0.0 + j0.25$	$0.0 + j0.1$		
Generator B	$0.0 + j0.25$	$0.0 + j0.1$		
Generator C	$0.0 + j0.25$	$0.0 + j0.1$		
Line A–B	$0.03 + j0.13$	$0.08 + j0.45$		
Line B–$C(N)$	$0.05 + j0.22$	$0.13 + j0.75$	Line B–$C(S)$	$0.08 + j0.48$
Line B–$C(S)$	$0.05 + j0.22$	$0.13 + j0.75$		
Line C–D	$0.02 + j0.11$	$0.07 + j0.37$		

b. Selecting ground as reference, form the incidence matrices A, K, B, $\hat{B}$, C, and $\hat{C}$ and verify the relations:

i. $A_b K^t = U$

ii. $B_l = A_l K^t$

iii. $C_b = -B_l^t$

iv. $\hat{C}\hat{B}^t = U$

c. Neglecting resistance, form the positive sequence network matrices Y_{BUS}, Z_{BUS}, Y_{BR}, Z_{BR}, Z_{LOOP}, and Y_{LOOP} by singular transformations.

d. Neglecting resistance, form the zero sequence network matrices Y_{BUS}, Z_{BUS}, Y_{BR}, Z_{BR}, Z_{LOOP}, and Y_{LOOP} by singular transformations.

e. Repeat *c* and *d* using nonsingular transformations.

f. Repeat *c* including resistance.

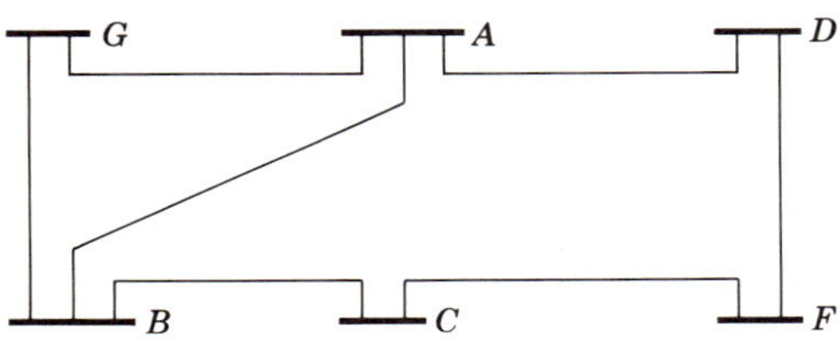

Fig. 3.15 Sample network for Prob. 3.3.

Table 3.6 Positive sequence reactances of sample network for Prob. 3.3

Element	*Positive sequence reactance*
G–A	0.04
G–B	0.05
A–B	0.04
B–C	0.03
A–D	0.02
C–F	0.07
D–F	0.10

3.3 The positive sequence reactances for the network shown in Fig. 3.15 are given in Table 3.6. Designate elements *A–B* and *D–F* as links and node *G* as the reference bus. Form:

a. The incidence matrices $\hat{A}$, A, K, B, $\hat{B}$, C, and $\hat{C}$

b. The network matrices Y_{BUS}, Y_{BR}, and Z_{LOOP} by singular transformations

c. The network matrices Y_{BUS}, Z_{BUS}, Z_{BR}, Z_{LOOP}, and Y_{LOOP} by nonsingular transformations

3.4 Prove that when there is no mutual coupling the diagonal and off-diagonal elements of the bus admittance matrix Y_{BUS} can be computed from

$$Y_{ii} = \sum_{j} y_{ij}$$

$$Y_{ij} = -\, y_{ij}$$

where y_{ij} is the sum of the admittances of all lines connecting buses i and j.

3.5 Using the bus impedance matrix Z_{BUS} computed in Prob. 3.2 and the internal generator voltages given in Table 3.7:

a. Compute the positive and zero sequence bus voltages of the network.

b. Compute the positive and zero sequence currents flowing in the line B–$C(N)$.

Table 3.7 Internal generator voltages for Prob. 3.5

	Internal per unit voltages	
Generator	*Positive sequence*	*Zero sequence*
A	$1.0\angle 0°$	0
B	$1.1\angle -10°$	0
C	$1.0\angle -10°$	$0.1\angle 0°$

3.6 Using the relations between interconnected and primitive network variables prove the following:

a. $A_b K^t = U$

b. $B_l = A_l K^t$

Bibliography

Byerly, R. T., R. W. Long, and C. W. King: Logic for Applying Topological Methods to Electric Networks, *Trans. AIEE*, vol. 77, pt. I, pp. 657–667, 1958.

Edelmann, H.: Numerical and Algebraic Generation of Mesh-impedance Matrices by Set-theoretical Intersection on a Digital Computer,

Trans. IEEE on Power Apparatus and Systems, vol. 83, pp. 397–402, 1964.

Guillemin, Ernst A.: "Introductory Circuit Theory," John Wiley & Sons, Inc., New York, 1953.

Hale, H. W.: A Logic for Identifying the Trees of a Graph, *Trans. AIEE*, vol. 80, pt. III, pp. 195–198, 1961.

Happ, H. H.: Orthogonal Networks, *Trans. IEEE on Power Apparatus and Systems*, vol. 85, pp. 281–294, 1966.

———: Special Cases of Orthogonal Networks—Tree and Link, *Trans. IEEE on Power Apparatus and Systems*, vol. 85, pp. 880–891, 1966.

Kron, Gabriel: "Tensor Analysis of Networks," John Wiley & Sons, Inc., New York, 1939.

———: "Diakoptics," Macdonald and Co., Ltd., London, 1963.

Le Corbeiller, P.: "Matrix Analysis of Electric Networks," John Wiley & Sons, Inc., New York, 1950.

Sato, N.: Digital Calculation of Network Inverse and Mesh Transformation Matrices, *Trans. AIEE*, vol. 79, pt. III, pp. 719–726, 1960.

Seshu, S., and M. B. Reed: "Linear Graphs and Electrical Networks," Addison-Wesley Publishing Company, Inc., Reading, Mass., 1961.

Tinney, W. F., and C. M. McIntyre: A Digital Method for Obtaining a Loop Connection Matrix, *Trans. AIEE*, vol. 79, pt. III, pp. 740–746, 1960.

chapter 4
Algorithms for formation of network matrices

4.1 Introduction

The methods presented in Secs. 3.5 and 3.6 require transformation and inversion of matrices to obtain network matrices. An alternative method based on an algorithm can be used to form the bus impedance matrix directly from system parameters and coded bus numbers. The underlying principle of the algorithm is the formation of the bus impedance matrix in steps, simulating the construction of the network by adding one element at a time (Brown, Person, Kirchmayer, and Stagg, 1960)†. A matrix is formed for the partial network represented after each element is connected to the network.

In addition, an algorithm is presented for deriving the loop admittance matrix from a given bus impedance matrix.

4.2 Algorithm for formation of bus impedance matrix

Performance equation of a partial network

Assume that the bus impedance matrix Z_{BUS} is known for a partial network of m buses and a reference node 0. The performance equation of this network, shown in Fig. 4.1, is

$$\bar{E}_{BUS} = Z_{BUS}\bar{I}_{BUS}$$

† Names in parentheses refer to the Bibliography at the end of each chapter.

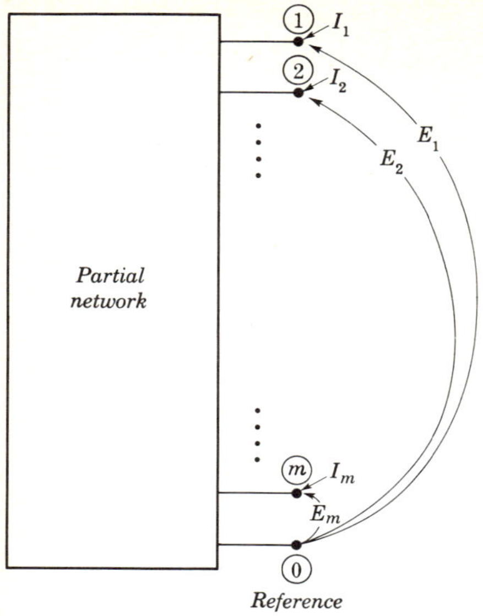

Fig. 4.1 Representation of a partial network.

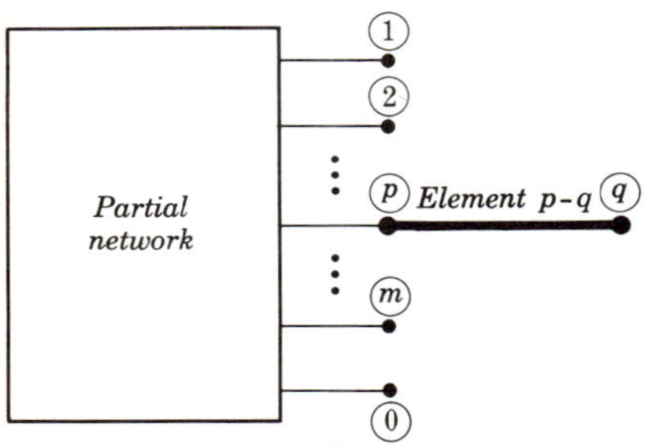

(a)

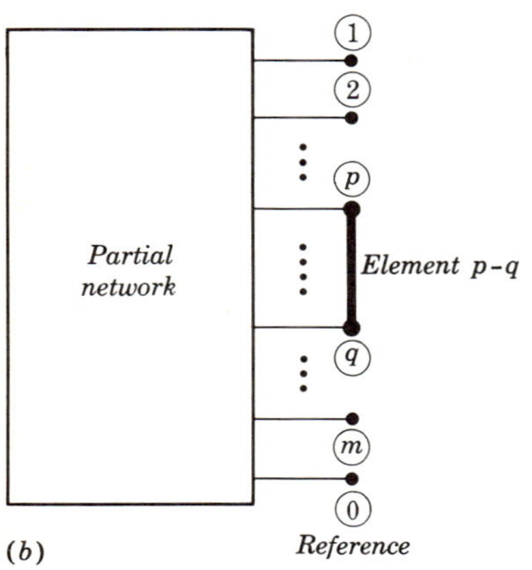

(b)

Fig. 4.2 Representations of a partial network with an added element. (a) Addition of a branch; (b) addition of a link.

where $\bar{E}_{BUS}$ = an $m \times 1$ vector of bus voltages measured with respect to the reference node

$\bar{I}_{BUS}$ = an $m \times 1$ vector of impressed bus currents

When an element p–q is added to the partial network it may be a branch or a link as shown in Fig. 4.2.

If p–q is a branch, a new bus q is added to the partial network and the resultant bus impedance matrix is of dimension $(m + 1) \times (m + 1)$. The new voltage and current vectors are of dimension $(m + 1) \times 1$. To determine the new bus impedance matrix requires only the calculation of the elements in the new row and column.

If p–q is a link, no new bus is added to the partial network. In this case, the dimensions of the matrices in the performance equation are unchanged, but all the elements of the bus impedance matrix must be recalculated to include the effect of the added link.

Addition of a branch

The performance equation for the partial network with an added branch p–q is

$$\begin{array}{c} \\ \left[\begin{array}{c} E_1 \\ E_2 \\ \cdots \\ E_p \\ \cdots \\ E_m \\ \hline E_q \end{array}\right] \end{array} = \begin{array}{c} \\ 1 \\ \\ \\ p \\ \\ m \\ q \end{array} \begin{array}{c} \begin{array}{ccccccc} 1 & & & p & & m & q \end{array} \\ \left[\begin{array}{cccccc|c} Z_{11} & Z_{12} & \cdots & Z_{1p} & \cdots & Z_{1m} & Z_{1q} \\ Z_{21} & Z_{22} & \cdots & Z_{2p} & \cdots & Z_{2m} & Z_{2q} \\ \cdots & \cdots & \cdots & \cdots & \cdots & \cdots & \cdots \\ Z_{p1} & Z_{p2} & \cdots & Z_{pp} & \cdots & Z_{pm} & Z_{pq} \\ \cdots & \cdots & \cdots & \cdots & \cdots & \cdots & \cdots \\ Z_{m1} & Z_{m2} & \cdots & Z_{mp} & \cdots & Z_{mm} & Z_{mq} \\ \hline Z_{q1} & Z_{q2} & \cdots & Z_{qp} & \cdots & Z_{qm} & Z_{qq} \end{array}\right] \end{array} \begin{array}{c} \\ \left[\begin{array}{c} I_1 \\ I_2 \\ \cdots \\ I_p \\ \cdots \\ I_m \\ \hline I_q \end{array}\right] \end{array} \qquad (4.2.1)$$

It is assumed that the network consists of bilateral passive elements. Hence $Z_{qi} = Z_{iq}$ where $i = 1, 2, \ldots, m$ and refers to the buses of the partial network, not including the new bus q. The added branch p–q is assumed to be mutually coupled with one or more elements of the partial network.

The elements Z_{qi} can be determined by injecting a current at the ith bus and calculating the voltage at the qth bus with respect to the reference

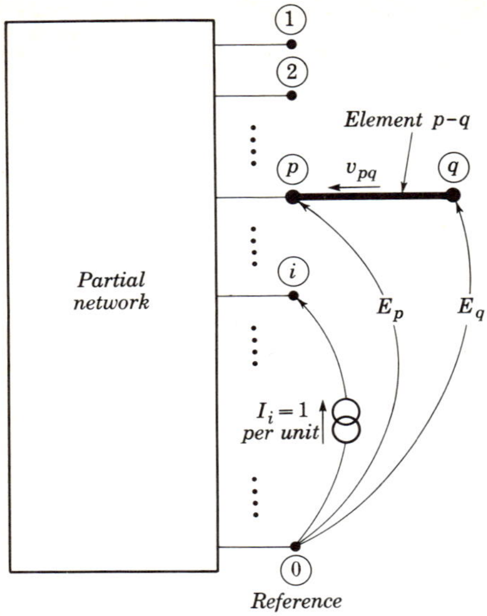

Fig. 4.3 Injected current and bus voltages for calculation of Z_{qi}.

node as shown in Fig. 4.3. Since all other bus currents equal zero, it follows from equation (4.2.1) that

$$\begin{aligned} E_1 &= Z_{1i}I_i \\ E_2 &= Z_{2i}I_i \\ &\cdots\cdots \\ E_p &= Z_{pi}I_i \\ &\cdots\cdots \\ E_m &= Z_{mi}I_i \\ E_q &= Z_{qi}I_i \end{aligned} \tag{4.2.2}$$

Letting $I_i = 1$ per unit in equations (4.2.2), Z_{qi} can be obtained directly by calculating E_q.

The bus voltages associated with the added element and the voltage across the element are related by

$$E_q = E_p - v_{pq} \tag{4.2.3}$$

The currents in the elements of the network in Fig. 4.3 are expressed in terms of the primitive admittances and the voltages across the elements by

$$\begin{bmatrix} i_{pq} \\ \hline i_{\rho\sigma} \end{bmatrix} = \begin{bmatrix} y_{pq,pq} & y_{pq,\rho\sigma} \\ \hline y_{\rho\sigma,pq} & y_{\rho\sigma,\rho\sigma} \end{bmatrix} \begin{bmatrix} v_{pq} \\ \hline v_{\rho\sigma} \end{bmatrix} \tag{4.2.4}$$

In equation (4.2.4) pq is a fixed subscript and refers to the added element and $\rho\sigma$ is a variable subscript and refers to all other elements. Then,

i_{pq} and v_{pq}	are, respectively, current through and voltage across the added element
$\bar{i}_{\rho\sigma}$ and $\bar{v}_{\rho\sigma}$	are the current and voltage vectors of the elements of the partial network
$y_{pq,pq}$	is the self-admittance of the added element
$\bar{y}_{pq,\rho\sigma}$	is the vector of mutual admittances between the added element p–q and the elements ρ–σ of the partial network
$\bar{y}_{\rho\sigma,pq}$	is the transpose of the vector $\bar{y}_{pq,\rho\sigma}$
$[y_{\rho\sigma,\rho\sigma}]$	is the primitive admittance matrix of the partial network

The current in the added branch, shown in Fig. 4.3, is

$$i_{pq} = 0 \tag{4.2.5}$$

However v_{pq} is not equal to zero since the added branch is mutually coupled to one or more of the elements of the partial network. Moreover,

$$\bar{v}_{\rho\sigma} = \bar{E}_{\rho} - \bar{E}_{\sigma} \tag{4.2.6}$$

where $\bar{E}_{\rho}$ and $\bar{E}_{\sigma}$ are the voltages at the buses in the partial network. From equations (4.2.4) and (4.2.5),

$$i_{pq} = y_{pq,pq}v_{pq} + \bar{y}_{pq,\rho\sigma}\bar{v}_{\rho\sigma} = 0$$

and therefore,

$$v_{pq} = -\frac{\bar{y}_{pq,\rho\sigma}\bar{v}_{\rho\sigma}}{y_{pq,pq}}$$

Substituting for $\bar{v}_{\rho\sigma}$ from equation (4.2.6),

$$v_{pq} = -\frac{\bar{y}_{pq,\rho\sigma}(\bar{E}_{\rho} - \bar{E}_{\sigma})}{y_{pq,pq}} \tag{4.2.7}$$

Substituting for v_{pq} in equation (4.2.3) from (4.2.7),

$$E_q = E_p + \frac{\bar{y}_{pq,\rho\sigma}(\bar{E}_{\rho} - \bar{E}_{\sigma})}{y_{pq,pq}}$$

Finally, substituting for E_q, E_p, $\bar{E}_{\rho}$, and $\bar{E}_{\sigma}$ from equation (4.2.2) with $I_i = 1$,

$$Z_{qi} = Z_{pi} + \frac{\bar{y}_{pq,\rho\sigma}(\bar{Z}_{\rho i} - \bar{Z}_{\sigma i})}{y_{pq,pq}} \qquad \begin{matrix} i = 1, 2, \ldots, m \\ i \neq q \end{matrix} \tag{4.2.8}$$

The element Z_{qq} can be calculated by injecting a current at the qth bus and calculating the voltage at that bus. Since all other bus currents equal zero, it follows from equation (4.2.1) that

$$\begin{aligned} E_1 &= Z_{1q}I_q \\ E_2 &= Z_{2q}I_q \\ &\cdots\cdots \\ E_p &= Z_{pq}I_q \\ &\cdots\cdots \\ E_m &= Z_{mq}I_q \\ E_q &= Z_{qq}I_q \end{aligned} \tag{4.2.9}$$

Letting $I_q = 1$ per unit in equations (4.2.9), Z_{qq} can be obtained directly by calculating E_q.

The voltages at buses p and q are related by equation (4.2.3), and the current through the added element is

$$i_{pq} = -I_q = -1 \tag{4.2.10}$$

The voltages across the elements of the partial network are given by equation (4.2.6) and the currents through these elements by (4.2.4). From equations (4.2.4) and (4.2.10),

$$i_{pq} = y_{pq,pq}v_{pq} + \bar{y}_{pq,\rho\sigma}\bar{v}_{\rho\sigma} = -1$$

and therefore,

$$v_{pq} = -\frac{1 + \bar{y}_{pq,\rho\sigma}\bar{v}_{\rho\sigma}}{y_{pq,pq}}$$

Substituting for $\bar{v}_{\rho\sigma}$ from equation (4.2.6),

$$v_{pq} = -\frac{1 + \bar{y}_{pq,\rho\sigma}(\bar{E}_\rho - \bar{E}_\sigma)}{y_{pq,pq}} \tag{4.2.11}$$

Substituting for v_{pq} in equation (4.2.3) from (4.2.11),

$$E_q = E_p + \frac{1 + \bar{y}_{pq,\rho\sigma}(\bar{E}_\rho - \bar{E}_\sigma)}{y_{pq,pq}}$$

Finally, substituting for E_q, E_p, $\bar{E}_\rho$, and $\bar{E}_\sigma$ from equation (4.2.9) with $I_q = 1$,

$$Z_{qq} = Z_{pq} + \frac{1 + \bar{y}_{pq,\rho\sigma}(\bar{Z}_{\rho q} - \bar{Z}_{\sigma q})}{y_{pq,pq}} \tag{4.2.12}$$

If there is no mutual coupling between the added branch and other elements of the partial network, then the elements of $\bar{y}_{pq,\rho\sigma}$ are zero and

$$z_{pq,pq} = \frac{1}{y_{pq,pq}}$$

It follows from equation (4.2.8) that

$$Z_{qi} = Z_{pi} \qquad \begin{matrix} i = 1, 2, \ldots, m \\ i \neq q \end{matrix}$$

and from equation (4.2.12) that

$$Z_{qq} = Z_{pq} + z_{pq,pq}$$

Furthermore, if there is no mutual coupling and p is the reference node,

$$Z_{pi} = 0 \qquad \begin{matrix} i = 1, 2, \ldots, m \\ i \neq q \end{matrix}$$

and

$$Z_{qi} = 0 \qquad \begin{matrix} i = 1, 2, \ldots, m \\ i \neq q \end{matrix}$$

Also

$$Z_{pq} = 0$$

and therefore,

$$Z_{qq} = z_{pq,pq}$$

Addition of a link

If the added element p–q is a link, the procedure for recalculating the elements of the bus impedance matrix is to connect in series with the added element a voltage source e_l as shown in Fig. 4.4. This creates a

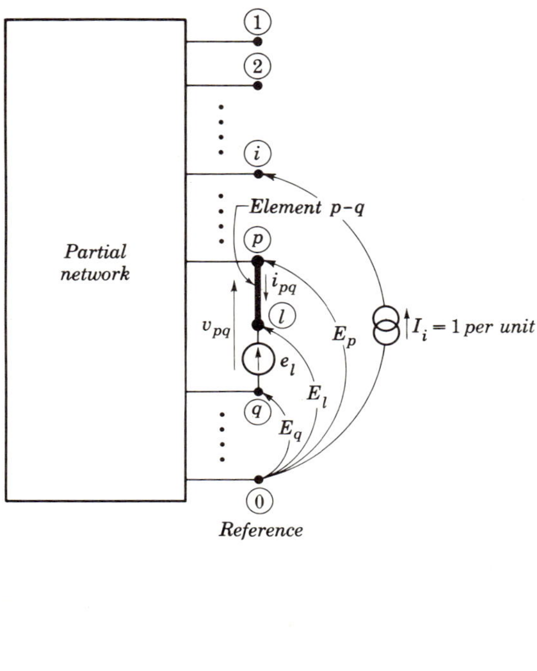

Fig. 4.4 ***Injected current voltage source in series with added link and bus voltages for calculation of*** Z_{li}.

fictitious node l which will be eliminated later. The voltage source e_l is selected such that the current through the added link is zero.

The performance equation for the partial network with the added element p–l and the series voltage source e_l is

$$\begin{array}{c} \\ \begin{bmatrix} E_1 \\ E_2 \\ \cdots \\ E_p \\ \cdots \\ E_m \\ \hline e_l \end{bmatrix} \end{array} = \begin{array}{c} \\ 1 \\ \\ \\ p \\ \\ m \\ l \end{array} \begin{array}{c} \begin{array}{ccccccc} 1 & & & p & & m & l \end{array} \\ \left[\begin{array}{cccccc|c} Z_{11} & Z_{12} & \cdots & Z_{1p} & \cdots & Z_{1m} & Z_{1l} \\ Z_{21} & Z_{22} & \cdots & Z_{2p} & \cdots & Z_{2m} & Z_{2l} \\ \cdots & \cdots & \cdots & \cdots & \cdots & \cdots & \cdots \\ Z_{p1} & Z_{p2} & \cdots & Z_{pp} & \cdots & Z_{pm} & Z_{pl} \\ \cdots & \cdots & \cdots & \cdots & \cdots & \cdots & \cdots \\ Z_{m1} & Z_{m2} & \cdots & Z_{mp} & \cdots & Z_{mm} & Z_{ml} \\ \hline Z_{l1} & Z_{l2} & \cdots & Z_{lp} & \cdots & Z_{lm} & Z_{ll} \end{array}\right] \end{array} \begin{array}{c} \\ \begin{bmatrix} I_1 \\ I_2 \\ \cdots \\ I_p \\ \cdots \\ I_m \\ \hline I_l \end{bmatrix} \end{array} \tag{4.2.13}$$

Since

$$e_l = E_l - E_q$$

the element Z_{li} can be determined by injecting a current at the ith bus and calculating the voltage at the lth node with respect to bus q. Since all other bus currents equal zero, it follows from equation (4.2.13) that

$$\begin{aligned} E_k &= Z_{ki}I_i \qquad k = 1, 2, \ldots, m \\ e_l &= Z_{li}I_i \end{aligned} \tag{4.2.14}$$

Letting $I_i = 1$ per unit in equations (4.2.14), Z_{li} can be obtained directly by calculating e_l.

The series voltage source is

$$e_l = E_p - E_q - v_{pl} \tag{4.2.15}$$

Since the current through the added link is

$$i_{pq} = 0$$

the element p–l can be treated as a branch. The current in this element in terms of primitive admittances and the voltages across the elements is

$$i_{pl} = y_{pl,pl}v_{pl} + \bar{y}_{pl,\rho\sigma}\bar{v}_{\rho\sigma}$$

where

$$i_{pl} = i_{pq} = 0$$

Therefore

$$v_{pl} = -\frac{\bar{y}_{pl,\rho\sigma}\bar{v}_{\rho\sigma}}{y_{pl,pl}}$$

Since

$$\bar{y}_{pl,\rho\sigma} = \bar{y}_{pq,\rho\sigma} \qquad \text{and} \qquad y_{pl,pl} = y_{pq,pq}$$

then

$$v_{pl} = -\frac{\bar{y}_{pq,\rho\sigma}\bar{v}_{\rho\sigma}}{y_{pq,pq}} \tag{4.2.16}$$

Substituting in order from equations (4.2.16), (4.2.6), and (4.2.14) with $I_i = 1$ into equation (4.2.15) yields

$$Z_{li} = Z_{pi} - Z_{qi} + \frac{\bar{y}_{pq,\rho\sigma}(\bar{Z}_{\rho i} - \bar{Z}_{\sigma i})}{y_{pq,pq}} \qquad \begin{array}{l} i = 1, 2, \ldots, m \\ i \neq l \end{array} \tag{4.2.17}$$

The element Z_{ll} can be calculated by injecting a current at the lth bus with bus q as reference and calculating the voltage at the lth bus with respect to bus q. Since all other bus currents equal zero, it follows from equation (4.2.13) that

$$\begin{aligned} E_k &= Z_{kl}I_l \qquad k = 1, 2, \ldots, m \\ e_l &= Z_{ll}I_l \end{aligned} \tag{4.2.18}$$

Letting $I_l = 1$ per unit in equation (4.2.18), Z_{ll} can be obtained directly by calculating e_l.

The current in the element p–l is

$$i_{pl} = -I_l = -1$$

This current in terms of primitive admittances and the voltages across the elements is

$$i_{pl} = y_{pl,pl}v_{pl} + \bar{y}_{pl,\rho\sigma}\bar{v}_{\rho\sigma} = -1$$

Again, since

$$\bar{y}_{pl,\rho\sigma} = \bar{y}_{pq,\rho\sigma} \qquad \text{and} \qquad y_{pl,pl} = y_{pq,pq}$$

then

$$v_{pl} = -\frac{1 + \bar{y}_{pq,\rho\sigma}\bar{v}_{\rho\sigma}}{y_{pq,pq}} \tag{4.2.19}$$

Substituting in order from equations (4.2.19), (4.2.6), and (4.2.18) with $I_l = 1$ into (4.2.15) yields

$$Z_{ll} = Z_{pl} - Z_{ql} + \frac{1 + \bar{y}_{pq,\rho\sigma}(\bar{Z}_{\rho l} - \bar{Z}_{\sigma l})}{y_{pq,pq}} \tag{4.2.20}$$

If there is no mutual coupling between the added element and other elements of the partial network, the elements of $\bar{y}_{pq,\rho\sigma}$ are zero and

$$z_{pq,pq} = \frac{1}{y_{pq,pq}}$$

It follows from equation (4.2.17) that

$$Z_{li} = Z_{pi} - Z_{qi} \qquad \begin{array}{l} i = 1, 2, \ldots, m \\ i \neq l \end{array}$$

and from equation (4.2.20),

$$Z_{ll} = Z_{pl} - Z_{ql} + z_{pq,pq}$$

Furthermore, if there is no mutual coupling and p is the reference node,

$$Z_{pi} = 0 \qquad \begin{array}{l} i = 1, 2, \ldots, m \\ i \neq l \end{array}$$

and

$$Z_{li} = -Z_{qi} \qquad \begin{array}{l} i = 1, 2, \ldots, m \\ i \neq l \end{array}$$

Also

$$Z_{pl} = 0$$

and therefore,

$$Z_{ll} = -Z_{ql} + z_{pq,pq}$$

The elements in the lth row and column of the bus impedance matrix for the augmented partial network are found from equations (4.2.17) and (4.2.20). It remains to calculate the required bus impedance matrix to include the effect of the added link. This can be accomplished by modifying the elements Z_{ij}, where $i, j = 1, 2, \ldots, m$, and eliminating the lth row and column corresponding to the fictitious node.

The fictitious node l is eliminated by short circuiting the series voltage source e_l. From equation (4.2.13),

$$\bar{E}_{BUS} = Z_{BUS}\bar{I}_{BUS} + \bar{Z}_{il}I_l \tag{4.2.21}$$

and

$$e_l = \bar{Z}_{lj}\bar{I}_{BUS} + Z_{ll}I_l = 0 \tag{4.2.22}$$

where $i, j = 1, 2, \ldots, m$. Solving for I_l from equation (4.2.22) and substituting into (4.2.21),

$$\bar{E}_{BUS} = \left(Z_{BUS} - \frac{\bar{Z}_{il}\bar{Z}_{lj}}{Z_{ll}}\right) \bar{I}_{BUS}$$

which is the performance equation of the partial network including the link p–q. It follows that the required bus impedance matrix is

$$Z_{BUS\,(\text{modified})} = Z_{BUS\,(\text{before elimination})} - \frac{\bar{Z}_{il}\bar{Z}_{lj}}{Z_{ll}}$$

where any element of $Z_{BUS\,(\text{modified})}$ is

$$Z_{ij\,(\text{modified})} = Z_{ij\,(\text{before elimination})} - \frac{Z_{il}Z_{lj}}{Z_{ll}}$$

A summary of the equations for the formation of the bus impedance matrix is given in Table 4.1.

4.3 Modification of the bus impedance matrix for changes in the network

The bus impedance matrix Z_{BUS} can be modified to reflect changes in the network. These changes may be addition of elements, removal of elements, or changes in the impedances of elements.

The method described in Sec. 4.2 based on the algorithm for forming a bus impedance matrix can be applied if elements are added to the network. Then Z_{BUS} is considered the matrix of the partial network at that stage and the new elements are added one at a time to produce the new bus impedance matrix Z'_{BUS}.

The procedure to remove elements or to change the impedances of elements is the same. If an element is removed which is not mutually coupled to any other element, the modified bus impedance matrix can be obtained by adding, in parallel with the element, a link whose impedance is equal to the negative of the impedance of the element to be removed. If the impedance of an uncoupled element is changed, the modified bus impedance matrix can be obtained by adding a link in parallel with the element such that the equivalent impedance of the two elements is the desired value.

When mutually coupled elements are removed or their impedances changed, the modified bus impedance matrix can not be obtained by adding a link and using the procedure described in Sec. 4.2. However, an equation can be derived for modifying the elements of Z_{BUS} by introducing appropriate changes in the bus currents of the original net-

Table 4.1 Summary of equations for formation of bus impedance matrix

Add p–q	Mutual coupling: p is not the reference bus	Mutual coupling: p is the reference bus	No mutual coupling: p is not the reference bus	No mutual coupling: p is the reference bus
Branch	$Z_{qi} = Z_{pi} + \dfrac{\bar{y}_{pq,\rho\sigma}(\bar{Z}_{\rho i} - \bar{Z}_{\sigma i})}{y_{pq,pq}}$ $i = 1, 2, \ldots, m$ $i \neq q$	$Z_{qi} = \dfrac{\bar{y}_{pq,\rho\sigma}(\bar{Z}_{\rho i} - \bar{Z}_{\sigma i})}{y_{pq,pq}}$ $i = 1, 2, \ldots, m$ $i \neq q$	$Z_{qi} = Z_{pi}$ $i = 1, 2, \ldots, m$ $i \neq q$	$Z_{qi} = 0$ $i = 1, 2, \ldots, m$ $i \neq q$
	$Z_{qq} = Z_{pq} + \dfrac{1 + \bar{y}_{pq,\rho\sigma}(\bar{Z}_{\rho q} - \bar{Z}_{\sigma q})}{y_{pq,pq}}$	$Z_{qq} = \dfrac{1 + \bar{y}_{pq,\rho\sigma}(\bar{Z}_{\rho q} - \bar{Z}_{\sigma q})}{y_{pq,pq}}$	$Z_{qq} = Z_{pq} + z_{pq,pq}$	$Z_{qq} = z_{pq,pq}$
Link	$Z_{li} = Z_{pi} - Z_{qi} + \dfrac{\bar{y}_{pq,\rho\sigma}(\bar{Z}_{\rho i} - \bar{Z}_{\sigma i})}{y_{pq,pq}}$ $i = 1, 2, \ldots, m$ $i \neq l$	$Z_{li} = -Z_{qi} + \dfrac{\bar{y}_{pq,\rho\sigma}(\bar{Z}_{\rho i} - \bar{Z}_{\sigma i})}{y_{pq,pq}}$ $i = 1, 2, \ldots, m$ $i \neq l$	$Z_{li} = Z_{pi} - Z_{qi}$ $i = 1, 2, \ldots, m$ $i \neq l$	$Z_{li} = -Z_{qi}$ $i = 1, 2, \ldots, m$ $i \neq l$
	$Z_{ll} = Z_{pl} - Z_{ql} + \dfrac{1 + \bar{y}_{pq,\rho\sigma}(\bar{Z}_{\rho l} - \bar{Z}_{\sigma l})}{y_{pq,pq}}$	$Z_{ll} = -Z_{ql} + \dfrac{1 + \bar{y}_{pq,\rho\sigma}(\bar{Z}_{\rho l} - \bar{Z}_{\sigma l})}{y_{pq,pq}}$	$Z_{ll} = Z_{pl} - Z_{ql} + z_{pq,pq}$	$Z_{ll} = -Z_{ql} + z_{pq,pq}$

Modification of the elements for elimination of lth node

$$Z_{ij\,(\text{modified})} = Z_{ij\,(\text{before elimination})} - \frac{Z_{il}Z_{lj}}{Z_{ll}} \qquad i, j = 1, 2, \ldots, m$$

work to simulate the removal of elements or changes in their impedances. The performance equation in terms of the new bus currents is

$$\bar{E}'_{BUS} = Z_{BUS}(\bar{I}_{BUS} + \overline{\Delta I}_{BUS}) \tag{4.3.1}$$

where $\overline{\Delta I}_{BUS}$ is a vector of bus current changes such that $\bar{E}'_{BUS}$ will reflect the desired changes in the network.

An element Z'_{ij} of the modified bus impedance matrix can be obtained by calculating for the modified network the voltage at bus i with a current injected at bus j. This is equivalent to calculating for the original network the voltage at bus i with the same value of current injected at bus j and appropriate changes in currents at the buses which are terminals of the elements being changed.

If the elements μ–ν coupled to elements ρ–σ are removed or their impedances are changed, the corresponding changes in the bus currents are

$$\begin{aligned}
\Delta I_k &= \Delta i_{\mu\nu} & k &= \mu \\
\Delta I_k &= -\Delta i_{\mu\nu} & k &= \nu \\
\Delta I_k &= \Delta i_{\rho\sigma} & k &= \rho \\
\Delta I_k &= -\Delta i_{\rho\sigma} & k &= \sigma \\
\Delta I_k &= 0 & &\text{for all other } k
\end{aligned} \tag{4.3.2}$$

Letting the injected current at the jth bus equal one per unit,

$$\begin{aligned}
I_j &= 1 \\
I_k &= 0 \qquad \begin{matrix} k = 1, 2, \ldots, n \\ k \neq j \end{matrix}
\end{aligned} \tag{4.3.3}$$

From the performance equation (4.3.1),

$$E'_i = \sum_{k=1}^{n} Z_{ik}(I_k + \Delta I_k) \qquad i = 1, 2, \ldots, n$$

Substituting for ΔI_k and I_k from equations (4.3.2) and (4.3.3),

$$E'_i = Z_{ij} + \bar{Z}_{i\mu}\overline{\Delta i}_{\mu\nu} - \bar{Z}_{i\nu}\overline{\Delta i}_{\mu\nu} + \bar{Z}_{i\rho}\overline{\Delta i}_{\rho\sigma} - \bar{Z}_{i\sigma}\overline{\Delta i}_{\rho\sigma}$$
$$E'_i = Z_{ij} + (\bar{Z}_{i\mu} - \bar{Z}_{i\nu})\overline{\Delta i}_{\mu\nu} + (\bar{Z}_{i\rho} - \bar{Z}_{i\sigma})\overline{\Delta i}_{\rho\sigma}$$

Using the subscript $\alpha\beta$ for network elements μ–ν and ρ–σ,

$$E'_i = Z_{ij} + (\bar{Z}_{i\alpha} - \bar{Z}_{i\beta})\overline{\Delta i}_{\alpha\beta} \qquad i = 1, 2, \ldots, n \tag{4.3.4}$$

From the performance equation of the primitive network,

$$\overline{\Delta i}_{\alpha\beta} = ([y_s] - [y'_s])\bar{v}'_{\gamma\delta} \tag{4.3.5}$$

where $[y_s]$ and $[y'_s]$ are respectively the square submatrices of the original and modified primitive admittance matrices. The rows and columns of the submatrices correspond to the network elements μ–ν and ρ–σ. The subscripts of the elements of $([y_s] - [y_s'])$ are $\alpha\beta$, $\gamma\delta$. The voltage vector in equation (4.3.5) is

$$\bar{v}'_{\gamma\delta} = \bar{E}'_\gamma - \bar{E}'_\delta$$

Substituting for $\bar{E}'_\gamma$ and $\bar{E}'_\delta$ from equation (4.3.4),

$$\bar{v}'_{\gamma\delta} = \bar{Z}_{\gamma j} - \bar{Z}_{\delta j} + ([Z_{\gamma\alpha}] - [Z_{\delta\alpha}] - [Z_{\gamma\beta}] + [Z_{\delta\beta}])\overline{\Delta i_{\alpha\beta}} \tag{4.3.6}$$

Substituting from equation (4.3.6) for $\bar{v}'_{\gamma\delta}$ into (4.3.5),

$$\overline{\Delta i_{\alpha\beta}} = ([y_s] - [y_s'])\{\bar{Z}_{\gamma j} - \bar{Z}_{\delta j} + ([Z_{\gamma\alpha}] - [Z_{\delta\alpha}] - [Z_{\gamma\beta}] + [Z_{\delta\beta}])\overline{\Delta i_{\alpha\beta}}\} \tag{4.3.7}$$

Solving equation (4.3.7) for $\overline{\Delta i_{\alpha\beta}}$,

$$\overline{\Delta i_{\alpha\beta}} = \{U - ([y_s] - [y_s'])([Z_{\gamma\alpha}] - [Z_{\delta\alpha}] - [Z_{\gamma\beta}] + [Z_{\delta\beta}])\}^{-1} ([y_s] - [y_s'])(\bar{Z}_{\gamma j} - \bar{Z}_{\delta j}) \tag{4.3.8}$$

Designating

$$[\Delta y_s] = [y_s] - [y_s']$$

and

$$[M] = \{U - [\Delta y_s]([Z_{\gamma\alpha}] - [Z_{\delta\alpha}] - [Z_{\gamma\beta}] + [Z_{\delta\beta}])\}$$

equation (4.3.8) becomes

$$\overline{\Delta i_{\alpha\beta}} = [M]^{-1}[\Delta y_s](\bar{Z}_{\gamma j} - \bar{Z}_{\delta j}) \tag{4.3.9}$$

Substituting from equation (4.3.9) for $\overline{\Delta i_{\alpha\beta}}$ into (4.3.4),

$$E_i' = Z_{ij} + (\bar{Z}_{i\alpha} - \bar{Z}_{i\beta})[M]^{-1}[\Delta y_s](\bar{Z}_{\gamma j} - \bar{Z}_{\delta j})$$

This equation gives, for the original network, the voltage at bus i as a result of injecting one per unit current at bus j and the appropriate current changes at buses μ, ν, ρ, and σ to simulate the effect of changes in the elements μ–ν. Thus, from the definition of the bus impedance matrix, the ijth element of the modified bus impedance matrix is

$$Z_{ij}' = Z_{ij} + (\bar{Z}_{i\alpha} - \bar{Z}_{i\beta})[M]^{-1}[\Delta y_s](\bar{Z}_{\gamma j} - \bar{Z}_{\delta j}) \qquad i = 1, 2, \ldots, n$$

The process is repeated for each $j = 1, 2, \ldots, n$ to obtain all elements of Z'_{BUS}.

4.4 Example of formation and modification of bus impedance matrix

The method based on the algorithm for forming the bus impedance matrix will be illustrated using the sample network given in Fig. 3.10.

Examples of the modification of this bus impedance matrix will also be given.

Problem

a. Form the bus impedance matrix Z_{BUS} of the network shown in Fig. 4.5.

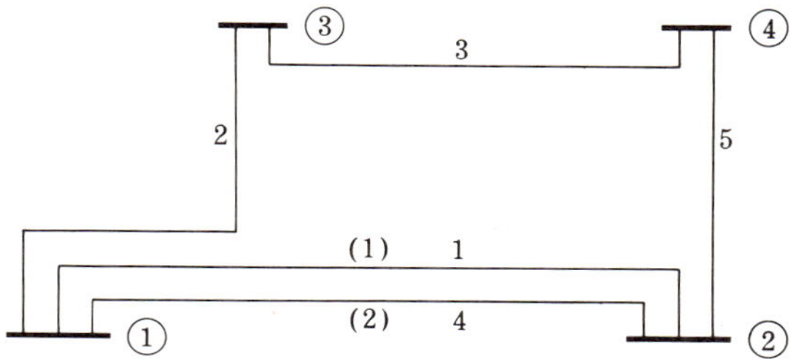

Fig. 4.5 Sample network.

b. Modify the bus impedance matrix obtained in part *a* to include the addition of an element from bus 2 to bus 4 with an impedance of 0.3 and coupled to element 5 with a mutual impedance of 0.1.

c. Modify the bus impedance matrix obtained in part *b* to remove the new element from bus 2 to bus 4.

Solution

The data for the network is given in Table 4.2. The bus impedance matrix will be formed by adding elements of the network in the order indicated in the first column of this table. Node 1 is selected as reference.

Table 4.2 Impedances for sample network

	Self		*Mutual*	
Element number	*Bus code* p–q	*Impedance* $z_{pq,pq}$	*Bus code* r–s	*Impedance* $z_{pq,rs}$
1	1–2(1)	0.6		
4	1–2(2)	0.4	1–2(1)	0.2
2	1–3	0.5	1–2(1)	0.1
3	3–4	0.5		
5	2–4	0.2		

a. Step 1. Start with element 1 which is a branch from $p = 1$ to $q = 2$. The elements of the bus impedance matrix for the partial network containing the single branch are

$Z_{BUS} =$

	①	②
①	0	0
②	0	0.6

Since node 1 is the reference, the elements of the first row and column are zero and need not be written. Thus

$Z_{BUS} =$

	②
②	0.6

Step 2. Add element 4, which is a link, from $p = 1$ (reference) to $q = 2$, mutually coupled with element 1. The augmented impedance matrix with the fictitious node l will be

	②	l
②	0.6	Z_{2l}
l	Z_{l2}	Z_{ll}

where

$$Z_{2l} = Z_{l2} = -Z_{22} + \frac{y_{12(2),12(1)}(Z_{12} - Z_{22})}{y_{12(2),12(2)}}$$

$$Z_{ll} = -Z_{2l} + \frac{1 + y_{12(2),12(1)}(Z_{1l} - Z_{2l})}{y_{12(2),12(2)}}$$

and $Z_{12} = Z_{1l} = 0$. Invert the primitive impedance matrix of the partial network to obtain the primitive admittance matrix.

$[z_{p\sigma,p\sigma}] =$

	1–2(1)	1–2(2)
1–2(1)	0.6	0.2
1–2(2)	0.2	0.4

$[z_{p\sigma,p\sigma}]^{-1} = [y_{p\sigma,p\sigma}] =$

	1–2(1)	1–2(2)
1–2(1)	2	−1
1–2(2)	−1	3

Then,

$$Z_{2l} = Z_{l2} = -0.6 + \frac{(-1)(-0.6)}{3} = -0.4$$

$$Z_{ll} = 0.4 + \frac{1\text{–}1(0.4)}{3} = 0.6$$

and the augmented matrix is

	②	l
②	0.6	−0.4
l	−0.4	0.6

Eliminating the lth row and column,

$$Z'_{22} = Z_{22} - \frac{Z_{2l}Z_{l2}}{Z_{ll}} = 0.6 - \frac{(-0.4)(-0.4)}{0.6} = 0.3333$$

and

	②
Z_{BUS} = ②	0.3333

Step 3. Add element 2, which is a branch, from $p = 1$ (reference) to $q = 3$, mutually coupled with element 1. This adds a new bus and the bus impedance matrix is

Z_{BUS} =	②	③
②	0.3333	Z_{23}
③	Z_{32}	Z_{33}

where

$$Z_{32} = Z_{23} = \frac{\begin{bmatrix} y_{13,12(1)} & y_{13,12(2)} \end{bmatrix} \begin{bmatrix} Z_{12} - Z_{22} \\ Z_{12} - Z_{22} \end{bmatrix}}{y_{13,13}}$$

$$Z_{33} = \frac{1 + \begin{bmatrix} y_{13,12(1)} & y_{13,12(2)} \end{bmatrix} \begin{bmatrix} Z_{13} - Z_{23} \\ Z_{13} - Z_{23} \end{bmatrix}}{y_{13,13}}$$

and $Z_{12} = Z_{13} = 0$. Invert the primitive impedance matrix to obtain the primitive admittance matrix.

$[z_{p\sigma,p\sigma}] =$

	1–2(1)	1–2(2)	1–3
1–2(1)	0.6	0.2	0.1
1–2(2)	0.2	0.4	
1–3	0.1		0.5

$[z_{p\sigma,p\sigma}]^{-1} = [y_{p\sigma,p\sigma}] =$

	1–2(1)	1–2(2)	1–3
1–2(1)	2.0833	−1.0417	−0.4167
1–2(2)	−1.0417	3.0208	0.2083
1–3	−0.4167	0.2083	2.0833

Then,

$$Z_{32} = Z_{23} = \frac{\begin{bmatrix} -0.4167 & 0.2083 \end{bmatrix} \begin{bmatrix} -0.3333 \\ -0.3333 \end{bmatrix}}{2.0833} = 0.0333$$

$$Z_{33} = \frac{1 + \begin{bmatrix} -0.4167 & 0.2083 \end{bmatrix} \begin{bmatrix} -0.0333 \\ -0.0333 \end{bmatrix}}{2.0833} = 0.4833$$

and

$Z_{BUS} =$

	②	③
②	0.3333	0.0333
③	0.0333	0.4833

Step 4. Add element 3, which is a branch, from $p = 3$ to $q = 4$, not mutually coupled. This adds a new bus.

$Z_{24} = Z_{42} = Z_{32} = 0.0333$
$Z_{34} = Z_{43} = Z_{33} = 0.4833$
$Z_{44} = Z_{34} + z_{34,34} = 0.4833 + 0.5 = 0.9833$

Thus,

	②	③	④
$Z_{BUS} =$ ②	0.3333	0.0333	0.0333
③	0.0333	0.4833	0.4833
④	0.0333	0.4833	0.9833

Step 5. Add element 5, which is a link, from $p = 2$ to $q = 4$, not mutually coupled. The elements of the lth row and column of the augmented matrix are

$$Z_{2l} = Z_{l2} = Z_{22} - Z_{42} = 0.3333 - 0.0333 = 0.3000$$
$$Z_{3l} = Z_{l3} = Z_{23} - Z_{43} = 0.0333 - 0.4833 = -0.4500$$
$$Z_{4l} = Z_{l4} = Z_{24} - Z_{44} = 0.0333 - 0.9833 = -0.9500$$
$$Z_{ll} = Z_{2l} - Z_{4l} + z_{24,24} = 0.3000 + 0.9500 + 0.2 = 1.4500$$

The augmented matrix is

	②	③	④	l
②	0.3333	0.0333	0.0333	0.3000
③	0.0333	0.4833	0.4833	−0.4500
④	0.0333	0.4833	0.9833	−0.9500
l	0.3000	−0.4500	−0.9500	1.4500

Eliminating the lth row and column,

$$Z'_{22} = Z_{22} - \frac{Z_{2l}Z_{l2}}{Z_{ll}} = 0.3333 - \frac{(0.3000)(0.3000)}{1.4500} = 0.2712$$

$$Z'_{23} = Z'_{32} = Z_{23} - \frac{Z_{2l}Z_{l3}}{Z_{ll}} = 0.0333 - \frac{(0.3000)(-0.4500)}{1.4500} = 0.1263$$

$$Z'_{24} = Z'_{42} = Z_{24} - \frac{Z_{2l}Z_{l4}}{Z_{ll}} = 0.0333 - \frac{(0.3000)(-0.9500)}{1.4500} = 0.2298$$

$$Z'_{33} = Z_{33} - \frac{Z_{3l}Z_{l3}}{Z_{ll}} = 0.4833 - \frac{(-0.4500)(-0.4500)}{1.4500} = 0.3436$$

$$Z'_{34} = Z'_{43} = Z_{34} - \frac{Z_{3l}Z_{l4}}{Z_{ll}} = 0.4833 - \frac{(-0.4500)(-0.9500)}{1.4500} = 0.1885$$

$$Z'_{44} = Z_{44} - \frac{Z_{4l}Z_{l4}}{Z_{ll}} = 0.9833 - \frac{(-0.9500)(-0.9500)}{1.4500} = 0.3609$$

and

$Z_{BUS} =$

	②	③	④
②	0.2712	0.1263	0.2298
③	0.1263	0.3436	0.1885
④	0.2298	0.1885	0.3609

b. Adding a new element, which is a link, from $p = 2$ to $q = 4$, mutually coupled with element 5 results in the augmented matrix

	②	③	④	l
②	0.2712	0.1263	0.2298	Z_{2l}
③	0.1263	0.3436	0.1885	Z_{3l}
④	0.2298	0.1885	0.3609	Z_{4l}
l	Z_{l2}	Z_{l3}	Z_{l4}	Z_{ll}

where

$$Z_{li} = Z_{pi} - Z_{qi} + \frac{\bar{y}_{pq,\rho\sigma}(\bar{Z}_{\rho i} - \bar{Z}_{\sigma i})}{y_{pq,pq}} \qquad i = 2, 3, 4$$

and

$$Z_{ll} = Z_{pl} - Z_{ql} + \frac{1 + \bar{y}_{pq,\rho\sigma}(\bar{Z}_{\rho l} - \bar{Z}_{\sigma l})}{y_{pq,pq}}$$

The primitive impedance matrix is

$[z] =$

	1–2(1)	1–2(2)	1–3	3–4	2–4(1)	2–4(2)
1–2(1)	0.6	0.2	0.1			
1–2(2)	0.2	0.4				
1–3	0.1		0.5			
3–4				0.5		
2–4(1)					0.2	0.1
2–4(2)					0.1	0.3

Since the new element is coupled to only one other element, it is sufficient to invert the submatrix for the coupled elements, which is

$[z_{p\sigma,p\sigma}] =$

	2–4(1)	2–4(2)
2–4(1)	0.2	0.1
2–4(2)	0.1	0.3

Thus

$[y_{p\sigma,p\sigma}] =$

	2–4(1)	2–4(2)
2–4(1)	6	−2
2–4(2)	−2	4

and

$$Z_{2l} = Z_{l2} = 0.2712 - 0.2298 + \frac{-2.0(0.2712 - 0.2298)}{4.0} = 0.0207$$

$$Z_{3l} = Z_{l3} = 0.1263 - 0.1885 + \frac{-2.0(0.1263 - 0.1885)}{4.0} = -0.0311$$

$$Z_{4l} = Z_{l4} = 0.2298 - 0.3609 + \frac{-2.0(0.2298 - 0.3609)}{4.0} = -0.0656$$

$$Z_{ll} = 0.0207 + 0.0656 + \frac{1 - 2.0(0.0207 + 0.0656)}{4.0} = 0.2931$$

and the augmented matrix is

	②	③	④	l
②	0.2712	0.1263	0.2298	0.0207
③	0.1263	0.3436	0.1885	−0.0311
④	0.2298	0.1885	0.3609	−0.0656
l	0.0207	−0.0311	−0.0656	0.2931

Eliminating the lth row and column,

$$Z'_{22} = 0.2712 - \frac{(0.0207)(0.0207)}{0.2931} = 0.2697$$

$$Z'_{23} = Z'_{32} = 0.1263 - \frac{(0.0207)(-0.0311)}{0.2931} = 0.1285$$

$$Z'_{24} = Z'_{42} = 0.2298 - \frac{(0.0207)(-0.0656)}{0.2931} = 0.2344$$

$$Z'_{33} = 0.3436 - \frac{(-0.0311)(-0.0311)}{0.2931} = 0.3403$$

$$Z'_{34} = Z'_{43} = 0.1885 - \frac{(-0.0311)(-0.0656)}{0.2931} = 0.1816$$

$$Z'_{44} = 0.3609 - \frac{(-0.0656)(-0.0656)}{0.2931} = 0.3462$$

Finally,

$Z_{BUS} =$

	②	③	④
②	0.2697	0.1285	0.2344
③	0.1285	0.3403	0.1816
④	0.2344	0.1816	0.3462

c. The modified elements of this bus impedance matrix for the removal of the network element 2–4(2) mutually coupled to network element 2–4(1) are obtained from

$$Z'_{ij} = Z_{ij} + (\bar{Z}_{i\alpha} - \bar{Z}_{i\beta})[M]^{-1}[\Delta y_s](\bar{Z}_{\gamma j} - \bar{Z}_{\delta j}) \qquad i, j = 2, 3, 4$$

where μ–ν is 2–4 and ρ–σ is also 2–4 and the indices $\alpha, \gamma = 2, 2$ and $\beta, \delta = 4, 4$.

The original primitive admittance submatrix is

$[y_s] =$

	2–4(1)	2–4(2)
2–4(1)	6	−2
2–4(2)	−2	4

and the modified primitive admittance submatrix is

$[y_s'] =$

	2–4(1)	2–4(2)
2–4(1)	5	
2–4(2)		

Thus

$[y_s] - [y_s'] = [\Delta y_s] =$

	2–4(1)	2–4(2)
2–4(1)	1	−2
2–4(2)	−2	4

Also

$$[M] = \{U - [\Delta y_s]([Z_{\gamma\alpha}] - [Z_{\delta\alpha}] - [Z_{\gamma\beta}] + [Z_{\delta\beta}])\}$$

where

$[Z_{\gamma\alpha}] =$

	②	②
②	0.2697	0.2697
②	0.2697	0.2697

$[Z_{\delta\alpha}] =$

	②	②
④	0.2344	0.2344
④	0.2344	0.2344

$[Z_{\gamma\beta}] =$

	④	④
②	0.2344	0.2344
②	0.2344	0.2344

$[Z_{\delta\beta}] =$

	④	④
④	0.3462	0.3462
④	0.3462	0.3462

Substituting in the above equation,

$$[M] = \begin{bmatrix} 1.1471 & 0.1471 \\ -0.2942 & 0.7058 \end{bmatrix}$$

$$[M]^{-1} = \begin{bmatrix} 0.82753 & -0.17247 \\ 0.34494 & 1.34494 \end{bmatrix}$$

and

$$[M]^{-1}[\Delta y_s] = \begin{bmatrix} 1.17247 & -2.34494 \\ -2.34494 & 4.68988 \end{bmatrix}$$

For $i = 2$ and $j = 2$,

$$Z'_{22} = Z_{22} + \left(\begin{bmatrix} Z_{22} & Z_{22} \end{bmatrix} - \begin{bmatrix} Z_{24} & Z_{24} \end{bmatrix} \right) [M]^{-1}[\Delta y_s] \left(\begin{bmatrix} Z_{22} \\ Z_{22} \end{bmatrix} - \begin{bmatrix} Z_{42} \\ Z_{42} \end{bmatrix} \right)$$

$$Z'_{22} = 0.2697 + \begin{bmatrix} 0.0353 & 0.0353 \end{bmatrix} \begin{bmatrix} 1.17247 & -2.34494 \\ -2.34494 & 4.68988 \end{bmatrix} \begin{bmatrix} 0.0353 \\ 0.0353 \end{bmatrix}$$

$$Z'_{22} = 0.2697 + 0.0015 = 0.2712$$

For $i = 2$ and $j = 3$,

$$Z'_{23} = Z_{23} + \left(\begin{bmatrix} Z_{22} & Z_{22} \end{bmatrix} - \begin{bmatrix} Z_{24} & Z_{24} \end{bmatrix} \right) [M]^{-1}[\Delta y_s] \left(\begin{bmatrix} Z_{23} \\ Z_{23} \end{bmatrix} - \begin{bmatrix} Z_{43} \\ Z_{43} \end{bmatrix} \right)$$

$$Z'_{23} = 0.1285 + \begin{bmatrix} 0.0353 & 0.0353 \end{bmatrix} \begin{bmatrix} 1.17247 & -2.34494 \\ -2.34494 & 4.68988 \end{bmatrix} \begin{bmatrix} -0.0531 \\ -0.0531 \end{bmatrix}$$

$$Z'_{23} = 0.1285 - 0.0022 = 0.1263$$

For $i = 2$ and $j = 4$,

$$Z'_{24} = Z_{24} + \left(\begin{bmatrix} Z_{22} & Z_{22} \end{bmatrix} - \begin{bmatrix} Z_{24} & Z_{24} \end{bmatrix} \right) [M]^{-1} [\Delta y_s] \left(\begin{bmatrix} Z_{24} \\ Z_{24} \end{bmatrix} - \begin{bmatrix} Z_{44} \\ Z_{44} \end{bmatrix} \right)$$

$$Z'_{24} = 0.2344 + \begin{bmatrix} 0.0353 & 0.0353 \end{bmatrix} \begin{bmatrix} 1.17247 & -2.34494 \\ -2.34494 & 4.68988 \end{bmatrix} \begin{bmatrix} -0.1118 \\ -0.1118 \end{bmatrix}$$

$$Z'_{24} = 0.2344 - 0.0046 = 0.2298$$

For $i = 3$ and $j = 3$,

$$Z'_{33} = Z_{33} + \left(\begin{bmatrix} Z_{32} & Z_{32} \end{bmatrix} - \begin{bmatrix} Z_{34} & Z_{34} \end{bmatrix} \right) [M]^{-1} [\Delta y_s] \left(\begin{bmatrix} Z_{23} \\ Z_{23} \end{bmatrix} - \begin{bmatrix} Z_{43} \\ Z_{43} \end{bmatrix} \right)$$

$$Z'_{33} = 0.3403 + \begin{bmatrix} -0.0531 & -0.0531 \end{bmatrix} \begin{bmatrix} 1.17247 & -2.34494 \\ -2.34494 & 4.68988 \end{bmatrix} \begin{bmatrix} -0.0531 \\ -0.0531 \end{bmatrix}$$

$$Z'_{33} = 0.3403 + 0.0033 = 0.3436$$

For $i = 3$ and $j = 4$,

$$Z'_{34} = Z_{34} + \left(\begin{bmatrix} Z_{32} & Z_{32} \end{bmatrix} - \begin{bmatrix} Z_{34} & Z_{34} \end{bmatrix} \right) [M]^{-1} [\Delta y_s] \left(\begin{bmatrix} Z_{24} \\ Z_{24} \end{bmatrix} - \begin{bmatrix} Z_{44} \\ Z_{44} \end{bmatrix} \right)$$

$$Z'_{34} = 0.1816 + \begin{bmatrix} -0.0531 & -0.0531 \end{bmatrix} \begin{bmatrix} 1.17247 & -2.34494 \\ -2.34494 & 4.68988 \end{bmatrix} \begin{bmatrix} -0.1118 \\ -0.1118 \end{bmatrix}$$

$$Z'_{34} = 0.1816 + 0.0069 = 0.1885$$

For $i = 4$ and $j = 4$,

$$Z'_{44} = Z_{44} + \left(\begin{bmatrix} Z_{42} & Z_{42} \end{bmatrix} - \begin{bmatrix} Z_{44} & Z_{44} \end{bmatrix} \right) [M]^{-1}[\Delta y_s] \left(\begin{bmatrix} Z_{24} \\ Z_{24} \end{bmatrix} - \begin{bmatrix} Z_{44} \\ Z_{44} \end{bmatrix} \right)$$

$$Z'_{44} = 0.3462 + \begin{bmatrix} -0.1118 & -0.1118 \end{bmatrix} \begin{bmatrix} 1.17247 & -2.34494 \\ -2.34494 & 4.68988 \end{bmatrix} \begin{bmatrix} -0.1118 \\ -0.1118 \end{bmatrix}$$

$$Z'_{44} = 0.3462 + 0.0147 = 0.3609$$

Thus

$Z_{BUS} =$

	②	③	④
②	0.2712	0.1263	0.2298
③	0.1263	0.3436	0.1885
④	0.2298	0.1885	0.3609

which is equal to the matrix obtained in part *a*.

4.5 Derivation of loop admittance matrix from bus impedance matrix

Derivation of node-pair impedance matrix from bus impedance matrix

An element of the node-pair impedance matrix Z_{NP} is designated by $Z_{ij,pq}$. If there is a current source between p and q only, an element of this matrix is defined as

$$Z_{ij,pq} = \frac{E_i - E_j}{I_{pq}} \qquad \begin{matrix} i, j = 1, 2, \ldots, n \\ i \neq j \end{matrix} \tag{4.5.1}$$

Letting the current $I_{pq} = 1$ per unit as shown in Fig. 4.6,

$$Z_{ij,pq} = E_i - E_j$$

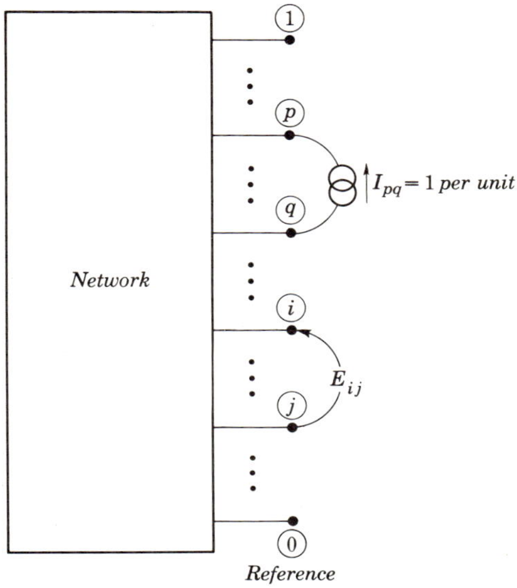

Fig. 4.6 Injected current for calculation of $\mathbf{Z}_{ij,pq}$.

The performance equation of the network written in terms of Z_{BUS} is

$$\begin{bmatrix} E_1 \\ \cdots \\ E_p \\ \cdots \\ E_q \\ \cdots \\ E_n \end{bmatrix} = \begin{array}{c|ccccccc|} & 1 & & p & & q & & n \\ 1 & Z_{11} & \cdots & Z_{1p} & \cdots & Z_{1q} & \cdots & Z_{1n} \\ & \cdots & \cdots & \cdots & \cdots & \cdots & \cdots & \cdots \\ p & Z_{p1} & \cdots & Z_{pp} & \cdots & Z_{pq} & \cdots & Z_{pn} \\ & \cdots & \cdots & \cdots & \cdots & \cdots & \cdots & \cdots \\ q & Z_{q1} & \cdots & Z_{qp} & \cdots & Z_{qq} & \cdots & Z_{qn} \\ & \cdots & \cdots & \cdots & \cdots & \cdots & \cdots & \cdots \\ n & Z_{n1} & \cdots & Z_{np} & \cdots & Z_{nq} & \cdots & Z_{nn} \end{array} \begin{bmatrix} I_1 \\ \cdots \\ I_p \\ \cdots \\ I_q \\ \cdots \\ I_n \end{bmatrix} \quad (4.5.2)$$

Since $I_{pq} = 1$, then $I_p = 1$ and $I_q = -1$. From equation (4.5.2) it follows that

$$E_i = Z_{ip} - Z_{iq} \qquad \text{and} \qquad E_j = Z_{jp} - Z_{jq}$$

From equation (4.5.1),

$$Z_{ij,pq} = Z_{ip} - Z_{jp} - Z_{iq} + Z_{jq} \qquad \begin{matrix} i, j = 1, 2, \ldots, n \\ i \neq j \end{matrix} \tag{4.5.3}$$

Using all node-pair combinations for p–q, all elements of Z_{NP} can be calculated from equation (4.5.3). The matrix Z_{NP} has dimension

$$\frac{n(n-1)}{2} \times \frac{n(n-1)}{2}$$

Derivation of element-pair admittance matrix from node-pair impedance matrix

The element-pair admittance matrix is designated by Y_{EP} and the element of the matrix by $Y_{ij,pq}$. If there is a voltage source only in series with p–q, an element of this matrix is defined as

$$Y_{ij,pq} = \frac{i_{ij}}{e_{pq}}$$

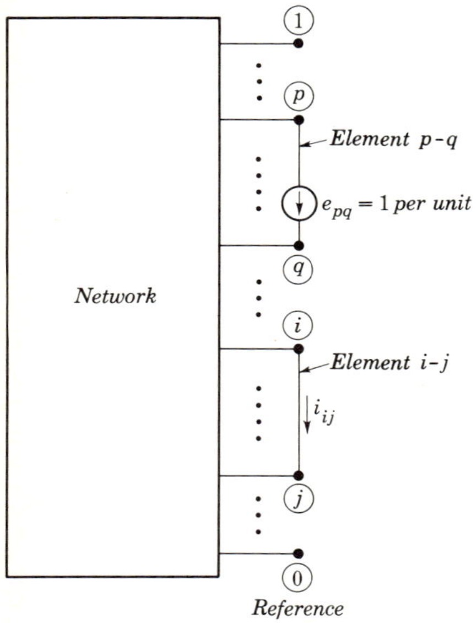

Fig. 4.7 Series voltage source for calculation of $Y_{ij,pq}$.

where i_{ij} = current through the element i–j
e_{pq} = voltage source in series with the element p–q
Let $e_{pq} = 1$ per unit, as shown in Fig. 4.7, then

$$Y_{ij,pq} = i_{ij} \tag{4.5.4}$$

It remains therefore to calculate the current i_{ij}.

The performance equation in admittance form for the primitive network is

$$\bar{\imath} + \bar{\jmath} = [y]\bar{v}$$

The current through the element i–j is

$$i_{ij} = -j_{ij} + \bar{y}_{ij,\rho\sigma}\bar{v}_{\rho\sigma} \tag{4.5.5}$$

where $\rho\sigma$ refers to all elements of the network. The voltage source in series with p–q induces currents in the elements mutually coupled with p–q. This voltage source can be replaced by equivalent current sources in parallel with each element, as shown in Fig. 4.8. The equivalent

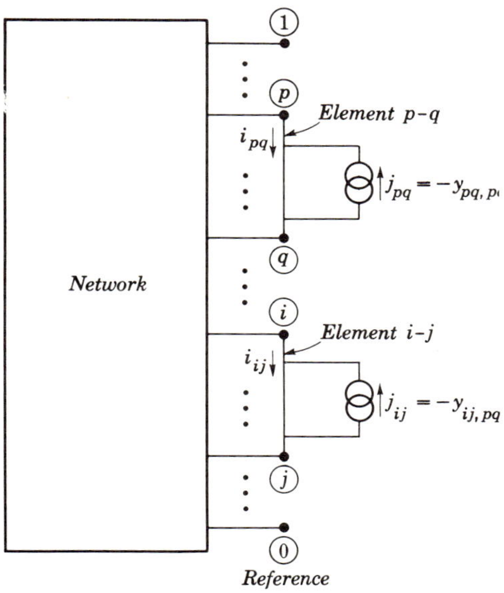

Fig. 4.8 Equivalent source currents for calculation of $Y_{ij,pq}$.

current source for the element i–j is

$$j_{ij} = -y_{ij,pq} \tag{4.5.6}$$

The voltages $\bar{v}_{\rho\sigma}$ can be obtained from the performance equation of the network using the node-pair impedance matrix

$$\bar{E}_{NP} = Z_{NP}\bar{I}_{NP} \tag{4.5.7}$$

From equation (4.5.7), the voltages across the elements ρ–σ of the network are

$$\bar{E}_\rho - \bar{E}_\sigma = [Z_{\rho\sigma,\mu\nu}]\bar{I}_{\mu\nu} \tag{4.5.8}$$

where

$$\bar{v}_{\rho\sigma} = \bar{E}_\rho - \bar{E}_\sigma \tag{4.5.9}$$

and the indices $\rho\sigma$ and $\mu\nu$ refer to the node-pairs corresponding to the terminals of the network elements. The elements of $Z_{\rho\sigma,\mu\nu}$ are obtained directly from Z_{NP}, and $Z_{\rho\sigma,\mu\nu}$ has dimension $e \times e$ where e equals the number of elements. The elements of the vector $\bar{I}_{\mu\nu}$ are equal to the shunt source currents which replace the series source voltage. Therefore,

$$\bar{I}_{\mu\nu} = -\bar{y}_{\mu\nu,pq} \tag{4.5.10}$$

Substituting from equations (4.5.9) and (4.5.10) into (4.5.8),

$$\bar{v}_{\rho\sigma} = -[Z_{\rho\sigma,\mu\nu}]\bar{y}_{\mu\nu,pq} \tag{4.5.11}$$

Substituting from equations (4.5.6) and (4.5.11) into (4.5.5) yields

$$i_{ij} = y_{ij,pq} - \bar{y}_{ij,\rho\sigma}[Z_{\rho\sigma,\mu\nu}]\bar{y}_{\mu\nu,pq}$$

Hence from equation (4.5.4),

$$Y_{ij,pq} = y_{ij,pq} - \bar{y}_{ij,\rho\sigma}[Z_{\rho\sigma,\mu\nu}]\bar{y}_{\mu\nu,pq} \tag{4.5.12}$$

and using all combinations of element pairs the matrix Y_{EP} can be obtained. The matrix Y_{EP} has dimension $e \times e$.

If there is no mutual coupling in the network, equation (4.5.12) reduces to

$$Y_{ij,pq} = -y_{ij,ij}Z_{ij,pq}y_{pq,pq} \qquad ij \neq pq$$

and

$$Y_{ij,ij} = y_{ij,ij} - y_{ij,ij}Z_{ij,ij}y_{ij,ij}$$

Equation (4.5.12) can be written in terms of the elements of the bus impedance matrix since

$$[Z_{\rho\sigma,\mu\nu}] = [Z_{\rho\mu}] - [Z_{\sigma\mu}] - [Z_{\rho\nu}] + [Z_{\sigma\nu}]$$

Then

$$Y_{ij,pq} = y_{ij,pq} - \bar{y}_{ij,\rho\sigma}([Z_{\rho\mu}] - [Z_{\sigma\mu}] - [Z_{\rho\nu}] + [Z_{\sigma\nu}])\bar{y}_{\mu\nu,pq}$$

If it is desired to derive Z_{NP} from a given Y_{EP}, the elements of Z_{NP} can be obtained in a manner similar to that described in this section. An element of Z_{NP}, in terms of the element-pair admittance matrix, can be expressed by

$$Z_{ij,pq} = z_{ij,pq} - \bar{z}_{ij,\rho\sigma}[Y_{\rho\sigma,\mu\nu}]\bar{z}_{\mu\nu,pq} \tag{4.5.13}$$

If there is no mutual coupling, equation (4.5.13) reduces to

$$Z_{ij,pq} = -z_{ij,ij}Y_{ij,pq}z_{pq,pq} \qquad ij \neq pq$$

and

$$Z_{ij,ij} = z_{ij,ij} - z_{ij,ij}Y_{ij,ij}z_{ij,ij}$$

Derivation of loop admittance matrix from element-pair admittance matrix

Given the element-pair admittance matrix Y_{EP}, the elements of the loop admittance matrix Y_{LOOP} can be obtained directly. Y_{EP} is partitioned as follows:

$Y_{EP} =$

	Branches	Links
Branches	Y_1	Y_2
Links	Y_3	Y_4

where the submatrix Y_4 is associated with the basic loops of the interconnected network since each link corresponds to a basic loop. Therefore

$$Y_{LOOP} = Y_4$$

It is possible to derive Y_{EP} from a given Y_{LOOP} since the elements of the submatrices Y_1, Y_2, and Y_3 can be determined from the elements of Y_4. Let i–j be a branch common to loops A, B, and C and let p–q be a branch common to loops G and H. If a voltage source $e_{pq} = 1$ per unit is applied in series with the branch p–q, then by definition of an element of Y_{EP},

$$Y_{ij,pq} = i_{ij}$$

Moreover, the voltages in the loops G and H are equal to one per unit. Hence, the currents in the loops A, B, and C are

$$I_A = Y_{AG} + Y_{AH}$$
$$I_B = Y_{BG} + Y_{BH}$$
$$I_C = Y_{CG} + Y_{CH}$$

where the admittances are obtained from the loop admittance matrix. Since the current in branch i–j is the algebraic sum of the currents in loops A, B, and C,

$$i_{ij} = I_A + I_B + I_C$$

Therefore,

$$Y_{ij,pq} = Y_{AG} + Y_{AH} + Y_{BG} + Y_{BH} + Y_{CG} + Y_{CH}$$

The signs of the loop admittance terms are determined by the orientation of the branches with respect to the loops.

4.6 Example of derivation of loop admittance matrix from bus impedance matrix

The method of deriving the loop admittance matrix from the bus impedance matrix will be illustrated for the sample network shown in Fig. 3.10.

Problem

Derive the loop admittance matrix Y_{LOOP} from the bus impedance matrix Z_{BUS} of the network shown in Fig. 3.10.

Solution

The primitive admittance matrix is

$[y] =$	1–2(1)	1–2(2)	1–3	2–4	3–4
1–2(1)	2.083	−1.042	−0.417		
1–2(2)	−1.042	3.021	0.208		
1–3	−0.417	0.208	2.083		
2–4				5.0	
3–4					2.0

The bus impedance matrix of the network obtained by nonsingular transformation is

$Z_{BUS} =$	②	③	④
②	0.271	0.126	0.230
③	0.126	0.344	0.189
④	0.230	0.189	0.362

First, form the node-pair impedance matrix Z_{NP}. The elements of the first row from the equation

$$Z_{ij,pq} = Z_{ip} - Z_{jp} - Z_{iq} + Z_{jq}$$

are

$$\begin{aligned}
Z_{12,12} &= Z_{11} - Z_{21} - Z_{12} + Z_{22} \\
&= 0 - 0 - 0 + 0.271 = 0.271 \\
Z_{12,13} &= Z_{11} - Z_{21} - Z_{13} + Z_{23} \\
&= 0 - 0 - 0 + 0.126 = 0.126 \\
Z_{12,24} &= Z_{12} - Z_{22} - Z_{14} + Z_{24} \\
&= 0 - 0.271 - 0 + 0.230 = -0.041 \\
Z_{12,34} &= Z_{13} - Z_{23} - Z_{14} + Z_{24} \\
&= 0 - 0.126 - 0 + 0.230 = 0.104 \\
Z_{12,14} &= Z_{11} - Z_{21} - Z_{14} + Z_{24} \\
&= 0 - 0 - 0 + 0.230 = 0.230 \\
Z_{12,23} &= Z_{12} - Z_{22} - Z_{13} + Z_{23} \\
&= 0 - 0.271 - 0 + 0.126 = -0.145
\end{aligned}$$

The elements of the remaining rows are obtained in a similar manner. The node-pair impedance matrix is

$Z_{NP} =$

	1–2	1–3	2–4	3–4	1–4	2–3
1–2	0.271	0.126	−0.041	0.104	0.230	−0.145
1–3	0.126	0.344	0.063	−0.155	0.189	0.218
2–4	−0.041	0.063	0.173	0.069	0.132	0.104
3–4	0.104	−0.155	0.069	0.328	0.173	−0.259
1–4	0.230	0.189	0.132	0.173	0.362	−0.041
2–3	−0.145	0.218	0.104	−0.259	−0.041	0.363

Then

$Z_{p\sigma,\mu\nu} =$

	1–2(1)	1–2(2)	1–3	2–4	3–4
1–2(1)	0.271	0.271	0.126	−0.041	0.104
1–2(2)	0.271	0.271	0.126	−0.041	0.104
1–3	0.126	0.126	0.344	0.063	−0.155
2–4	−0.041	−0.041	0.063	0.173	0.069
3–4	0.104	0.104	−0.155	0.069	0.328

Second, form the submatrix Y_4 of the element-pair admittance matrix Y_{EP}. The elements of the submatrix are obtained from the equation

$$Y_{ij,pq} = y_{ij,pq} - \bar{y}_{ij,p\sigma}[Z_{p\sigma,\mu\nu}]\bar{y}_{\mu\nu,pq}$$

The element $Y_{12(2),12(2)}$ of Y_4 is

$$Y_{12(2),12(2)} = y_{12(2),12(2)} - \begin{bmatrix} y_{12(2),12(1)} & y_{12(2),12(2)} & y_{12(2),13} & y_{12(2),24} & y_{12(2),34} \end{bmatrix} [Z_{\rho\sigma,\mu\nu}] \begin{bmatrix} y_{12(1),12(2)} \\ y_{12(2),12(2)} \\ y_{13,12(2)} \\ y_{24,12(2)} \\ y_{34,12(2)} \end{bmatrix}$$

$$= 3.021 - \begin{bmatrix} -1.042 & 3.021 & 0.208 & & \end{bmatrix} \begin{bmatrix} 0.271 & 0.271 & 0.126 & -0.041 & 0.104 \\ 0.271 & 0.271 & 0.126 & -0.041 & 0.104 \\ 0.126 & 0.126 & 0.344 & 0.063 & -0.155 \\ -0.041 & -0.041 & 0.063 & 0.173 & 0.069 \\ 0.104 & 0.104 & -0.155 & 0.069 & 0.328 \end{bmatrix} \begin{bmatrix} -1.042 \\ 3.021 \\ 0.208 \\ \\ \end{bmatrix}$$

$$= 3.021 - \begin{bmatrix} 0.5625 & 0.5625 & 0.3209 & -0.0680 & 0.1736 \end{bmatrix} \begin{bmatrix} -1.042 \\ 3.021 \\ 0.208 \\ \\ \end{bmatrix}$$

$$= 3.021 - 1.180 = 1.841$$

The element $Y_{12(2),24}$ is

$$Y_{12(2),24} = y_{12(2),24} - \begin{bmatrix} y_{12(2),12(1)} & y_{12(2),12(2)} & y_{12(2),13} & y_{12(2),24} & y_{12(2),34} \end{bmatrix} [Z_{\rho\sigma,\mu\nu}] \begin{bmatrix} y_{12(1),24} \\ y_{12(2),24} \\ y_{13,24} \\ y_{24,24} \\ y_{34,24} \end{bmatrix}$$

$$= 0 - \begin{bmatrix} -1.042 & 3.021 & 0.208 & & \end{bmatrix} \begin{bmatrix} 0.271 & 0.271 & 0.126 & -0.041 & 0.104 \\ 0.271 & 0.271 & 0.126 & -0.041 & 0.104 \\ 0.126 & 0.126 & 0.344 & 0.063 & -0.155 \\ -0.041 & -0.041 & 0.063 & 0.173 & 0.069 \\ 0.104 & 0.104 & -0.155 & 0.069 & 0.328 \end{bmatrix} \begin{bmatrix} \\ \\ \\ 5.0 \\ \\ \end{bmatrix}$$

$$= 0 - (-0.340) = 0.340$$

The elements $Y_{24,12(2)}$ and $Y_{24,24}$ are calculated in a similar manner. The submatrix Y_4 is the loop admittance matrix

$Y_{LOOP} =$

	D	E
D	1.841	0.340
E	0.340	0.675

Problems

4.1 Using the data for the sample power system given in Prob. 3.2 and neglecting resistance, form the following positive sequence matrices:
a. The bus impedance matrix using the algorithm
b. The node-pair impedance matrix Z_{NP}
c. The element-pair admittance matrix Y_{EP}
d. The loop admittance matrix Y_{LOOP} from Y_{EP}

4.2 Repeat Prob. 4.1 using the zero sequence network data and neglecting resistance.

4.3 Modify the positive and zero sequence bus impedance matrices obtained in Probs. 4.1 and 4.2 to reflect the opening of the north circuit N between buses B and C.

4.4 Derive the equation

$$Z_{ij,pq} = z_{ij,pq} - \bar{z}_{ij,\rho\sigma}[Y_{\rho\sigma,\mu\nu}]\bar{z}_{\mu\nu,pq}$$

for obtaining the elements of the node-pair impedance matrix Z_{NP} using the element-pair admittance matrix Y_{EP}.

4.5 Form the node-pair impedance matrix Z_{NP} using the loop admittance matrix Y_{LOOP} obtained from Prob. 4.1, part *d*.

4.6 Show that the branch impedance matrix Z_{BR} can be derived from the node-pair impedance matrix Z_{NP}.

Bibliography

Brown, H. E., C. E. Person, L. K. Kirchmayer, and G. W. Stagg: Digital Calculation of Three-phase Short Circuits by Matrix Method, *Trans. AIEE*, vol. 79, pt. III, pp. 1277–1281, 1960.

El-Abiad, Ahmed H.: Algorithms for Direct Computation and Modification of Solution (Inverted) Matrices of Networks Including Mutual Impedances, *IEEE Proc. Power Ind. Computer Appl. Conf.*, *Phoenix, Ariz.*, pp. 150–166, 1963.

———: Digital Calculation of Line to Ground Short Circuits by Matrix Method, *Trans. AIEE*, vol. 79, pt. III, pp. 323–332, 1960.

———: Digital Computer Analysis of Large Linear Systems, *Proc. First Ann. Allerton Conf. Circuits and System Theory, University of Illinois, Urbana*, 1963.

Glimn, A. F., R. Habermann, Jr., J. M. Henderson, and L. K. Kirchmayer: Digital Calculation of Network Impedances, *Trans. AIEE*, vol. 74, pt. III, pp. 1285–1297, 1955.

Hale, H. W., and J. B. Ward: Digital Computation of Driving Point and Transfer Impedances, *Trans. AIEE*, vol. 76, pt. III, pp. 476–481, 1957.

Shipley, R. Bruce: Tensors Extend Network Analyzer Fault Studies, Electric Light and Power, Chicago, December, 1955.

chapter 5
Three-phase networks

5.1 Introduction

Power systems are operated usually with balanced three-phase generation and loads. A balanced network is obtained by the transposing of transmission lines. This makes possible the treatment of many three-phase power system problems on a single-phase basis. If there is unbalanced excitation on a balanced network the solution of network problems can be obtained by one of two methods. The first method analyzes the network in terms of actual phase quantities. The second method involves the transformation of unbalanced phase quantities into balanced sequence quantities. Two important types of sequence quantities are symmetrical components and Clarke's components. For symmetrical components, the balanced sequence impedances are uncoupled for both stationary and rotating elements. For Clarke's components the balanced sequence impedances are uncoupled only for stationary elements. Transformations for unbalanced networks, in general, do not yield uncoupled sequence impedances.

5.2 Three-phase network elements

A three-phase network component represented in impedance form is shown in Fig. 5.1. This component represented in admittance form is shown in Fig. 5.2. The variables and parameters are:

$v^a_{pq}, v^b_{pq}, v^c_{pq}$	are the voltages across the element p–q for phases a, b, and c, respectively
$e^a_{pq}, e^b_{pq}, e^c_{pq}$	are the source voltages in series with phases a, b, and c, respectively, of the element p–q

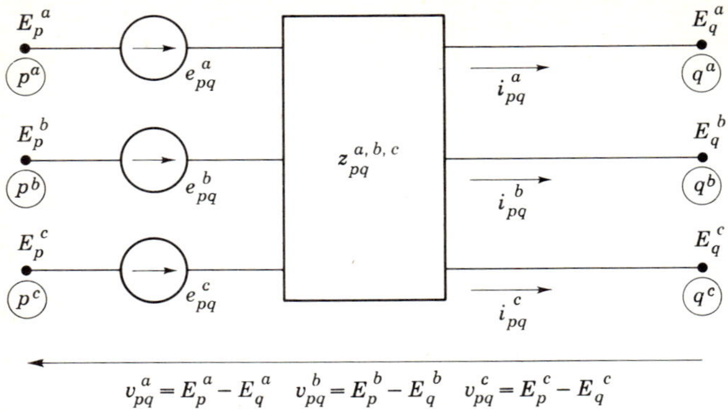

Fig. 5.1 ***Representation of three-phase network component in impedance form.***

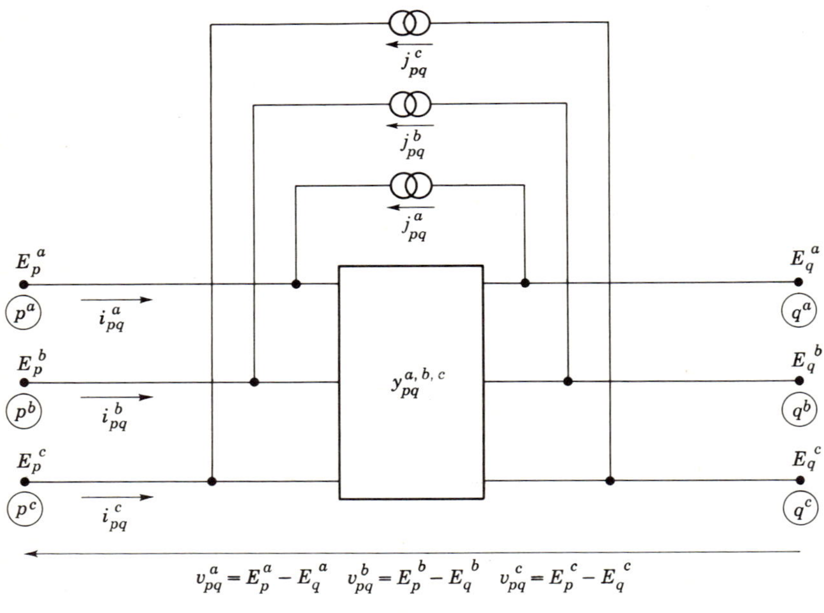

Fig. 5.2 ***Representation of three-phase network component in admittance form.***

$i_{pq}^{a}, i_{pq}^{b}, i_{pq}^{c}$ are the currents through the element p–q for phases a, b, and c, respectively

$j_{pq}^{a}, j_{pq}^{b}, j_{pq}^{c}$ are the source currents in parallel with phases a, b, and c, respectively, of the element p–q

$z_{pq}^{a,b,c}$ is the three-phase impedance matrix for the element p–q

$y_{pq}^{a,b,c}$ is the three-phase admittance matrix for the element p–q

The performance equation of a three-phase element in impedance form is

$$\begin{bmatrix} v_{pq}^{a} \\ v_{pq}^{b} \\ v_{pq}^{c} \end{bmatrix} + \begin{bmatrix} e_{pq}^{a} \\ e_{pq}^{b} \\ e_{pq}^{c} \end{bmatrix} = \begin{bmatrix} z_{pq}^{aa} & z_{pq}^{ab} & z_{pq}^{ac} \\ z_{pq}^{ba} & z_{pq}^{bb} & z_{pq}^{bc} \\ z_{pq}^{ca} & z_{pq}^{cb} & z_{pq}^{cc} \end{bmatrix} \begin{bmatrix} i_{pq}^{a} \\ i_{pq}^{b} \\ i_{pq}^{c} \end{bmatrix} \tag{5.2.1}$$

where z_{pq}^{aa} = self-impedance of phase a of the three-phase element connecting nodes p and q

z_{pq}^{ab} = mutual impedance between phases a and b

z_{pq}^{ac} = mutual impedance between phases a and c

and so forth.

Equation (5.2.1) can be written more concisely as

$$v_{pq}^{a,b,c} + e_{pq}^{a,b,c} = z_{pq}^{a,b,c} i_{pq}^{a,b,c} \tag{5.2.2}$$

The performance equation in admittance form is

$$\begin{bmatrix} i_{pq}^{a} \\ i_{pq}^{b} \\ i_{pq}^{c} \end{bmatrix} + \begin{bmatrix} j_{pq}^{a} \\ j_{pq}^{b} \\ j_{pq}^{c} \end{bmatrix} = \begin{bmatrix} y_{pq}^{aa} & y_{pq}^{ab} & y_{pq}^{ac} \\ y_{pq}^{ba} & y_{pq}^{bb} & y_{pq}^{bc} \\ y_{pq}^{ca} & y_{pq}^{cb} & y_{pq}^{cc} \end{bmatrix} \begin{bmatrix} v_{pq}^{a} \\ v_{pq}^{b} \\ v_{pq}^{c} \end{bmatrix}$$

which can be written

$$i_{pq}^{a,b,c} + j_{pq}^{a,b,c} = y_{pq}^{a,b,c} v_{pq}^{a,b,c}$$

where

$$y_{pq}^{a,b,c} = (z_{pq}^{a,b,c})^{-1}$$

The parallel three-phase source current in admittance form and the three-phase series source voltage in impedance form have the relationship, as is the case in single-phase representation,

$$j_{pq}^{a,b,c} = -y_{pq}^{a,b,c} e_{pq}^{a,b,c}$$

The impedance matrix $z_{pq}^{a,b,c}$ and the admittance matrix $y_{pq}^{a,b,c}$ of a stationary bilateral element are symmetric. If, in addition, the three-phase element is balanced, then the diagonal elements of $z_{pq}^{a,b,c}$, designated by z_{pq}^{s}, are equal and the off-diagonal elements, designated by z_{pq}^{m}, are equal, that is,

$$z_{pq}^{aa} = z_{pq}^{bb} = z_{pq}^{cc} = z_{pq}^{s}$$

and

$$z_{pq}^{ab} = z_{pq}^{ac} = z_{pq}^{ba} = z_{pq}^{bc} = z_{pq}^{ca} = z_{pq}^{cb} = z_{pq}^{m}$$

The corresponding relations are true in the admittance matrix $y_{pq}^{a,b,c}$.

The impedance and admittance matrices of balanced three-phase rotating elements are not symmetric. However, the mutual coupling from phase a to phase b, b to c, and c to a for the phase sequence a, b, c are identical, that is,

$$z_{pq}^{ab} = z_{pq}^{bc} = z_{pq}^{ca} = z_{pq}^{m1}$$

Similarly,

$$z_{pq}^{ac} = z_{pq}^{ba} = z_{pq}^{cb} = z_{pq}^{m2}$$

The performance equation of the three-phase primitive network in impedance form is

$$\bar{v}^{a,b,c} + \bar{e}^{a,b,c} = [z^{a,b,c}]\bar{\imath}^{a,b,c}$$

or in the admittance form is

$$\bar{\imath}^{a,b,c} + \bar{\jmath}^{a,b,c} = [y^{a,b,c}]\bar{v}^{a,b,c}$$

The vectors representing the variables are composed of 3×1 submatrices corresponding to the variables of a particular three-phase network element. The parameter matrices are composed of 3×3 submatrices. These submatrices correspond to the self and mutual three-phase impedance or admittance matrices of the network elements.

5.3 Three-phase balanced network elements

Balanced excitation

The excitation of any three-phase element is balanced when the source voltages or source currents of all phases are equal in magnitude and displaced from each other by 120°. For balanced excitation,

$$e_{pq}^{a,b,c} = \begin{bmatrix} e_{pq}^{a} \\ e_{pq}^{b} \\ e_{pq}^{c} \end{bmatrix} = \begin{bmatrix} 1 \\ a^2 \\ a \end{bmatrix} e_{pq}^{a} \qquad \text{and} \qquad j_{pq}^{a,b,c} = \begin{bmatrix} j_{pq}^{a} \\ j_{pq}^{b} \\ j_{pq}^{c} \end{bmatrix} = \begin{bmatrix} 1 \\ a^2 \\ a \end{bmatrix} j_{pq}^{a}$$

where

$$a = \epsilon^{j(2\pi/3)} = -\tfrac{1}{2} + j\tfrac{1}{2}\sqrt{3}$$

It follows that $a^3 = 1$, $a^2 + a + 1 = 0$, and $a^* = a^2$. The phase voltages and phase currents are balanced if the excitation of a balanced three-phase element is balanced. Then, the performance equation, in impedance form, for a stationary element is

$$\begin{bmatrix} 1 \\ a^2 \\ a \end{bmatrix} v_{pq}^{a} + \begin{bmatrix} 1 \\ a^2 \\ a \end{bmatrix} e_{pq}^{a} = \begin{bmatrix} z_{pq}^{s} & z_{pq}^{m} & z_{pq}^{m} \\ z_{pq}^{m} & z_{pq}^{s} & z_{pq}^{m} \\ z_{pq}^{m} & z_{pq}^{m} & z_{pq}^{s} \end{bmatrix} \begin{bmatrix} 1 \\ a^2 \\ a \end{bmatrix} i_{pq}^{a} \tag{5.3.1}$$

and for a rotating element is

$$\begin{bmatrix} 1 \\ a^2 \\ a \end{bmatrix} v_{pq}^{a} + \begin{bmatrix} 1 \\ a^2 \\ a \end{bmatrix} e_{pq}^{a} = \begin{bmatrix} z_{pq}^{s} & z_{pq}^{m1} & z_{pq}^{m2} \\ z_{pq}^{m2} & z_{pq}^{s} & z_{pq}^{m1} \\ z_{pq}^{m1} & z_{pq}^{m2} & z_{pq}^{s} \end{bmatrix} \begin{bmatrix} 1 \\ a^2 \\ a \end{bmatrix} i_{pq}^{a} \tag{5.3.2}$$

Both sides of equation (5.3.1) can be premultiplied by the conjugate transpose of

$$\begin{bmatrix} 1 \\ a^2 \\ a \end{bmatrix}$$

that is,

1	a	a^2

to obtain

$$3v_{pq}^a + 3e_{pq}^a = 3(z_{pq}^s - z_{pq}^m)i_{pq}^a \tag{5.3.3}$$

Dividing by 3, equation (5.3.3) becomes

$$v_{pq}^a + e_{pq}^a = (z_{pq}^s - z_{pq}^m)i_{pq}^a$$

where $(z_{pq}^s - z_{pq}^m)$ is the positive sequence impedance, which is designated by $z_{pq}^{(1)}$. Thus, a balanced three-phase element with balanced excitation can be treated as a single-phase element in network problems. The power in the element is equal to three times the power per phase.

In a similar manner, equation (5.3.2) can be reduced to

$$v_{pq}^a + e_{pq}^a = (z_{pq}^s + a^2 z_{pq}^{m1} + a z_{pq}^{m2})i_{pq}^a$$

where $z_{pq}^s + a^2 z_{pq}^{m1} + a z_{pq}^{m2}$ is the positive sequence impedance.

The performance equation, in admittance form, for a stationary element is

$$i_{pq}^a + j_{pq}^a = (y_{pq}^s - y_{pq}^m)v_{pq}^a$$

and for a rotating element is

$$i_{pq}^a + j_{pq}^a = (y_{pq}^s + a^2 y_{pq}^{m1} + a y_{pq}^{m2})v_{pq}^a$$

Unbalanced excitation

When the excitation is unbalanced, the performance equation of a three-phase element can be reduced to three independent equations by diagonalizing the impedance matrix $z_{pq}^{a,b,c}$. Using a complex transformation matrix T then the phase variables are expressed in terms of a new set of variables as follows:

$$\begin{aligned} v_{pq}^{a,b,c} &= T v_{pq}^{i,j,k} \\ e_{pq}^{a,b,c} &= T e_{pq}^{i,j,k} \\ i_{pq}^{a,b,c} &= T i_{pq}^{i,j,k} \end{aligned} \tag{5.3.4}$$

The complex power in the element is

$$S_{pq} = P_{pq} + jQ_{pq} = \{(i_{pq}^{a,b,c})^*\}^t e_{pq}^{a,b,c}$$

Substituting from equations (5.3.4),

$$S_{pq} = \{(i_{pq}^{i,j,k})^*\}^t (T^*)^t T e_{pq}^{i,j,k} \tag{5.3.5}$$

The complex power in terms of the i, j, k sequence variables is

$$S'_{pq} = \{(i_{pq}^{i,j,k})^*\}^t e_{pq}^{i,j,k} \tag{5.3.6}$$

If the complex powers S_{pq} and S'_{pq} are equal, that is, the selected transformation T is power-invariant, then from equations (5.3.5) and (5.3.6),

$$(T^*)^t T = U = T(T^*)^t$$

Thus T is a unitary matrix.

Substituting from equations (5.3.4) the performance equation (5.2.2) becomes

$$T(v_{pq}^{i,j,k} + e_{pq}^{i,j,k}) = z_{pq}^{a,b,c} T i_{pq}^{i,j,k} \tag{5.3.7}$$

Both sides of equation (5.3.7) can be premultiplied by $(T^*)^t$ to obtain

$$v_{pq}^{i,j,k} + e_{pq}^{i,j,k} = (T^*)^t z_{pq}^{a,b,c} T i_{pq}^{i,j,k}$$

It follows that

$$z_{pq}^{i,j,k} = (T^*)^t z_{pq}^{a,b,c} T \tag{5.3.8}$$

A similar transformation can be obtained for the performance equation in its admittance form.

5.4 Transformation matrices

Symmetrical components

Two particular transformations for three-phase balanced elements are of interest. One of these transforms the three-phase quantities into zero, positive, and negative sequence quantities, known as symmetrical components. The matrix for this transformation is

$$T_s = \frac{1}{\sqrt{3}} \begin{bmatrix} 1 & 1 & 1 \\ 1 & a^2 & a \\ 1 & a & a^2 \end{bmatrix}$$

which is a unitary matrix, that is, $(T_s^*)^t T_s = U$; and furthermore, because T_s is symmetric, $T_s^* = T_s^{-1}$. Using this transformation the impedance matrix for a stationary element $z_{pq}^{i,j,k}$ from equation (5.3.8) becomes

$$z_{pq}^{0,1,2} = \frac{1}{\sqrt{3}} \begin{bmatrix} 1 & 1 & 1 \\ 1 & a & a^2 \\ 1 & a^2 & a \end{bmatrix} \begin{bmatrix} z_{pq}^s & z_{pq}^m & z_{pq}^m \\ z_{pq}^m & z_{pq}^s & z_{pq}^m \\ z_{pq}^m & z_{pq}^m & z_{pq}^s \end{bmatrix} \frac{1}{\sqrt{3}} \begin{bmatrix} 1 & 1 & 1 \\ 1 & a^2 & a \\ 1 & a & a^2 \end{bmatrix}$$

That is,

$$z_{pq}^{0,1,2} = \begin{bmatrix} z_{pq}^s + 2z_{pq}^m & & \\ & z_{pq}^s - z_{pq}^m & \\ & & z_{pq}^s - z_{pq}^m \end{bmatrix}$$

where the zero sequence impedance is

$$z_{pq}^{(0)} = z_{pq}^s + 2z_{pq}^m$$

the positive sequence impedance is

$$z_{pq}^{(1)} = z_{pq}^s - z_{pq}^m$$

the negative sequence impedance is

$$z_{pq}^{(2)} = z_{pq}^s - z_{pq}^m$$

and $z_{pq}^{0,1,2}$ refers to the transformed impedance matrix, which is diagonal for a balanced three-phase element.

The transformation matrix T_s also diagonalizes the impedance matrix for a rotating element, even though $z_{pq}^{a,b,c}$ is not symmetric. This diagonalized matrix is

$$z_{pq}^{0,1,2} = \begin{bmatrix} z_{pq}^s + z_{pq}^{m1} + z_{pq}^{m2} & & \\ & z_{pq}^s + a^2 z_{pq}^{m1} + a z_{pq}^{m2} & \\ & & z_{pq}^s + a z_{pq}^{m1} + a^2 z_{pq}^{m2} \end{bmatrix}$$

where

$$\begin{aligned} z_{pq}^{(0)} &= z_{pq}^s + z_{pq}^{m1} + z_{pq}^{m2} \\ z_{pq}^{(1)} &= z_{pq}^s + a^2 z_{pq}^{m1} + a z_{pq}^{m2} \\ z_{pq}^{(2)} &= z_{pq}^s + a z_{pq}^{m1} + a^2 z_{pq}^{m2} \end{aligned}$$

Clarke's components

Another transformation matrix transforms the three-phase quantities into zero, alpha, and beta sequence quantities, known as Clarke's components. The matrix for this transformation is

$$T_c = \frac{1}{\sqrt{3}} \begin{bmatrix} 1 & \sqrt{2} & 0 \\ 1 & -\sqrt{1/2} & \sqrt{3/2} \\ 1 & -\sqrt{1/2} & -\sqrt{3/2} \end{bmatrix}$$

which is an orthogonal matrix, that is, $T_c^t T_c = U$. Therefore $T_c^t = T_c^{-1}$. Using this transformation the impedance matrix for a stationary element $z_{pq}^{i,j,k}$ from equation (5.3.8) becomes

$$z_{pq}^{0,\alpha,\beta} = \begin{bmatrix} z_{pq}^s + 2z_{pq}^m & & \\ & z_{pq}^s - z_{pq}^m & \\ & & z_{pq}^s - z_{pq}^m \end{bmatrix}$$

where the diagonal elements are the zero, alpha, and beta impedance components, respectively, and $z_{pq}^{0,\alpha,\beta}$ refers to the transformed impedance matrix which is diagonal for a balanced three-phase element.

The transformation matrix T_c does not diagonalize the nonsymmetric impedance matrix $z_{pq}^{a,b,c}$ for a rotating element. The following is obtained by this transformation.

$$z_{pq}^{0,\alpha,\beta} = \begin{bmatrix} z_{pq}^s + z_{pq}^{m1} + z_{pq}^{m2} & & \\ & z_{pq}^s - \tfrac{1}{2}(z_{pq}^{m1} + z_{pq}^{m2}) & \frac{\sqrt{3}}{2}(z_{pq}^{m1} - z_{pq}^{m2}) \\ & \frac{\sqrt{3}}{2}(z_{pq}^{m2} - z_{pq}^{m1}) & z^s - \tfrac{1}{2}(z_{pq}^{m1} + z_{pq}^{m2}) \end{bmatrix}$$

5.5 Three-phase unbalanced network elements

When a three-phase element is unbalanced, the transformation T_s, or T_c, on $z_{pq}^{a,b,c}$ does not yield uncoupled sequence impedances. Even though it is possible to diagonalize $z_{pq}^{a,b,c}$, no single transformation exists for diagonalizing the impedance matrices for all elements of a network because the unbalance of the different elements, in general, is not related. Therefore, it may be desirable to maintain the original three-phase quantities for the solution of network problems. When the transformation T_s is used the sequence networks cannot be treated independently.

5.6 Incidence and network matrices for three-phase networks

Incidence and network matrices for a three-phase balanced or unbalanced network can be formed by the same procedures as those described in Chap. 3 for single-phase networks. The entries 1, -1, and 0 in the incidence matrices for a single-phase network, however, will be replaced by the 3×3 matrices, U, $-U$, and null, respectively. Also, the impedance or admittance of a network element will be a 3×3 matrix. The rows and columns of this matrix refer to the phases a, b, and c or to the appropriate sequence components. The network matrices will be composed of 3×3 submatrices whose elements also refer to the phase or sequence components.

5.7 Algorithm for formation of three-phase bus impedance matrix

Performance equation of a partial three-phase network

The performance equation for a three-phase network representation in the bus frame of reference and impedance form is

$$\bar{E}_{BUS}^{a,b,c} = Z_{BUS}^{a,b,c} \bar{I}_{BUS}^{a,b,c}$$

where $\bar{E}_{BUS}^{a,b,c}$ = vector of the three-phase bus voltages measured with respect to the reference bus

$\bar{I}_{BUS}^{a,b,c}$ = vector of impressed three-phase bus currents

$Z_{BUS}^{a,b,c}$ = three-phase bus impedance matrix

When the three-phase elements of the network are balanced, their impedance or admittance matrices can be diagonalized by the transformation matrix T_s or T_c. In this case, the three sequence networks can be treated independently. The procedures based on the algorithm described in Chap. 4 can be applied to form the independent sequence network matrices.

When the three-phase elements of the network are unbalanced, the 3×3 submatrices $Z_{ij}^{a,b,c}$ and $Z_{ji}^{a,b,c}$ of the bus impedance matrix are not equal. The equations for the formation of the three-phase bus impedance matrix by the algorithm can be derived in a manner similar to that for single-phase networks.

Addition of a branch

The performance equation of the partial network with an added branch p–q, in terms of three-phase quantities, is

$$
\begin{array}{c}
\begin{bmatrix} E_1^{a,b,c} \\ E_2^{a,b,c} \\ \cdots \\ E_p^{a,b,c} \\ \cdots \\ E_m^{a,b,c} \\ \hline E_q^{a,b,c} \end{bmatrix}
\end{array}
=
\begin{array}{c|cccccccc|c}
 & 1 & 2 & & p & & m & q \\
1 & Z_{11}^{a,b,c} & Z_{12}^{a,b,c} & \cdots & Z_{1p}^{a,b,c} & \cdots & Z_{1m}^{a,b,c} & Z_{1q}^{a,b,c} \\
2 & Z_{21}^{a,b,c} & Z_{22}^{a,b,c} & \cdots & Z_{2p}^{a,b,c} & \cdots & Z_{2m}^{a,b,c} & Z_{2q}^{a,b,c} \\
 & \cdots & \cdots & \cdots & \cdots & \cdots & \cdots & \cdots \\
p & Z_{p1}^{a,b,c} & Z_{p2}^{a,b,c} & \cdots & Z_{pp}^{a,b,c} & \cdots & Z_{pm}^{a,b,c} & Z_{pq}^{a,b,c} \\
 & \cdots & \cdots & \cdots & \cdots & \cdots & \cdots & \cdots \\
m & Z_{m1}^{a,b,c} & Z_{m2}^{a,b,c} & \cdots & Z_{mp}^{a,b,c} & \cdots & Z_{mm}^{a,b,c} & Z_{mq}^{a,b,c} \\
\hline
q & Z_{q1}^{a,b,c} & Z_{q2}^{a,b,c} & \cdots & Z_{qp}^{a,b,c} & \cdots & Z_{qm}^{a,b,c} & Z_{qq}^{a,b,c}
\end{array}
\begin{bmatrix} I_1^{a,b,c} \\ I_2^{a,b,c} \\ \cdots \\ I_p^{a,b,c} \\ \cdots \\ I_m^{a,b,c} \\ \hline I_q^{a,b,c} \end{bmatrix}
\qquad (5.7.1)
$$

The elements $Z_{qi}^{a,b,c}$ can be determined by injecting a three-phase current at the ith bus, as shown in Fig. 5.3, and measuring the voltage at the qth bus with respect to the reference node. Similarly, the elements $Z_{iq}^{a,b,c}$ can be determined by injecting a three-phase current at the qth bus, as

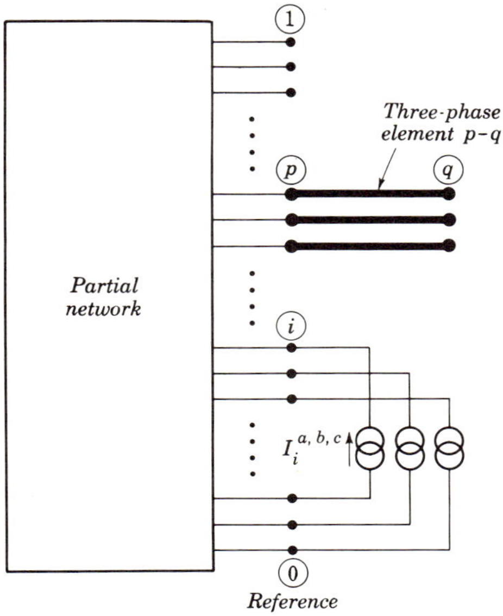

Fig. 5.3 Injected three-phase current for calculation of $Z_{qi}^{a,b,c}$.

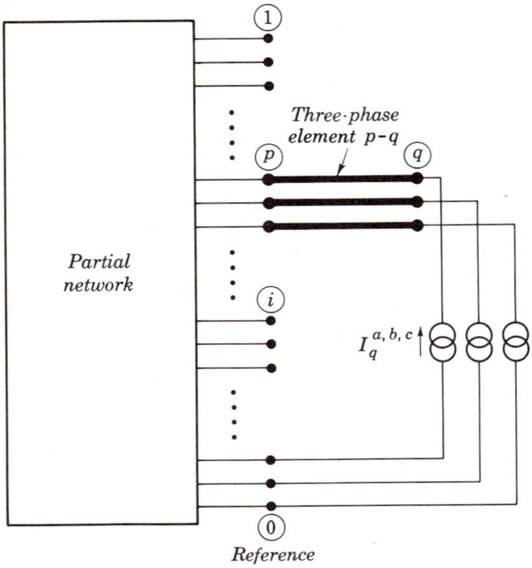

Fig. 5.4 Injected three-phase current for calculation of $Z_{iq}^{a,b,c}$ and $Z_{qq}^{a,b,c}$.

shown in Fig. 5.4, and measuring the voltage at the ith bus with respect to the reference node.

To calculate $Z_{qi}^{a,b,c}$ let the current at the ith bus be $I_i^{a,b,c}$ and all other bus currents equal zero. The voltage across the added element p–q is

$$v_{pq}^{a,b,c} = E_p^{a,b,c} - E_q^{a,b,c} \tag{5.7.2}$$

The vector of voltages across the elements ρ–σ of the partial network is

$$\bar{v}_{\rho\sigma}^{a,b,c} = \bar{E}_\rho^{a,b,c} - \bar{E}_\sigma^{a,b,c} \tag{5.7.3}$$

The current in the element p–q, in terms of the primitive admittances and the voltages across the elements, is

$$i_{pq}^{a,b,c} = y_{pq,pq}^{a,b,c} v_{pq}^{a,b,c} + \bar{y}_{pq,\rho\sigma}^{a,b,c} \bar{v}_{\rho\sigma}^{a,b,c} \tag{5.7.4}$$

Since $i_{pq}^{a,b,c} = 0$, from equation (5.7.4),

$$v_{pq}^{a,b,c} = -(y_{pq,pq}^{a,b,c})^{-1} \bar{y}_{pq,\rho\sigma}^{a,b,c} \bar{v}_{\rho\sigma}^{a,b,c} \tag{5.7.5}$$

Substituting from equations (5.7.2) and (5.7.3) into (5.7.5),

$$E_p^{a,b,c} - E_q^{a,b,c} = -(y_{pq,pq}^{a,b,c})^{-1} \bar{y}_{pq,\rho\sigma}^{a,b,c} (\bar{E}_\rho^{a,b,c} - \bar{E}_\sigma^{a,b,c}) \tag{5.7.6}$$

From equation (5.7.1),

$$\begin{aligned} E_p^{a,b,c} &= Z_{pi}^{a,b,c} I_i^{a,b,c} \\ E_q^{a,b,c} &= Z_{qi}^{a,b,c} I_i^{a,b,c} \end{aligned} \tag{5.7.7}$$

and at any bus k,

$$E_k^{a,b,c} = Z_{ki}^{a,b,c} I_i^{a,b,c}$$

Using the relationships from equation (5.7.7) in (5.7.6) and solving for $Z_{qi}^{a,b,c} I_i^{a,b,c}$,

$$Z_{qi}^{a,b,c} I_i^{a,b,c} = Z_{pi}^{a,b,c} I_i^{a,b,c} + \{(y_{pq,pq}^{a,b,c})^{-1} \bar{y}_{pq,\rho\sigma}^{a,b,c} (\bar{Z}_{\rho i}^{a,b,c} - \bar{Z}_{\sigma i}^{a,b,c})\} I_i^{a,b,c} \tag{5.7.8}$$

Since equation (5.7.8) is valid for all values of $I_i^{a,b,c}$, it follows that

$$Z_{qi}^{a,b,c} = Z_{pi}^{a,b,c} + (y_{pq,pq}^{a,b,c})^{-1} \bar{y}_{pq,\rho\sigma}^{a,b,c} (\bar{Z}_{\rho i}^{a,b,c} - \bar{Z}_{\sigma i}^{a,b,c}) \tag{5.7.9}$$

To calculate $Z_{iq}^{a,b,c}$, let the current at the qth bus be $I_q^{a,b,c}$ and all other bus currents equal zero. If the added element p–q were not mutually coupled to the elements of the partial network, the voltage at the ith bus would be the same whether the current $I_q^{a,b,c}$ is injected at bus p or q, that is, $I_p^{a,b,c} = I_q^{a,b,c}$, and therefore

$$E_i^{a,b,c} = Z_{iq}^{a,b,c} I_q^{a,b,c} = Z_{ip}^{a,b,c} I_q^{a,b,c}$$

However, the element p–q is assumed to be mutually coupled to one or more elements of the partial network; therefore,

$$E_i^{a,b,c} = Z_{iq}^{a,b,c} I_q^{a,b,c} = Z_{ip}^{a,b,c} I_q^{a,b,c} + \Delta E_i^{a,b,c} \tag{5.7.10}$$

where $\Delta E_i^{a,b,c}$ is the change in voltage at bus i due to the effect of mutual coupling. The vector of voltages induced in the elements ρ–σ is

$$\bar{e}_{\rho\sigma}^{a,b,c} = \bar{z}_{\rho\sigma,pq}^{a,b,c} I_q^{a,b,c} \tag{5.7.11}$$

The series source voltages $\bar{e}_{\rho\sigma}^{a,b,c}$ and parallel source currents $\bar{j}_{\rho\sigma}^{a,b,c}$ are related by

$$\bar{j}_{\rho\sigma}^{a,b,c} = -[z_{\rho\sigma,\rho\sigma}^{a,b,c}]^{-1} \bar{e}_{\rho\sigma}^{a,b,c} \tag{5.7.12}$$

Substituting from equation (5.7.11) into (5.7.12),

$$\bar{j}_{\rho\sigma}^{a,b,c} = -[z_{\rho\sigma,\rho\sigma}^{a,b,c}]^{-1} \bar{z}_{\rho\sigma,pq}^{a,b,c} I_q^{a,b,c} \tag{5.7.13}$$

However, in terms of bus currents,

$$\bar{j}_{\rho\sigma}^{a,b,c} = \bar{I}_\rho^{a,b,c} = -\bar{I}_\sigma^{a,b,c} \tag{5.7.14}$$

and the change in voltage at the ith bus due to mutual coupling is

$$\Delta E_i^{a,b,c} = \bar{Z}_{i\rho}^{a,b,c} \bar{I}_\rho^{a,b,c} + \bar{Z}_{i\sigma}^{a,b,c} \bar{I}_\sigma^{a,b,c} \tag{5.7.15}$$

Substituting for $\bar{I}_\rho^{a,b,c}$ and $\bar{I}_\sigma^{a,b,c}$, from equations (5.7.13) and (5.7.14), equation (5.7.15) becomes

$$\Delta E_i^{a,b,c} = -(\bar{Z}_{i\rho}^{a,b,c} - \bar{Z}_{i\sigma}^{a,b,c})[z_{\rho\sigma,\rho\sigma}^{a,b,c}]^{-1} \bar{z}_{\rho\sigma,pq}^{a,b,c} I_q^{a,b,c} \tag{5.7.16}$$

Substituting from equation (5.7.16) into (5.7.10), it follows that

$$Z_{iq}^{a,b,c} = Z_{ip}^{a,b,c} - (\bar{Z}_{ip}^{a,b,c} - \bar{Z}_{i\sigma}^{a,b,c})[z_{\rho\sigma,\rho\sigma}^{a,b,c}]^{-1}\bar{z}_{\rho\sigma,pq}^{a,b,c} \tag{5.7.17}$$

From the matrix equation,

$$\left[\begin{array}{c|c} z_{pq,pq}^{a,b,c} & z_{pq,\rho\sigma}^{a,b,c} \\ \hline z_{\rho\sigma,pq}^{a,b,c} & z_{\rho\sigma,\rho\sigma}^{a,b,c} \end{array}\right] \left[\begin{array}{c|c} y_{pq,pq}^{a,b,c} & y_{pq,\rho\sigma}^{a,b,c} \\ \hline y_{\rho\sigma,pq}^{a,b,c} & y_{\rho\sigma,\rho\sigma}^{a,b,c} \end{array}\right] = \left[\begin{array}{c|c} U & \\ \hline & U \end{array}\right]$$

then

$$\bar{z}_{\rho\sigma,pq}^{a,b,c} y_{pq,pq}^{a,b,c} = -[z_{\rho\sigma,\rho\sigma}^{a,b,c}]\bar{y}_{\rho\sigma,pq}^{a,b,c} \tag{5.7.18}$$

Premultiplying by $[z_{\rho\sigma,\rho\sigma}^{a,b,c}]^{-1}$ and postmultiplying by $(y_{pq,pq}^{a,b,c})^{-1}$, equation (5.7.18) becomes

$$[z_{\rho\sigma,\rho\sigma}^{a,b,c}]^{-1}\bar{z}_{\rho\sigma,pq}^{a,b,c} = -\bar{y}_{\rho\sigma,pq}^{a,b,c}(y_{pq,pq}^{a,b,c})^{-1} \tag{5.7.19}$$

Substituting from equation (5.7.19) into (5.7.17),

$$Z_{iq}^{a,b,c} = Z_{ip}^{a,b,c} + (\bar{Z}_{ip}^{a,b,c} - \bar{Z}_{i\sigma}^{a,b,c})\bar{y}_{\rho\sigma,pq}^{a,b,c}(y_{pq,pq}^{a,b,c})^{-1} \tag{5.7.20}$$

The element $Z_{qq}^{a,b,c}$ can be determined by injecting a three-phase current at the qth bus and measuring the voltage at that bus with respect to the reference node. Let the current at the qth bus be $I_q^{a,b,c}$ and all other bus currents equal zero. Since $i_{pq}^{a,b,c} = -I_q^{a,b,c}$, substituting in equation (5.7.4) for $i_{pq}^{a,b,c}$ and solving for $v_{pq}^{a,b,c}$,

$$v_{pq}^{a,b,c} = -(y_{pq,pq}^{a,b,c})^{-1}(I_q^{a,b,c} + \bar{y}_{pq,\rho\sigma}^{a,b,c}\bar{v}_{\rho\sigma}^{a,b,c}) \tag{5.7.21}$$

Substituting from equations (5.7.2) and (5.7.3) into (5.7.21),

$$E_p^{a,b,c} - E_q^{a,b,c} = -(y_{pq,pq}^{a,b,c})^{-1}\{I_q^{a,b,c} + \bar{y}_{pq,\rho\sigma}^{a,b,c}(\bar{E}_\rho^{a,b,c} - \bar{E}_\sigma^{a,b,c})\} \tag{5.7.22}$$

From equation (5.7.1),

$$\begin{aligned} E_p^{a,b,c} &= Z_{pq}^{a,b,c} I_q^{a,b,c} \\ E_q^{a,b,c} &= Z_{qq}^{a,b,c} I_q^{a,b,c} \end{aligned} \tag{5.7.23}$$

and at any bus k,

$$E_k^{a,b,c} = Z_{kq}^{a,b,c} I_q^{a,b,c}$$

Substituting from equation (5.7.23) into (5.7.22) and solving for $Z_{qq}^{a,b,c} I_q^{a,b,c}$, it follows, since the resulting equation is valid for all values of $I_q^{a,b,c}$, that

$$Z_{qq}^{a,b,c} = Z_{pq}^{a,b,c} + (y_{pq,pq}^{a,b,c})^{-1}\{U + \bar{y}_{pq,\rho\sigma}^{a,b,c}(\bar{Z}_{\rho q}^{a,b,c} - \bar{Z}_{\sigma q}^{a,b,c})\} \tag{5.7.24}$$

If there is no mutual coupling between the added branch and the elements of the partial network, the elements of $\bar{y}^{a,b,c}_{pq,\rho\sigma}$ are zero and $(y^{a,b,c}_{pq,pq})^{-1} = z^{a,b,c}_{pq,pq}$. Then, equations (5.7.9), (5.7.20), and (5.7.24) reduce to

$$
\begin{aligned}
Z^{a,b,c}_{qi} &= Z^{a,b,c}_{pi} \\
Z^{a,b,c}_{iq} &= Z^{a,b,c}_{ip} \\
Z^{a,b,c}_{qq} &= Z^{a,b,c}_{pq} + z^{a,b,c}_{pq,pq}
\end{aligned}
$$

If, in addition, p is the reference node, the elements of $Z^{a,b,c}_{qi}$ and $Z^{a,b,c}_{iq}$ are zero. Also

$$Z^{a,b,c}_{qq} = z^{a,b,c}_{pq,pq}$$

If the network elements are balanced, then $Z^{a,b,c}_{qi} = Z^{a,b,c}_{iq}$ and either equation (5.7.9) or (5.7.20) can be used.

Addition of a link

As in the case of single-phase networks, when the new element is a link it is connected in series with a voltage source as shown in Fig. 5.5. The three-phase voltage source $e^{a,b,c}_l$ is selected such that the current through the added link is zero. Then the element p–l, where l is a fictitious node,

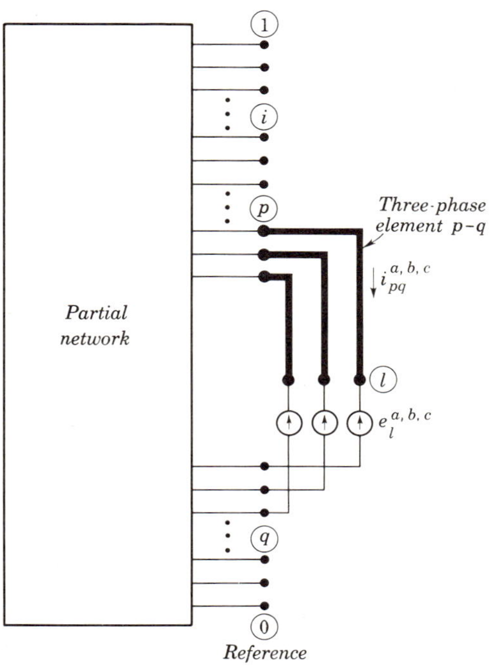

***Fig. 5.5** Three-phase voltage source in series with added link for calculation of $Z^{a,b,c}_{li}$, $Z^{a,b,c}_{il}$, and $Z^{a,b,c}_{ll}$.*

can be treated as a branch. The performance equation of the partial network with the added branch p–l, in terms of three-phase quantities, is

$$\begin{array}{c} \\ \left[\begin{array}{c} E_1^{a,b,c} \\ E_2^{a,b,c} \\ \cdots \\ E_p^{a,b,c} \\ \cdots \\ E_m^{a,b,c} \\ \hline e_l^{a,b,c} \end{array}\right] \end{array} = \begin{array}{c} \\ 1 \\ 2 \\ \\ p \\ \\ m \\ l \end{array} \begin{array}{c} \begin{array}{ccccccc} 1 & 2 & & p & & m & l \end{array} \\ \left[\begin{array}{cccccc|c} Z_{11}^{a,b,c} & Z_{12}^{a,b,c} & \cdots & Z_{1p}^{a,b,c} & \cdots & Z_{1m}^{a,b,c} & Z_{1l}^{a,b,c} \\ Z_{21}^{a,b,c} & Z_{22}^{a,b,c} & \cdots & Z_{2p}^{a,b,c} & \cdots & Z_{2m}^{a,b,c} & Z_{2l}^{a,b,c} \\ \cdots & \cdots & \cdots & \cdots & \cdots & \cdots & \cdots \\ Z_{p1}^{a,b,c} & Z_{p2}^{a,b,c} & \cdots & Z_{pp}^{a,b,c} & \cdots & Z_{pm}^{a,b,c} & Z_{pl}^{a,b,c} \\ \cdots & \cdots & \cdots & \cdots & \cdots & \cdots & \cdots \\ Z_{m1}^{a,b,c} & Z_{m2}^{a,b,c} & \cdots & Z_{mp}^{a,b,c} & \cdots & Z_{mm}^{a,b,c} & Z_{ml}^{a,b,c} \\ \hline Z_{l1}^{a,b,c} & Z_{l2}^{a,b,c} & \cdots & Z_{lp}^{a,b,c} & \cdots & Z_{lm}^{a,b,c} & Z_{ll}^{a,b,c} \end{array}\right] \end{array} \begin{array}{c} \\ \left[\begin{array}{c} I_1^{a,b,c} \\ I_2^{a,b,c} \\ \cdots \\ I_p^{a,b,c} \\ \cdots \\ I_m^{a,b,c} \\ \hline I_l^{a,b,c} \end{array}\right] \end{array} \qquad (5.7.25)$$

The elements $Z_{li}^{a,b,c}$ can be determined by injecting a three-phase current at the ith bus and measuring the voltage at the fictitious node l with respect to bus q. Let the current at the ith bus be $I_i^{a,b,c}$ and all other bus currents equal zero. Then, from equation (5.7.25),

$$e_l^{a,b,c} = Z_{li}^{a,b,c} I_i^{a,b,c} \qquad (5.7.26)$$

Also, as shown in Fig. 5.5,

$$e_l^{a,b,c} = v_{pq}^{a,b,c} - v_{pl}^{a,b,c} \qquad (5.7.27)$$

The current $i_{pq}^{a,b,c}$ in terms of the primitive admittances and the voltages across the elements is

$$i_{pq}^{a,b,c} = y_{pl,pl}^{a,b,c} v_{pl}^{a,b,c} + \bar{y}_{pl,\rho\sigma}^{a,b,c} \bar{v}_{\rho\sigma}^{a,b,c} \qquad (5.7.28)$$

Since

$$\begin{aligned} \bar{y}_{pl,\rho\sigma}^{a,b,c} &= \bar{y}_{pq,\rho\sigma}^{a,b,c} \\ y_{pl,pl}^{a,b,c} &= y_{pq,pq}^{a,b,c} \end{aligned}$$

and the elements of $i_{pq}^{a,b,c}$ are zero, then the voltage $v_{pl}^{a,b,c}$, from equation (5.7.28), is

$$v_{pl}^{a,b,c} = -(y_{pq,pq}^{a,b,c})^{-1} \bar{y}_{pq,\rho\sigma}^{a,b,c} \bar{v}_{\rho\sigma}^{a,b,c} \qquad (5.7.29)$$

Substituting from equation (5.7.29) into (5.7.27),

$$e_l^{a,b,c} = v_{pq}^{a,b,c} + (y_{pq,pq}^{a,b,c})^{-1} \bar{y}_{pq,\rho\sigma}^{a,b,c} \bar{v}_{\rho\sigma}^{a,b,c} \qquad (5.7.30)$$

Substituting for $v_{pq}^{a,b,c}$ and $\bar{v}_{\rho\sigma}^{a,b,c}$ from equations (5.7.2), (5.7.3), and (5.7.7) into (5.7.30),

$$e_l^{a,b,c} = (Z_{pi}^{a,b,c} - Z_{qi}^{a,b,c})I_i^{a,b,c} + (y_{pq,pq}^{a,b,c})^{-1}\bar{y}_{pq,\rho\sigma}^{a,b,c}(\bar{Z}_{\rho i}^{a,b,c} - \bar{Z}_{\sigma i}^{a,b,c})I_i^{a,b,c} \tag{5.7.31}$$

Substituting for $e_l^{a,b,c}$ from equation (5.7.26) into (5.7.31), it follows, since the resulting equation is valid for all values of $I_i^{a,b,c}$, that

$$Z_{li}^{a,b,c} = Z_{pi}^{a,b,c} - Z_{qi}^{a,b,c} + (y_{pq,pq}^{a,b,c})^{-1}\bar{y}_{pq,\rho\sigma}^{a,b,c}(\bar{Z}_{\rho i}^{a,b,c} - \bar{Z}_{\sigma i}^{a,b,c}) \tag{5.7.32}$$

The elements $Z_{il}^{a,b,c}$ can be determined by injecting a three-phase current between q and l and measuring the voltage at bus i. Let the current between q and l be $I_l^{a,b,c}$ and all other bus currents equal zero. If the added element were not mutually coupled to the elements of the partial network, then

$$E_i^{a,b,c} = (Z_{ip}^{a,b,c} - Z_{iq}^{a,b,c})I_l^{a,b,c}$$

However, because of the effect of mutual coupling,

$$E_i^{a,b,c} = Z_{il}^{a,b,c}I_l^{a,b,c} = (Z_{ip}^{a,b,c} - Z_{iq}^{a,b,c})I_l^{a,b,c} + \Delta E_i^{a,b,c} \tag{5.7.33}$$

Following the same procedure as in the case where the added element is a branch, then

$$\Delta E_i^{a,b,c} = -(\bar{Z}_{i\rho}^{a,b,c} - \bar{Z}_{i\sigma}^{a,b,c})[z_{\rho\sigma,\rho\sigma}^{a,b,c}]^{-1}\bar{z}_{\rho\sigma,pq}^{a,b,c}I_l^{a,b,c} \tag{5.7.34}$$

and from equation (5.7.19),

$$[z_{\rho\sigma,\rho\sigma}^{a,b,c}]^{-1}\bar{z}_{\rho\sigma,pq}^{a,b,c} = -\bar{y}_{\rho\sigma,pq}^{a,b,c}(y_{pq,pq}^{a,b,c})^{-1}$$

Substituting from equation (5.7.34) into equation (5.7.33) it follows, since the resultant equation is valid for all values of $I_l^{a,b,c}$, that

$$Z_{il}^{a,b,c} = Z_{ip}^{a,b,c} - Z_{iq}^{a,b,c} + (\bar{Z}_{i\rho}^{a,b,c} - \bar{Z}_{i\sigma}^{a,b,c})\bar{y}_{\rho\sigma,pq}^{a,b,c}(y_{pq,pq}^{a,b,c})^{-1} \tag{5.7.35}$$

The element $Z_{ll}^{a,b,c}$ can be determined by injecting a three-phase current between q and l and measuring the voltage at node l with respect to bus q. Let the current between q and l be $I_l^{a,b,c}$ and all other bus currents equal zero. Since $i_{pq}^{a,b,c} = -I_l^{a,b,c}$, substituting in equation (5.7.28) for $i_{pq}^{a,b,c}$ and solving for $v_{pl}^{a,b,c}$, then

$$v_{pl}^{a,b,c} = -(y_{pq,pq}^{a,b,c})^{-1}(I_l^{a,b,c} + \bar{y}_{pq,\rho\sigma}^{a,b,c}\bar{v}_{\rho\sigma}^{a,b,c}) \tag{5.7.36}$$

Substituting from equation (5.7.36) into (5.7.27),

$$e_l^{a,b,c} = v_{pq}^{a,b,c} + (y_{pq,pq}^{a,b,c})^{-1}(I_l^{a,b,c} + \bar{y}_{pq,\rho\sigma}^{a,b,c}\bar{v}_{\rho\sigma}^{a,b,c}) \tag{5.7.37}$$

Substituting for $v_{pq}^{a,b,c}$ and $\bar{v}_{\rho\sigma}^{a,b,c}$ from equations (5.7.2), (5.7.3), and (5.7.7) into (5.7.37),

$$e_l^{a,b,c} = (Z_{pl}^{a,b,c} - Z_{ql}^{a,b,c})I_l^{a,b,c} + (y_{pq,pq}^{a,b,c})^{-1}\{I_l^{a,b,c} + \bar{y}_{pq,\rho\sigma}^{a,b,c}(\bar{Z}_{\rho l}^{a,b,c} - \bar{Z}_{\sigma l}^{a,b,c})I_l^{a,b,c}\} \tag{5.7.38}$$

Table 5.1 Summary of equations for formation of three-phase bus impedance matrix

Add p–q	Mutual coupling: p is not the reference bus	Mutual coupling: p is the reference bus
Branch	$Z_{qi}^{a,b,c} = Z_{pi}^{a,b,c} + (y_{pq,pq}^{a,b,c})^{-1}\bar{y}_{pq,\rho\sigma}^{a,b,c}(\bar{Z}_{pi}^{a,b,c} - \bar{Z}_{\sigma i}^{a,b,c})$ $i = 1, 2, \ldots, m$ $i \neq q$	$Z_{qi}^{a,b,c} = (y_{pq,pq}^{a,b,c})^{-1}\bar{y}_{pq,\rho\sigma}^{a,b,c}(\bar{Z}_{pi}^{a,b,c} - \bar{Z}_{\sigma i}^{a,b,c})$ $i = 1, 2, \ldots, m$ $i \neq q$
	$Z_{iq}^{a,b,c} = Z_{ip}^{a,b,c} + (\bar{Z}_{ip}^{a,b,c} - \bar{Z}_{i\sigma}^{a,b,c})\bar{y}_{\rho\sigma,pq}^{a,b,c}(y_{pq,pq}^{a,b,c})^{-1}$ $i = 1, 2, \ldots, m$ $i \neq q$	$Z_{iq}^{a,b,c} = (\bar{Z}_{ip}^{a,b,c} - \bar{Z}_{i\sigma}^{a,b,c})\bar{y}_{\rho\sigma,pq}^{a,b,c}(y_{pq,pq}^{a,b,c})^{-1}$ $i = 1, 2, \ldots, m$ $i \neq q$
	$Z_{qq}^{a,b,c} = Z_{pq}^{a,b,c} + (y_{pq,pq}^{a,b,c})^{-1}\{U + \bar{y}_{pq,\rho\sigma}^{a,b,c}(\bar{Z}_{pq}^{a,b,c} - \bar{Z}_{\sigma q}^{a,b,c})\}$	$Z_{qq}^{a,b,c} = (y_{pq,pq}^{a,b,c})^{-1}\{U + \bar{y}_{pq,\rho\sigma}^{a,b,c}(\bar{Z}_{pq}^{a,b,c} - \bar{Z}_{\sigma q}^{a,b,c})\}$
Link	$Z_{li}^{a,b,c} = Z_{pi}^{a,b,c} - Z_{qi}^{a,b,c} + (y_{pq,pq}^{a,b,c})^{-1}\bar{y}_{pq,\rho\sigma}^{a,b,c}(\bar{Z}_{pi}^{a,b,c} - \bar{Z}_{\sigma i}^{a,b,c})$ $i = 1, 2, \ldots, m$ $i \neq l$	$Z_{li}^{a,b,c} = -Z_{qi}^{a,b,c} + (y_{pq,pq}^{a,b,c})^{-1}\bar{y}_{pq,\rho\sigma}^{a,b,c}(\bar{Z}_{pi}^{a,b,c} - \bar{Z}_{\sigma i}^{a,b,c})$ $i = 1, 2, \ldots, m$ $i \neq l$
	$Z_{il}^{a,b,c} = Z_{ip}^{a,b,c} - Z_{iq}^{a,b,c} + (\bar{Z}_{ip}^{a,b,c} - \bar{Z}_{i\sigma}^{a,b,c})\bar{y}_{\rho\sigma,pq}^{a,b,c}(y_{pq,pq}^{a,b,c})^{-1}$ $i = 1, 2, \ldots, m$ $i \neq l$	$Z_{il}^{a,b,c} = -Z_{iq}^{a,b,c} + (\bar{Z}_{ip}^{a,b,c} - \bar{Z}_{i\sigma}^{a,b,c})\bar{y}_{\rho\sigma,pq}^{a,b,c}(y_{pq,pq}^{a,b,c})^{-1}$ $i = 1, 2, \ldots, m$ $i \neq l$
	$Z_{ll}^{a,b,c} = Z_{pl}^{a,b,c} - Z_{ql}^{a,b,c} + (y_{pq,pq}^{a,b,c})^{-1}\{U + \bar{y}_{pq,\rho\sigma}^{a,b,c}(\bar{Z}_{pl}^{a,b,c} - \bar{Z}_{\sigma l}^{a,b,c})\}$	$Z_{ll}^{a,b,c} = -Z_{ql}^{a,b,c} + (y_{pq,pq}^{a,b,c})^{-1}\{U + \bar{y}_{pq,\rho\sigma}^{a,b,c}(\bar{Z}_{pl}^{a,b,c} - \bar{Z}_{\sigma l}^{a,b,c})\}$

Add p–q	No mutual coupling	
	p is not the reference bus	*p is the reference bus*
Branch	$Z_{qi}^{a,b,c} = Z_{pi}^{a,b,c}$ $i = 1, 2, \ldots, m$ $i \neq q$ $Z_{iq}^{a,b,c} = Z_{ip}^{a,b,c}$ $i = 1, 2, \ldots, m$ $i \neq q$ $Z_{qq}^{a,b,c} = Z_{pq}^{a,b,c} + z_{pq,pq}^{a,b,c}$	$Z_{qi}^{a,b,c} = 0$ $i = 1, 2, \ldots, m$ $i \neq q$ $Z_{iq}^{a,b,c} = 0$ $i = 1, 2, \ldots, m$ $i \neq q$ $Z_{qq}^{a,b,c} = z_{pq,pq}^{a,b,c}$
Link	$Z_{li}^{a,b,c} = Z_{pi}^{a,b,c} - Z_{qi}^{a,b,c}$ $i = 1, 2, \ldots, m$ $i \neq l$ $Z_{il}^{a,b,c} = Z_{ip}^{a,b,c} - Z_{iq}^{a,b,c}$ $i = 1, 2, \ldots, m$ $i \neq l$ $Z_{ll}^{a,b,c} = Z_{pl}^{a,b,c} - Z_{ql}^{a,b,c} + z_{pq,pq}^{a,b,c}$	$Z_{li}^{a,b,c} = -Z_{qi}^{a,b,c}$ $i = 1, 2, \ldots, m$ $i \neq l$ $Z_{il}^{a,b,c} = -Z_{iq}^{a,b,c}$ $i = 1, 2, \ldots, m$ $i \neq l$ $Z_{ll}^{a,b,c} = -Z_{ql}^{a,b,c} + z_{pq,pq}^{a,b,c}$

Modification of the elements for elimination of lth node

$$Z_{ij\,(\text{modified})}^{a,b,c} = Z_{ij\,(\text{before elimination})}^{a,b,c} - Z_{il}^{a,b,c}(Z_{ll}^{a,b,c})^{-1}Z_{lj}^{a,b,c} \qquad i,j = 1, 2, \ldots, m$$

From equation (5.7.25),

$$e_l^{a,b,c} = Z_{ll}^{a,b,c} I_l^{a,b,c} \tag{5.7.39}$$

Substituting from equation (5.7.39) into (5.7.38), it follows, since the resultant equation is valid for all values of $I_l^{a,b,c}$, that

$$Z_{ll} = (Z_{pl}^{a,b,c} - Z_{ql}^{a,b,c}) + (y_{pq,pq}^{a,b,c})^{-1}\{U + \bar{y}_{pq,\rho\sigma}^{a,b,c}(\bar{Z}_{\rho l}^{a,b,c} - \bar{Z}_{\sigma l}^{a,b,c})\} \tag{5.7.40}$$

If there is no mutual coupling between the added link and the elements of the partial network, equations (5.7.32), (5.7.35), and (5.7.40) reduce to

$$Z_{li}^{a,b,c} = Z_{pi}^{a,b,c} - Z_{qi}^{a,b,c}$$
$$Z_{il}^{a,b,c} = Z_{ip}^{a,b,c} - Z_{iq}^{a,b,c}$$
$$Z_{ll}^{a,b,c} = Z_{pl}^{a,b,c} - Z_{ql}^{a,b,c} + z_{pq,pq}^{a,b,c}$$

If, in addition, p is the reference node,

$$Z_{li}^{a,b,c} = -Z_{qi}^{a,b,c}$$
$$Z_{il}^{a,b,c} = -Z_{iq}^{a,b,c}$$
$$Z_{ll}^{a,b,c} = -Z_{ql}^{a,b,c} + z_{pq,pq}^{a,b,c}$$

Furthermore, if the elements are balanced,

$$Z_{li}^{a,b,c} = Z_{il}^{a,b,c}$$

The fictitious node l is eliminated by short circuiting the link voltage source $e_l^{a,b,c}$. From equation (5.7.25),

$$\bar{E}_{BUS}^{a,b,c} = Z_{BUS}^{a,b,c} \bar{I}_{BUS}^{a,b,c} + \bar{Z}_{il}^{a,b,c} I_l^{a,b,c} \tag{5.7.41}$$

and

$$e_l^{a,b,c} = \bar{Z}_{lj}^{a,b,c} \bar{I}_{BUS}^{a,b,c} + Z_{ll}^{a,b,c} I_l^{a,b,c} = 0 \tag{5.7.42}$$

where $i, j = 1, 2, \ldots, m$. Solving for $I_l^{a,b,c}$ from equation (5.7.42) and substituting into (5.7.41),

$$\bar{E}_{BUS}^{a,b,c} = \{Z_{BUS}^{a,b,c} - \bar{Z}_{il}^{a,b,c}(Z_{ll}^{a,b,c})^{-1}\bar{Z}_{lj}^{a,b,c}\}\bar{I}_{BUS}^{a,b,c}$$

Therefore,

$$Z_{ij\,(\text{modified})}^{a,b,c} = Z_{ij\,(\text{before elimination})}^{a,b,c} - Z_{il}^{a,b,c}(Z_{ll}^{a,b,c})^{-1}Z_{lj}^{a,b,c}$$

A summary of equations for the formation of the three-phase bus impedance matrix is given in Table 5.1. These equations can be written in terms of symmetrical or Clarke's components.

5.8 Modification of the three-phase bus impedance matrix for changes in the network

The formulas given in Table 5.1 can be used to modify a three-phase bus impedance matrix when an element is added to the network. These formulas can be used also when an element not mutually coupled to other elements of the network is removed or its impedance is changed. The procedures are similar to those used for single-phase networks. When an element is removed, the modified three-phase bus impedance matrix can be obtained by adding a parallel element whose three-phase impedance is equal to the negative of the impedance of the element to be removed. When the impedance of an element is to be changed, the modified three-phase bus impedance matrix can be obtained by adding a parallel element such that the equivalent three-phase impedance of the two elements is the desired value.

The same procedures as those used for single-phase networks can be employed to derive an equation for modifying the submatrices of the three-phase bus impedance matrix when mutually coupled elements are removed or their impedances are changed. This equation is

$$Z'^{a,b,c}_{ij} = Z^{a,b,c}_{ij} + (\bar{Z}^{a,b,c}_{i\alpha} - \bar{Z}^{a,b,c}_{i\beta})[M^{a,b,c}]^{-1}[\Delta y^{a,b,c}_{s}](\bar{Z}^{a,b,c}_{\gamma j} - \bar{Z}^{a,b,c}_{\delta j})$$

where

$$[\Delta y^{a,b,c}_{s}] = [y^{a,b,c}_{s}] - [y'^{a,b,c}_{s}]$$
$$[M^{a,b,c}] = \{U - [\Delta y^{a,b,c}_{s}]([Z^{a,b,c}_{\gamma\alpha}] - [Z^{a,b,c}_{\delta\alpha}] - [Z^{a,b,c}_{\gamma\beta}] + [Z^{a,b,c}_{\delta\beta}])\}$$

5.9 Example of formation and modification of three-phase network matrices

The methods of forming three-phase network matrices by transformation and by algorithm will be illustrated using the sample system shown in Fig. 5.6*a*.

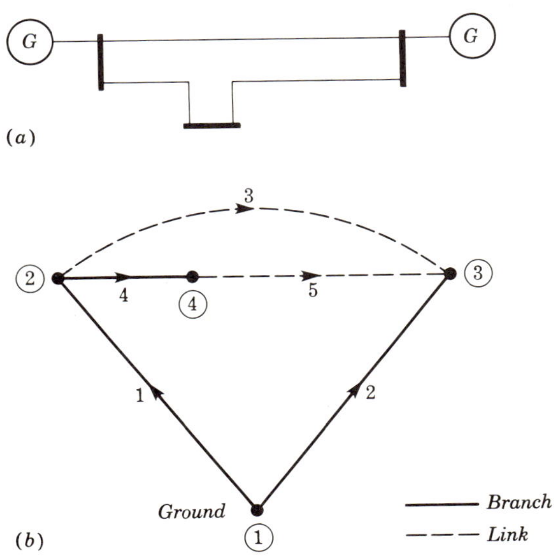

***Fig. 5.6** Sample three-phase system. (a) Single line diagram; (b) tree and cotree of oriented connected graph.*

Problem

a. Form the bus incidence matrix A with ground as reference.
b. Form the bus admittance matrix Y_{BUS} by transformation.
c. Form the bus impedance matrix Z_{BUS} using the algorithm.
d. Modify the three-phase bus impedance matrix obtained in part *c* to remove element 3 between bus 2 and bus 3.

Table 5.2 Three-phase impedances for sample system

	Self				Mutual			
Element number	Bus code p–q	Impedance $z^{a,b,c}_{pq,pq}$			Bus code r–s	Impedance $z^{a,b,c}_{pq,rs}$		
		0.080	−0.025	−0.020				
1	1–2	−0.020	0.080	−0.025				
		−0.025	−0.020	0.080				
		0.080	−0.025	−0.020				
2	1–3	−0.020	0.080	−0.025				
		−0.025	−0.020	0.080				
		1.50	0.50	0.50				
3	2–3	0.50	1.50	0.50				
		0.50	0.50	1.50				
		0.60	0.20	0.20		0.20	0.20	0.20
4	2–4	0.20	0.60	0.20	2–3	0.20	0.20	0.20
		0.20	0.20	0.60		0.20	0.20	0.20
		0.90	0.30	0.30		0.30	0.30	0.30
5	4–3	0.30	0.90	0.30	2–3	0.30	0.30	0.30
		0.30	0.30	0.90		0.30	0.30	0.30

Solution

The data for the sample three-phase system is given in Table 5.2. The impedances for this system are represented by real numbers equal to the generator and line reactances. The branches and links of the oriented connected graph for the single-line representation of the system are shown in Fig. 5.6*b*.

a. The bus incidence matrix is

$A =$

e \ bus	②	③	④
1	$-U$		
2		$-U$	
4	U		$-U$
3	U	$-U$	
5		$-U$	U

where U is a 3×3 unit matrix.

b. The primitive impedance matrix of the three-phase system from Table 5.2 is

$[z^{a,b,c}] =$

e \ e	1 a	1 b	1 c	2 a	2 b	2 c	4 a	4 b	4 c	3 a	3 b	3 c	5 a	5 b	5 c
1 a	.080	−.025	−.020												
1 b	−.020	.080	−.025												
1 c	−.025	−.020	.080												
2 a				.080	−.025	−.020									
2 b				−.020	.080	−.025									
2 c				−.025	−.020	.080									
4 a							.600	.200	.200	.200	.200	.200			
4 b							.200	.600	.200	.200	.200	.200			
4 c							.200	.200	.600	.200	.200	.200			
3 a							.200	.200	.200	1.500	.500	.500	.300	.300	.300
3 b							.200	.200	.200	.500	1.500	.500	.300	.300	.300
3 c							.200	.200	.200	.500	.500	1.500	.300	.300	.300
5 a										.300	.300	.300	.900	.300	.300
5 b										.300	.300	.300	.300	.900	.300
5 c										.300	.300	.300	.300	.300	.900

The primitive admittance matrix is the inverse of $[z^{a,b,c}]$.

The bus admittance matrix obtained by singular transformation is

$$Y_{BUS}^{a,b,c} = A^t[y^{a,b,c}]A$$

		1 a	1 b	1 c	2 a	2 b	2 c	4 a	4 b	4 c	3 a	3 b	3 c	5 a	5 b	5 c
	a	−1						1			1					
②	b		−1						1			1				
	c			−1						1			1			
	a				−1						−1			−1		
= ③	b					−1						−1			−1	
	c						−1						−1			−1
	a							−1						1		
④	b								−1						1	
	c									−1						1

		1 a	1 b	1 c	2 a	2 b	2 c	4 a	4 b	4 c	3 a	3 b	3 c	5 a	5 b	5 c
	a	16.0163	6.5151	6.0400												
1	b	6.0400	16.0163	6.5151												
	c	6.5151	6.0400	16.0163												
	a				16.0163	6.5151	6.0400									
2	b				6.0400	16.0163	6.5151									
	c				6.5151	6.0400	16.0163									
	a							2.0750	−.4250	−.4250	−.1250	−.1250	−.1250	.0750	.0750	.0750
4	b							−.4250	2.0750	−.4250	−.1250	−.1250	−.1250	.0750	.0750	.0750
	c							−.4250	−.4250	2.0750	−.1250	−.1250	−.1250	.0750	.0750	.0750
	a							−.1250	−.1250	−.1250	.8750	−.1250	−.1250	−.1250	−.1250	−.1250
3	b							−.1250	−.1250	−.1250	−.1250	.8750	−.1250	−.1250	−.1250	−.1250
	c							−.1250	−.1250	−.1250	−.1250	−.1250	.8750	−.1250	−.1250	−.1250
	a							.0750	.0750	.0750	−.1250	−.1250	−.1250	1.4083	−.2583	−.2583
5	b							.0750	.0750	.0750	−.1250	−.1250	−.1250	−.2583	1.4083	−.2583
	c							.0750	.0750	.0750	−.1250	−.1250	−.1250	−.2583	−.2583	1.4083

		② a	② b	② c	③ a	③ b	③ c	④ a	④ b	④ c
	a	−1								
1	b		−1							
	c			−1						
	a				−1					
2	b					−1				
	c						−1			
	a	1						−1		
4	b		1						−1	
	c			1						−1
	a	1			−1					
3	b		1			−1				
	c			1			−1			
	a				−1			1		
5	b					−1			1	
	c						−1			1

		② a	② b	② c	③ a	③ b	③ c	④ a	④ b	④ c
	a	18.7163	5.7151	5.2400	−.7000	.3000	.3000	−2.0000	.5000	.5000
②	b	5.2400	18.7163	5.7151	.3000	−.7000	.3000	.5000	−2.0000	.5000
	c	5.7151	5.2400	18.7163	.3000	.3000	−.7000	.5000	.5000	−2.0000
	a	−.7000	.3000	.3000	18.0496	5.8818	5.4067	−1.3333	.3333	.3333
= ③	b	.3000	−.7000	.3000	5.4067	18.0496	5.8818	.3333	−1.3333	.3333
	c	.3000	.3000	−.7000	5.8818	5.4067	18.0496	.3333	.3333	−1.3333
	a	−2.0000	.5000	.5000	−1.3333	.3333	.3333	3.3333	−.8333	−.8333
④	b	.5000	−2.0000	.5000	.3333	−1.3333	.3333	−.8333	3.3333	−.8333
	c	.5000	.5000	−2.0000	.3333	.3333	−1.3333	−.8333	−.8333	3.3333

c. The bus impedance matrix will be formed by first adding all branches and then adding the links.

Step 1. Start with element 1, the branch from $p = 1$ to $q = 2$. The elements of the bus impedance matrix of the partial network are

$Z^{a,b,c}_{BUS} =$

		② a	② b	② c
②	a	0.080	−0.025	−0.020
	b	−0.020	0.080	−0.025
	c	−0.025	−0.020	0.080

Step 2. Add element 2, the branch from $p = 1$ to $q = 3$. This adds a new bus and the bus impedance matrix is

$Z^{a,b,c}_{BUS} =$

		② a	② b	② c	③ a	③ b	③ c
②	a	0.080	−0.025	−0.020			
	b	−0.020	0.080	−0.025			
	c	−0.025	−0.020	0.080			
③	a				0.080	−0.025	−0.020
	b				−0.020	0.080	−0.025
	c				−0.025	−0.020	0.080

Step 3. Add element 4, the branch from $p = 2$ to $q = 4$. This element is not connected to the reference node and its addition creates a new bus. Using the formulas in Table 5.1

$$Z^{a,b,c}_{qi} = Z^{a,b,c}_{pi} \qquad i = 2, 3$$
$$Z^{a,b,c}_{iq} = Z^{a,b,c}_{ip} \qquad i = 2, 3$$

and

$$Z^{a,b,c}_{qq} = Z^{a,b,c}_{pq} + z^{a,b,c}_{pq,pq}$$

The bus impedance matrix is

$Z_{BUS}^{a,b,c} =$

		② a	② b	② c	③ a	③ b	③ c	④ a	④ b	④ c
②	a	0.080	−0.025	−0.020				0.080	−0.025	−0.020
	b	−0.020	0.080	−0.025				−0.020	0.080	−0.025
	c	−0.025	−0.020	0.080				−0.025	−0.020	0.080
③	a				0.080	−0.025	−0.020			
	b				−0.020	0.080	−0.025			
	c				−0.025	−0.020	0.080			
④	a	0.080	−0.025	−0.020				0.680	0.175	0.180
	b	−0.020	0.080	−0.025				0.180	0.680	0.175
	c	−0.025	−0.020	0.080				0.175	0.180	0.680

Step 4. Add element 5, the link from $p = 4$ to $q = 3$. This element is not connected to the reference node and is not mutually coupled to an existing element of the partial network. The elements of the rows and columns corresponding to the fictitious node l are obtained from

$$Z_{li}^{a,b,c} = Z_{pi}^{a,b,c} - Z_{qi}^{a,b,c} \qquad i = 2, 3, 4$$
$$Z_{il}^{a,b,c} = Z_{ip}^{a,b,c} - Z_{iq}^{a,b,c} \qquad i = 2, 3, 4$$

and

$$Z_{ll}^{a,b,c} = Z_{pl}^{a,b,c} - Z_{ql}^{a,b,c} + z_{pq,pq}^{a,b,c}$$

Thus, the augmented impedance matrix is

		②			③			④			l	
	a	b	c	a	b	c	a	b	c	a	b	c
② a	.080	−.025	−.020				.080	−.025	−.020	.080	−.025	−.020
② b	−.020	.080	−.025				−.020	.080	−.025	−.020	.080	−.025
② c	−.025	−.020	.080				−.025	−.020	.080	−.025	−.020	.080
③ a				.080	−.025	−.020				−.080	.025	.020
③ b				−.020	.080	−.025				.020	−.080	.025
③ c				−.025	−.020	.080				.025	.020	−.080
④ a	.080	−.025	−.020				.680	.175	.180	.680	.175	.180
④ b	−.020	.080	−.025				.180	.680	.175	.180	.680	.175
④ c	−.025	−.020	.080				.175	.180	.680	.175	.180	.680
l a	.080	−.025	−.020	−.080	.025	.020	.680	.175	.180	1.660	.450	.460
l b	−.020	.080	−.025	.020	−.080	.025	.180	.680	.175	.460	1.660	.450
l c	−.025	−.020	.080	.025	.020	−.080	.175	.180	.680	.450	.460	1.660

The rows and columns corresponding to the fictitious node l are eliminated using the formula

$$Z^{a,b,c}_{ij\,(\text{modified})} = Z^{a,b,c}_{ij\,(\text{before elimination})} - Z^{a,b,c}_{il}(Z^{a,b,c}_{ll})^{-1}Z^{a,b,c}_{lj}$$

Then, the bus impedance matrix is

$Z^{a,b,c}_{BUS} =$

		②			③			④	
	a	b	c	a	b	c	a	b	c
② a	.0740	−.0219	−.0176	.0060	−.0031	−.0024	.0468	−.0144	−.0115
② b	−.0176	.0740	−.0219	−.0024	.0060	−.0031	−.0115	.0468	−.0144
② c	−.0219	−.0176	.0740	−.0031	−.0024	.0060	−.0144	−.0115	.0468
③ a	.0060	−.0031	−.0024	.0740	−.0219	−.0176	.0332	−.0106	−.0085
③ b	−.0024	.0060	−.0031	−.0176	.0740	−.0219	−.0085	.0332	−.0106
③ c	−.0031	−.0024	.0060	−.0219	−.0176	.0740	−.0106	−.0085	.0032
④ a	.0468	−.0144	−.0115	.0332	−.0106	−.0085	.4014	.1071	.1097
④ b	−.0115	.0468	−.0144	−.0085	.0332	−.0106	.1097	.4014	.1071
④ c	−.0144	−.0115	.0468	−.0106	−.0085	.0332	.1071	.1097	.4014

Step 5. Add element 3, the link from $p = 2$ to $q = 3$ mutually coupled to elements 4 and 5. This element is not connected to the reference node. The elements of the rows and columns corresponding to the fictitious node l are obtained from

$$Z_{li}^{a,b,c} = Z_{pi}^{a,b,c} - Z_{qi}^{a,b,c} + (y_{pq,pq}^{a,b,c})^{-1}\bar{y}_{pq,\rho\sigma}^{a,b,c}(\bar{Z}_{\rho i}^{a,b,c} - \bar{Z}_{\sigma i}^{a,b,c}) \qquad i = 2, 3, 4$$

$$Z_{il}^{a,b,c} = Z_{ip}^{a,b,c} - Z_{iq}^{a,b,c} + (\bar{Z}_{i\rho}^{a,b,c} - \bar{Z}_{i\sigma}^{a,b,c})\bar{y}_{\rho\sigma,pq}^{a,b,c}(y_{pq,pq}^{a,b,c})^{-1} \qquad i = 2, 3, 4$$

and

$$Z_{ll}^{a,b,c} = Z_{pl}^{a,b,c} - Z_{ql}^{a,b,c} + (y_{pq,pq}^{a,b,c})^{-1}\{U + \bar{y}_{pq,\rho\sigma}^{a,b,c}(\bar{Z}_{\rho l}^{a,b,c} - \bar{Z}_{\sigma l}^{a,b,c})\}$$

The submatrix $Z_{l2}^{a,b,c}$ is

$$Z_{l2}^{a,b,c} = Z_{22}^{a,b,c} - Z_{32}^{a,b,c} + (y_{23,23}^{a,b,c})^{-1} \begin{bmatrix} y_{23,24}^{a,b,c} & y_{23,43}^{a,b,c} \end{bmatrix} \begin{bmatrix} Z_{22}^{a,b,c} - Z_{42}^{a,b,c} \\ Z_{42}^{a,b,c} - Z_{32}^{a,b,c} \end{bmatrix}$$

$$= \begin{bmatrix} .0740 & -.0219 & -.0176 \\ -.0176 & .0740 & -.0219 \\ -.0219 & -.0176 & .0740 \end{bmatrix} - \begin{bmatrix} .0060 & -.0031 & -.0024 \\ -.0024 & .0060 & -.0031 \\ -.0031 & -.0024 & .0060 \end{bmatrix}$$

$$+ \begin{bmatrix} 1.2 & .2 & .2 \\ .2 & 1.2 & .2 \\ .2 & .2 & 1.2 \end{bmatrix} \begin{bmatrix} -.1250 & -.1250 & -.1250 & -.1250 & -.1250 & -.1250 \\ -.1250 & -.1250 & -.1250 & -.1250 & -.1250 & -.1250 \\ -.1250 & -.1250 & -.1250 & -.1250 & -.1250 & -.1250 \end{bmatrix} \begin{bmatrix} .0272 & -.0075 & -.0061 \\ -.0061 & .0272 & -.0075 \\ -.0075 & -.0061 & .0272 \\ .0408 & -.0113 & -.0091 \\ -.0091 & .0408 & -.0113 \\ -.0113 & -.0091 & .0408 \end{bmatrix}$$

$$= \begin{bmatrix} .0612 & -.0256 & -.0220 \\ -.0220 & .0612 & -.0256 \\ -.0256 & -.0220 & .0612 \end{bmatrix}$$

The submatrix $Z_{2l}^{a,b,c}$ is

$$Z_{2l}^{a,b,c} = Z_{22}^{a,b,c} - Z_{23}^{a,b,c} + \left[\begin{array}{c|c} Z_{22}^{a,b,c} - Z_{24}^{a,b,c} & Z_{24}^{a,b,c} - Z_{23}^{a,b,c} \end{array}\right] \left[\begin{array}{c} y_{24,23}^{a,b,c} \\ \hline y_{43,23}^{a,b,c} \end{array}\right] (y_{23,23}^{a,b,c})^{-1}$$

$$= \left[\begin{array}{ccc} .0740 & -.0219 & -.0176 \\ -.0176 & .0740 & -.0219 \\ -.0219 & -.0176 & .0740 \end{array}\right] - \left[\begin{array}{ccc} .0060 & -.0031 & -.0024 \\ -.0024 & .0060 & -.0031 \\ -.0031 & -.0024 & .0060 \end{array}\right]$$

$$+ \left[\begin{array}{ccc|ccc} .0272 & -.0075 & -.0061 & .0408 & -.0113 & -.0091 \\ -.0061 & .0272 & -.0075 & -.0091 & .0408 & -.0113 \\ -.0075 & -.0061 & .0272 & -.0113 & -.0091 & .0408 \end{array}\right] \left[\begin{array}{ccc} -.1250 & -.1250 & -.1250 \\ -.1250 & -.1250 & -.1250 \\ -.1250 & -.1250 & -.1250 \\ \hline -.1250 & -.1250 & -.1250 \\ -.1250 & -.1250 & -.1250 \\ -.1250 & -.1250 & -.1250 \end{array}\right] \left[\begin{array}{ccc} 1.2 & .2 & .2 \\ .2 & 1.2 & .2 \\ .2 & .2 & 1.2 \end{array}\right]$$

$$= \left[\begin{array}{ccc} .0612 & -.0256 & -.0220 \\ -.0220 & .0612 & -.0256 \\ -.0256 & -.0220 & .0612 \end{array}\right]$$

In a similar manner the remaining submatrices $Z_{li}^{a,b,c}$ and $Z_{il}^{a,b,c}$ for $i = 3, 4$ can be calculated. In this problem

$$Z_{2l}^{a,b,c} = Z_{l2}^{a,b,c} \qquad Z_{3l}^{a,b,c} = Z_{l3}^{a,b,c} \qquad \text{and} \qquad Z_{4l}^{a,b,c} = Z_{l4}^{a,b,c}$$

since the elements of the network are balanced.

The submatrix $Z_{ll}^{a,b,c}$ is

$$Z_{ll}^{a,b,c} = Z_{2l}^{a,b,c} - Z_{3l}^{a,b,c} + (y_{23,23}^{a,b,c})^{-1}\left\{U + \begin{bmatrix} y_{23,24}^{a,b,c} & y_{23,43}^{a,b,c} \end{bmatrix} \begin{bmatrix} Z_{2l}^{a,b,c} - Z_{4l}^{a,b,c} \\ Z_{4l}^{a,b,c} - Z_{3l}^{a,b,c} \end{bmatrix}\right\}$$

$$= \begin{bmatrix} .0612 & -.0256 & -.0220 \\ -.0220 & .0612 & -.0256 \\ -.0256 & -.0220 & .0612 \end{bmatrix} - \begin{bmatrix} -.0612 & .0256 & .0220 \\ .0220 & -.0612 & .0256 \\ .0256 & .0220 & -.0612 \end{bmatrix}$$

$$+ \begin{bmatrix} 1.2 & .2 & .2 \\ .2 & 1.2 & .2 \\ .2 & .2 & 1.2 \end{bmatrix} \left\{ \begin{bmatrix} 1 & & \\ & 1 & \\ & & 1 \end{bmatrix} + \begin{bmatrix} -.1250 & -.1250 & -.1250 & -.1250 & -.1250 & -.1250 \\ -.1250 & -.1250 & -.1250 & -.1250 & -.1250 & -.1250 \\ -.1250 & -.1250 & -.1250 & -.1250 & -.1250 & -.1250 \end{bmatrix} \begin{bmatrix} .0490 & -.0204 & -.0176 \\ -.0176 & .0490 & -.0204 \\ -.0204 & -.0176 & .0490 \\ .0734 & -.0308 & -.0264 \\ -.0264 & .0734 & -.0308 \\ -.0308 & -.0264 & .0734 \end{bmatrix} \right\}$$

$$= \begin{bmatrix} 1.3170 & .1434 & .1506 \\ .1506 & 1.3170 & .1434 \\ .1434 & .1506 & 1.3170 \end{bmatrix}$$

The augmented impedance matrix is

		②			③			④			l	
	a	b	c	a	b	c	a	b	c	a	b	c
a	.0740	−.0219	−.0176	.0060	−.0031	−.0024	.0468	−.0144	−.0115	.0612	−.0256	−.0220
② b	−.0176	.0740	−.0219	−.0024	.0060	−.0031	−.0115	.0468	−.0144	−.0220	.0612	−.0256
c	−.0219	−.0176	.0740	−.0031	−.0024	.0060	−.0144	−.0115	.0468	−.0256	−.0220	.0612
a	.0060	−.0031	−.0024	.0740	−.0219	−.0176	.0332	−.0106	−.0085	−.0612	.0256	.0220
③ b	−.0024	.0060	−.0031	−.0176	.0740	−.0219	−.0085	.0332	−.0106	.0220	−.0612	.0256
c	−.0031	−.0024	.0060	−.0219	−.0176	.0740	−.0106	−.0085	.0332	.0256	.0220	−.0612
a	.0468	−.0144	−.0115	.0332	−.0106	−.0085	.4014	.1071	.1097	.0122	−.0052	−.0044
④ b	−.0115	.0468	−.0144	−.0085	.0332	−.0106	.1097	.4014	.1071	−.0044	.0122	−.0052
c	−.0144	−.0115	.0468	−.0106	−.0085	.0332	.1071	.1097	.4014	−.0052	−.0044	.0122
a	.0612	−.0256	−.0220	−.0612	.0256	.0220	.0122	−.0052	−.0044	1.3170	.1434	.1506
l b	−.0220	.0612	−.0256	.0220	−.0612	.0256	−.0044	.0122	−.0052	.1506	1.3170	.1434
c	−.0256	−.0220	.0612	.0256	.0220	−.0612	−.0052	−.0044	.0122	.1434	.1506	1.3170

Eliminating the rows and columns corresponding to the fictitious node l, the bus impedance matrix of the system is

$Z_{BUS}^{a,b,c} =$

		② a	② b	② c	③ a	③ b	③ c	④ a	④ b	④ c
②	a	.0699	−.0196	−.0159	.0101	−.0054	−.0041	.0460	−.0139	−.0112
②	b	−.0159	.0699	−.0196	−.0041	.0101	−.0054	−.0112	.0460	−.0139
②	c	−.0196	−.0159	.0699	−.0054	−.0041	.0101	−.0139	−.0112	.0460
③	a	.0101	−.0054	−.0041	.0699	−.0196	−.0159	.0340	−.0111	−.0088
③	b	−.0041	.0101	−.0054	−.0159	.0699	−.0196	−.0088	.0340	−.0111
③	c	−.0054	−.0041	.0101	−.0196	−.0159	.0699	−.0111	−.0088	.0340
④	a	.0460	−.0139	−.0112	.0340	−.0111	−.0088	.4012	.1072	.1098
④	b	−.0112	.0460	−.0139	−.0088	.0340	−.0111	.1098	.4012	.1072
④	c	−.0139	−.0112	.0460	−.0111	−.0088	.0340	.1072	.1098	.4012

This matrix can be checked by inverting the bus admittance matrix obtained in part b of this problem.

d. The modified elements of the bus impedance matrix for the removal of the network element 2–3 mutually coupled to the network elements 2–4 and 4–3 are obtained from

$$Z'^{a,b,c}_{ij} = Z^{a,b,c}_{ij} + (\bar{Z}^{a,b,c}_{i\alpha} - \bar{Z}^{a,b,c}_{i\beta})[M^{a,b,c}]^{-1}[\Delta y^{a,b,c}_s](\bar{Z}^{a,b,c}_{\gamma j} - \bar{Z}^{a,b,c}_{\delta j}) \qquad \begin{aligned} i, j &= 2, 3, 4 \\ \alpha, \gamma &= 2, 4, 2 \\ \beta, \delta &= 4, 3, 3 \end{aligned}$$

The original primitive admittance submatrix is

$[y^{a,b,c}_s] =$

	2–4 a	2–4 b	2–4 c	4–3 a	4–3 b	4–3 c	2–3 a	2–3 b	2–3 c
2–4 a	2.0750	−.4250	−.4250	.0750	.0750	.0750	−.1250	−.1250	−.1250
2–4 b	−.4250	2.0750	−.4250	.0750	.0750	.0750	−.1250	−.1250	−.1250
2–4 c	−.4250	−.4250	2.0750	.0750	.0750	.0750	−.1250	−.1250	−.1250
4–3 a	.0750	.0750	.0750	1.4083	−.2583	−.2583	−.1250	−.1250	−.1250
4–3 b	.0750	.0750	.0750	−.2583	1.4083	−.2583	−.1250	−.1250	−.1250
4–3 c	.0750	.0750	.0750	−.2583	−.2583	1.4083	−.1250	−.1250	−.1250
2–3 a	−.1250	−.1250	−.1250	−.1250	−.1250	−.1250	.8750	−.1250	−.1250
2–3 b	−.1250	−.1250	−.1250	−.1250	−.1250	−.1250	−.1250	.8750	−.1250
2–3 c	.1250	.1250	.1250	−.1250	−.1250	−.1250	−.1250	−.1250	.8750

The new primitive admittance submatrix is

$[y_s'^{a,b,c}] =$

		2–4			4–3			2–3		
		a	b	c	a	b	c	a	b	c
2–4	a	2.0000	−.5000	−.5000						
	b	−.5000	2.0000	−.5000						
	c	−.5000	−.5000	2.0000						
4–3	a				1.3333	−.3333	−.3333			
	b				−.3333	1.3333	−.3333			
	c				−.3333	−.3333	1.3333			
2–3	a									
	b									
	c									

Then, $[\Delta y_s^{a,b,c}] = [y_s^{a,b,c}] - [y_s'^{a,b,c}]$

$[\Delta y_s^{a,b,c}] =$

		2–4 a	2–4 b	2–4 c	4–3 a	4–3 b	4–3 c	2–3 a	2–3 b	2–3 c
2–4	a	.0750	.0750	.0750	.0750	.0750	.0750	−.1250	−.1250	−.1250
2–4	b	.0750	.0750	.0750	.0750	.0750	.0750	−.1250	−.1250	−.1250
2–4	c	.0750	.0750	.0750	.0750	.0750	.0750	−.1250	−.1250	−.1250
4–3	a	.0750	.0750	.0750	.0750	.0750	.0750	−.1250	−.1250	−.1250
4–3	b	.0750	.0750	.0750	.0750	.0750	.0750	−.1250	−.1250	−.1250
4–3	c	.0750	.0750	.0750	.0750	.0750	.0750	−.1250	−.1250	−.1250
2–3	a	−.1250	−.1250	−.1250	−.1250	−.1250	−.1250	.8750	−.1250	−.1250
2–3	b	−.1250	−.1250	−.1250	−.1250	−.1250	−.1250	−.1250	.8750	−.1250
2–3	c	−.1250	−.1250	−.1250	−.1250	−.1250	−.1250	−.1250	−.1250	.8750

Also

$$[M^{a,b,c}] = \{U - [\Delta y_s^{a,b,c}]([Z_{\gamma\alpha}^{a,b,c}] - [Z_{\delta\alpha}^{a,b,c}] - [Z_{\gamma\beta}^{a,b,c}] + [Z_{\delta\beta}^{a,b,c}])\}$$

$$\alpha, \gamma = 2, 4, 2$$
$$\beta, \delta = 4, 3, 3$$

where
$$[Z_{\gamma\alpha}^{a,b,c}] = \begin{bmatrix} Z_{22}^{a,b,c} & Z_{24}^{a,b,c} & Z_{22}^{a,b,c} \\ Z_{42}^{a,b,c} & Z_{44}^{a,b,c} & Z_{42}^{a,b,c} \\ Z_{22}^{a,b,c} & Z_{24}^{a,b,c} & Z_{22}^{a,b,c} \end{bmatrix}$$

$$[Z_{\delta\alpha}^{a,b,c}] = \begin{bmatrix} Z_{42}^{a,b,c} & Z_{44}^{a,b,c} & Z_{42}^{a,b,c} \\ Z_{32}^{a,b,c} & Z_{34}^{a,b,c} & Z_{32}^{a,b,c} \\ Z_{32}^{a,b,c} & Z_{34}^{a,b,c} & Z_{32}^{a,b,c} \end{bmatrix}$$

$$[Z_{\gamma\beta}^{a,b,c}] = \begin{bmatrix} Z_{24}^{a,b,c} & Z_{23}^{a,b,c} & Z_{23}^{a,b,c} \\ Z_{44}^{a,b,c} & Z_{43}^{a,b,c} & Z_{43}^{a,b,c} \\ Z_{24}^{a,b,c} & Z_{23}^{a,b,c} & Z_{23}^{a,b,c} \end{bmatrix}$$

$$[Z_{\delta\beta}^{a,b,c}] = \begin{bmatrix} Z_{44}^{a,b,c} & Z_{43}^{a,b,c} & Z_{43}^{a,b,c} \\ Z_{34}^{a,b,c} & Z_{33}^{a,b,c} & Z_{33}^{a,b,c} \\ Z_{34}^{a,b,c} & Z_{33}^{a,b,c} & Z_{33}^{a,b,c} \end{bmatrix}$$

and $Z_{22}^{a,b,c}$, $Z_{24}^{a,b,c}$, and so forth, are obtained from the original bus impedance matrix.

Substituting, then

$[M^{a,b,c}] =$

1.00135	.00135	.00135	.00203	.00203	.00203	.00338	.00338	.00338
.00135	1.00135	.00135	.00203	.00203	.00203	.00338	.00338	.00338
.00135	.00135	1.00135	.00203	.00203	.00203	.00338	.00338	.00338
.00135	.00135	.00135	1.00203	.00203	.00203	.00338	.00338	.00338
.00135	.00135	.00135	.00203	1.00203	.00203	.00338	.00338	.00338
.00135	.00135	.00135	.00203	.00203	1.00203	.00338	.00338	.00338
−.04105	.01815	.01615	−.06165	.02715	.02435	.89730	.04530	.04050
.01615	−.04105	.01815	.02435	−.06165	.02715	.04050	.89730	.04530
.01815	.01615	−.04105	.02715	.02435	−.06165	.04530	.04050	.89730

For the calculation of $Z'^{a,b,c}_{23}$, $i = 2$, $j = 3$, $\alpha, \gamma = 2, 4, 2$, and $\beta, \delta = 4, 3, 3$. Then

$$(\bar{Z}^{a,b,c}_{i\alpha} - \bar{Z}^{a,b,c}_{i\beta}) = \left[Z^{a,b,c}_{22} - Z^{a,b,c}_{24} \;\middle|\; Z^{a,b,c}_{24} - Z^{a,b,c}_{23} \;\middle|\; Z^{a,b,c}_{22} - Z^{a,b,c}_{23} \right]$$

.0239	−.0057	−.0047	.0359	−.0085	−.0071	.0598	−.0142	−.0118
−.0047	.0239	−.0057	−.0071	.0359	−.0085	−.0118	.0598	−.0142
−.0057	−.0047	.0239	−.0085	−.0071	.0359	−.0142	−.0118	.0598

and

$$(\bar{Z}^{a,b,c}_{\gamma j} - \bar{Z}^{a,b,c}_{\delta j}) = \begin{bmatrix} Z^{a,b,c}_{23} - Z^{a,b,c}_{43} \\ \hline Z^{a,b,c}_{43} - Z^{a,b,c}_{33} \\ \hline Z^{a,b,c}_{23} - Z^{a,b,c}_{33} \end{bmatrix}$$

=

−.0239	.0057	.0047
.0047	−.0239	.0057
.0057	.0047	−.0239
−.0359	.0085	.0071
.0071	−.0359	.0085
.0085	.0071	−.0359
−.0598	.0142	.0118
.0118	−.0598	.0142
.0142	.0118	−.0598

$$Z'^{a,b,c}_{23} = Z^{a,b,c}_{23} + \begin{bmatrix} Z^{a,b,c}_{22} - Z^{a,b,c}_{24} & Z^{a,b,c}_{24} - Z^{a,b,c}_{23} & Z^{a,b,c}_{22} - Z^{a,b,c}_{23} \end{bmatrix} [M^{a,b,c}]^{-1}[\Delta y^{a,b,c}_{s}] \begin{bmatrix} Z^{a,b,c}_{23} - Z^{a,b,c}_{43} \\ Z^{a,b,c}_{43} - Z^{a,b,c}_{33} \\ Z^{a,b,c}_{23} - Z^{a,b,c}_{33} \end{bmatrix}$$

$$= \begin{bmatrix} .0101 & -.0054 & -.0041 \\ -.0041 & .0101 & -.0054 \\ -.0054 & -.0041 & .0101 \end{bmatrix} + \begin{bmatrix} -.0042 & .0023 & .0018 \\ .0018 & -.0042 & .0023 \\ .0023 & .0018 & -.0042 \end{bmatrix}$$

$$= \begin{array}{c|ccc|} & a & b & c \\ \hline a & .0059 & -.0031 & -.0023 \\ b & -.0023 & .0059 & -.0031 \\ c & -.0031 & -.0023 & .0059 \\ \hline \end{array}$$

The remaining submatrices of $Z'^{a,b,c}_{BUS}$ are calculated in a similar manner to obtain the final modified matrix.

$Z'^{a,b,c}_{BUS}$ =

		② a	② b	② c	③ a	③ b	③ c	④ a	④ b	④ c
②	a	.0741	−.0218	−.0177	.0059	−.0031	−.0023	.0468	−.0143	−.0116
	b	−.0177	.0741	−.0218	−.0023	.0059	−.0031	−.0116	.0468	−.0143
	c	−.0218	−.0177	.0741	−.0031	−.0023	.0059	−.0143	−.0116	.0468
③	a	.0059	−.0031	−.0023	.0741	−.0218	−.0177	.0332	−.0107	−.0084
	b	−.0023	.0059	−.0031	−.0177	.0741	−.0218	−.0084	.0332	−.0107
	c	−.0031	−.0023	.0059	−.0218	−.0177	.0741	−.0107	−.0084	.0332
④	a	.0468	−.0143	−.0116	.0332	−.0107	−.0084	.4014	.1071	.1097
	b	−.0116	.0468	−.0143	−.0084	.0332	−.0107	.1097	.4014	.1071
	c	−.0143	−.0116	.0468	−.0107	−.0084	.0332	.1071	.1097	.4014

This matrix checks with that obtained after step 4 in part c except for a slight difference due to round-off error.

Problems

5.1 In Prob. 3.2, the positive and zero sequence data for the sample system shown in Fig. 3.14 is given in Table 3.5. For this system:

a. With ground as reference, form the three-phase incidence matrices A, K, B, $\hat{B}$, C, and $\hat{C}$ for the oriented connected graph selected for Prob. 3.2 and verify the relations:

i. $A_b K^t = U$

ii. $B_l = A_l K^t$

iii. $C_b = -B_l^t$

iv. $\hat{C}\hat{B}^t = U$

b. Neglecting resistance and assuming all negative sequence reactances are equal to the corresponding positive sequence reactances, form the three-phase network matrices $Y_{BUS}^{a,b,c}$ and $Z_{LOOP}^{a,b,c}$ by singular transformations.

c. Neglecting resistance and assuming the positive and negative sequence impedances are equal, form the three-phase network matrix $Z_{BUS}^{a,b,c}$ using the algorithm and ground as reference.

d. Transform $Z_{BUS}^{a,b,c}$ calculated in part c to $Z_{BUS}^{0,1,2}$. The submatrices for positive and zero sequences can be verified with those obtained in Prob. 3.2.

5.2 The sequence impedance data for the sample system shown in Fig. 5.7 is given in Table 5.3. Selecting ground as reference (node 0), compute $Z_{BUS}^{0,1,2}$ using the algorithm.

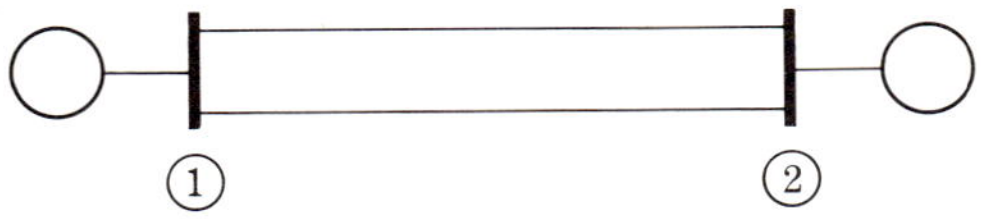

Fig. 5.7 Sample system for Prob. 5.2.

Table 5.3 Sequence impedance data of sample power system for Prob. 5.2

	Self				Mutual			
Element number	Bus code p–q	Impedance $z^{0,1,2}_{pq,pq}$			Bus code r–s	Impedance $z^{0,1,2}_{pq,rs}$		
		$0 + j0.04$						
1	0–1		$0 + j0.10$					
				$0 + j0.10$				
		$0 + j0.04$						
2	0–2		$0 + j0.10$					
				$0 + j0.10$				
		$0.35 + j0.65$	$0.02 - j0.01$	$-0.02 - j0.01$				
3	1–2(1)	$-0.02 - j0.01$	$0.15 + j0.60$	$-0.04 + j0.02$				
		$0.02 - j0.01$	$0.04 + j0.02$	$0.15 + j0.60$				
		$0.35 + j0.65$	$0.02 - j0.01$	$-0.02 - j0.01$		$0.20 + j1.20$		
4	1–2(2)	$-0.02 - j0.01$	$0.15 + j0.60$	$-0.04 + j0.02$	1–2(1)			
		$0.02 - j0.01$	$0.04 + j0.02$	$0.15 + j0.60$				

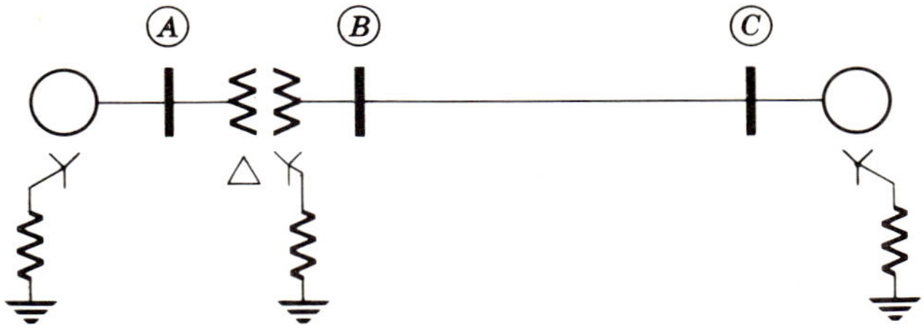

Fig. 5.8 Sample system for Prob. 5.3.

5.3 The reactance data for the three-phase system shown in Fig. 5.8 is Generators A and C:

$$x^{(1)} = x^{(2)} = 0.1$$
$$x^{(0)} = 0.04$$
$$x_g = 0.02$$

Transformer A–B:

$$x^{(1)} = x^{(2)} = x^{(0)} = 0.1$$
$$x_g = 0.05$$

Transmission line B–C:

$$x^{a,b,c} = \begin{bmatrix} 0.3 & 0.2 & 0 \\ 0.2 & 0.4 & 0.1 \\ 0 & 0.1 & 0.2 \end{bmatrix}$$

a. With ground as reference form $Y^{0,1,2}_{BUS}$.
b. Form $Z^{0,1,2}_{BUS}$ using the algorithm.
c. Determine $Z^{a,b,c}_{BUS}$ from $Z^{0,1,2}_{BUS}$ obtained in part b.

5.4 Assume that the transmission line B–C of Prob. 5.3 is balanced and its reactance is

$$x^{a,b,c} = \begin{bmatrix} 0.3 & 0.1 & 0.1 \\ 0.1 & 0.3 & 0.1 \\ 0.1 & 0.1 & 0.3 \end{bmatrix}$$

a. Compute $Z^{(0)}_{BUS}$, $Z^{(1)}_{BUS}$, and $Z^{(2)}_{BUS}$.

b. Determine $Z^{a,b,c}_{BUS}$ from $Z^{0,1,2}_{BUS}$ obtained in part *a* and compare the the results to those obtained for the unbalanced line in Prob. 5.3, part *c.*

5.5 The sequence impedance data for the sample system shown in Fig. 5.9 is given in Table 5.4. The mutual impedances $z^{0,1,2}_{12,13}$ and $z^{0,1,2}_{13,12}$ are not equal because of the circuit arrangement. For this system compute $Z^{0,1,2}_{BUS}$ using the ground (node 0) as reference.

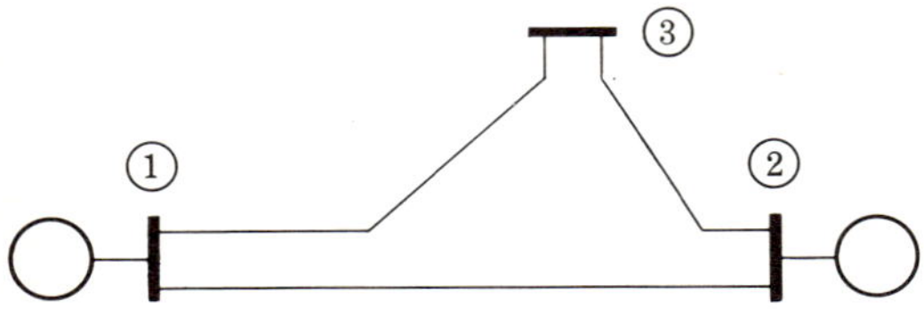

Fig. 5.9 Sample system for Prob. 5.5.

Table 5.4 Sequence impedance data of sample power system for Prob. 5.5

Element Number	Self: Bus code p–q	Self: Impedance $z^{0,1,2}_{pq,pq}$			Mutual: Bus code r–s	Mutual: Impedance $z^{0,1,2}_{pq,rs}$		
		$0 + j.0400$						
1	0–1		$0 + j.1000$					
				$0 + j.1000$				
		$0 + j.0400$						
2	0–2		$0 + j.1000$					
				$0 + j.1000$				
		$.3504 + j.9965$	$.0041 - j.0125$	$-.0065 - j.0022$		$.1317 + j.1545$	$-.0166 - j.0117$	$.0096 - j.0141$
3	1–2	$-.0065 - j.0022$	$.0644 + j.4510$	$-.0244 - j.0152$	1–3	$-.0018 + j.0202$	$-.0029 - j.0052$	$.0050 - j.0028$
		$.0041 - j.0125$	$.0257 + j.0136$	$.0644 + j.4510$		$.0074 + j.0153$	$-.0049 - j.0029$	$.0026 - j.0045$
		$.3504 + j.9965$	$.0052 - j.0043$	$-.0129 + j.0027$		$.1317 + j.1545$	$.0074 + j.0153$	$-.0018 + j.0202$
4	1–3	$-.0129 + j.0027$	$.0644 + j.4510$	$-.0247 + j.0156$	1–2	$.0096 - j.0141$	$.0026 - j.0045$	$.0050 - j.0028$
		$.0052 - j.0043$	$.0254 + j.0135$	$.0644 + j.4510$		$-.0166 - j.0117$	$.0049 - j.0029$	$-.0029 - j.0052$
		$.1776 + j.4936$	$.0018 - j.0044$	$-.0047 + j.0006$				
5	2–3	$-.0047 + j.0006$	$.0319 + j.2256$	$-.0124 + j.0076$				
		$.0018 - j.0044$	$.0128 + j.0069$	$.0319 + j.2256$				

Bibliography

Clarke, Edith: "Circuit Analysis of A-C Power Systems, Symmetrical and Related Components," vol. 1, John Wiley & Sons, Inc., New York, 1943.

Kron, Gabriel: "Tensor Analysis of Networks," John Wiley & Sons, Inc., New York, 1939.

Rothe, Frederick S.: "An Introduction to Power System Analysis," John Wiley & Sons, Inc., New York, 1953.

Stevenson, William D., Jr.: "Elements of Power System Analysis," McGraw-Hill Book Company, New York, 1962.

Wagner, C. F., and R. D. Evans: "Symmetrical Components," McGraw-Hill Book Company, New York, 1933.

chapter 6
Short circuit studies

6.1 Introduction

Short circuit calculations provide currents and voltages on a power system during fault conditions. This information is required to design an adequate protective relaying system and to determine interrupting requirements for circuit breakers at each switching location. Relaying systems must recognize the existence of a fault and initiate circuit breaker operation to disconnect faulted facilities. This action is required to assure minimum disruption of electrical service and to limit damage in the faulted equipment. The currents and voltages resulting from various types of faults occurring at many locations throughout the power system must be calculated to provide sufficient data to develop an effective relaying and switching system. To obtain the required information a special purpose analog computer, called a network analyzer, was used extensively for short circuit studies before digital techniques were available.

The bus frame of reference in admittance form was employed in the first application of digital computers to short circuit studies. This method, which was patterned after similar techniques employed for load flow calculations, used an iterative technique (Coombe and Lewis, 1956). This required a complete iterative solution for each fault type and location. The procedure was time-consuming, particularly if, as was usually the case, the currents and voltages were required for a large number of fault locations. Consequently, this method was not adopted generally.

The development of techniques for applying a digital computer to form the bus impedance matrix made it feasible to use Thevenin's theorem for short circuit calculations. This approach provided an efficient means of determining short circuit currents and voltages because these values can be obtained with few arithmetic operations involving only related portions of the bus impedance matrix.

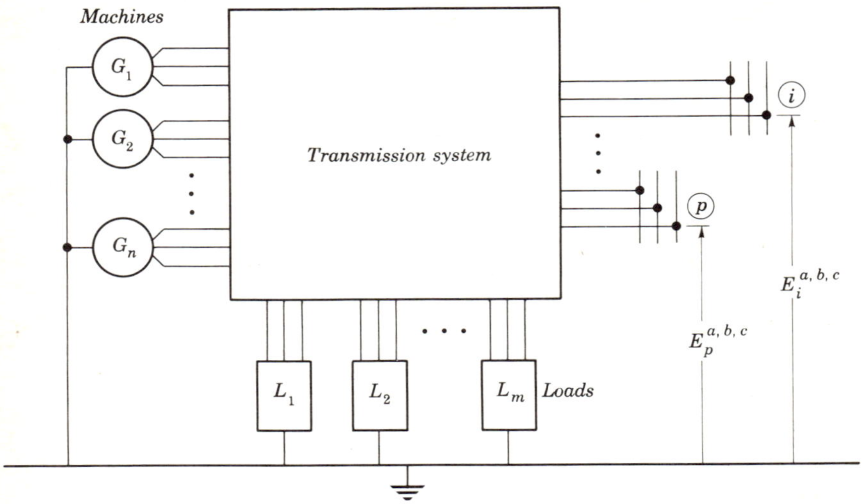

Fig. 6.1 Three-phase representation of a power system.

6.2 Short circuit calculations using Z_{BUS}

System representation

The three-phase representation of a power system under steady state condition is shown in Fig. 6.1. In general, sufficient accuracy in short circuit studies can be obtained with a simplified representation. The simplified three-phase representation is shown in Fig. 6.2 and is obtained by:

1. Representing each machine by a constant voltage behind the machine reactance, transient or subtransient
2. Neglecting shunt connections, e.g., loads, line charging, etc.
3. Setting all transformers at nominal taps

In many short circuit studies, particularly for high voltage systems, it is sufficient to represent transmission line and transformer impedances as real numbers equal to the corresponding reactances.

Fault currents and voltages

The use of the bus impedance matrix provides a convenient means of calculating short circuit currents and voltages when the ground is selected as reference. One of the distinct advantages is that, once the bus

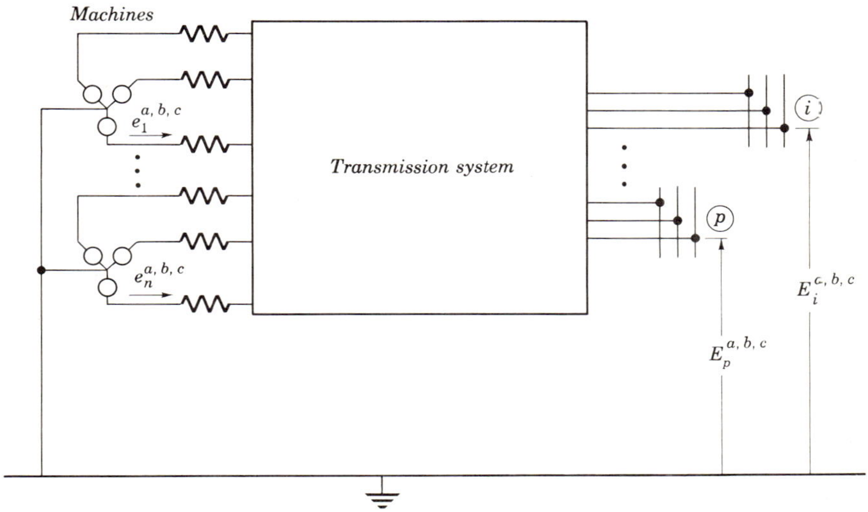

Fig. 6.2 Three-phase representation of a power system for short circuit studies.

impedance matrix is formed, the elements of this matrix can be used directly to calculate the currents and voltages associated with various types of faults and fault locations.

The representation of the system with a fault at bus p is shown in Fig. 6.3. In this representation, derived by means of Thevenin's theorem, the internal impedance is represented by the bus impedance matrix including machine reactances, and the open-circuited voltage is represented by the bus voltages prior to the fault.

The performance equation of the system during a fault is

$$\bar{E}^{a,b,c}_{BUS(F)} = \bar{E}^{a,b,c}_{BUS(0)} - Z^{a,b,c}_{BUS}\bar{I}^{a,b,c}_{BUS(F)} \tag{6.2.1}$$

The unknown voltage vector is

$$\bar{E}^{a,b,c}_{BUS(F)} = \begin{bmatrix} E^{a,b,c}_{1(F)} \\ \cdots \\ E^{a,b,c}_{p(F)} \\ \cdots \\ E^{a,b,c}_{n(F)} \end{bmatrix}$$

where the elements of $\bar{E}^{a,b,c}_{BUS(F)}$ are the three-phase voltage vectors $E^{a,b,c}_{i(F)} \qquad i = 1, 2, \ldots, n$

The known voltage vector prior to the fault is

$$\bar{E}^{a,b,c}_{BUS(0)} = \begin{bmatrix} E^{a,b,c}_{1(0)} \\ \cdots \\ E^{a,b,c}_{p(0)} \\ \cdots \\ E^{a,b,c}_{n(0)} \end{bmatrix}$$

The unknown bus current vector during a fault at bus p is

$$\bar{I}^{a,b,c}_{BUS(F)} = \begin{bmatrix} 0 \\ \cdots \\ 0 \\ I^{a,b,c}_{p(F)} \\ 0 \\ \cdots \\ 0 \end{bmatrix}$$

The three-phase bus impedance matrix is

$$Z^{a,b,c}_{BUS} = \begin{bmatrix} Z^{a,b,c}_{11} & \cdots & Z^{a,b,c}_{1p} & \cdots & Z^{a,b,c}_{1n} \\ \cdots & \cdots & \cdots & \cdots & \cdots \\ Z^{a,b,c}_{p1} & \cdots & Z^{a,b,c}_{pp} & \cdots & Z^{a,b,c}_{pn} \\ \cdots & \cdots & \cdots & \cdots & \cdots \\ Z^{a,b,c}_{n1} & \cdots & Z^{a,b,c}_{np} & \cdots & Z^{a,b,c}_{nn} \end{bmatrix}$$

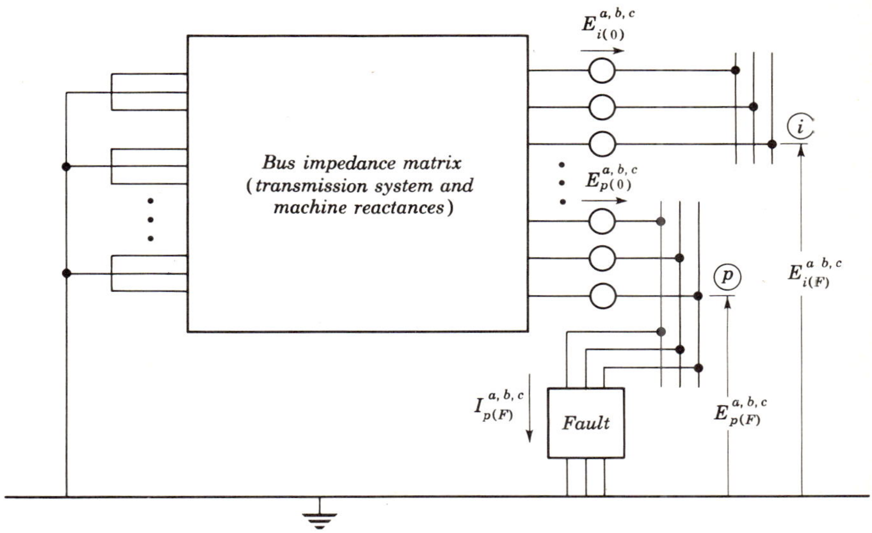

Fig. 6.3 Three-phase representation of a power system with a fault at bus p.

where the elements of $Z^{a,b,c}_{BUS}$ are matrices of dimension 3 × 3. Equation (6.2.1) can be written as follows:

$$
\begin{aligned}
E^{a,b,c}_{1(F)} &= E^{a,b,c}_{1(0)} - Z^{a,b,c}_{1p} I^{a,b,c}_{p(F)} \\
E^{a,b,c}_{2(F)} &= E^{a,b,c}_{2(0)} - Z^{a,b,c}_{2p} I^{a,b,c}_{p(F)} \\
&\cdots\cdots\cdots \\
E^{a,b,c}_{p(F)} &= E^{a,b,c}_{p(0)} - Z^{a,b,c}_{pp} I^{a,b,c}_{p(F)} \\
&\cdots\cdots\cdots \\
E^{a,b,c}_{n(F)} &= E^{a,b,c}_{n(0)} - Z^{a,b,c}_{np} I^{a,b,c}_{p(F)}
\end{aligned}
\tag{6.2.2}
$$

The three-phase voltage vector at the faulted bus p is, from Fig. 6.3,

$$E^{a,b,c}_{p(F)} = Z^{a,b,c}_{F} I^{a,b,c}_{p(F)} \tag{6.2.3}$$

where $Z^{a,b,c}_{F}$ is the three-phase impedance matrix for the fault. The elements of this 3 × 3 matrix depend on the type of fault and fault impedance. Substituting from equation (6.2.3) for $E^{a,b,c}_{p(F)}$, the pth equation of (6.2.2) becomes

$$Z^{a,b,c}_{F} I^{a,b,c}_{p(F)} = E^{a,b,c}_{p(0)} - Z^{a,b,c}_{pp} I^{a,b,c}_{p(F)} \tag{6.2.4}$$

Solving equation (6.2.4) for $I^{a,b,c}_{p(F)}$ yields

$$I^{a,b,c}_{p(F)} = (Z^{a,b,c}_{F} + Z^{a,b,c}_{pp})^{-1} E^{a,b,c}_{p(0)} \tag{6.2.5}$$

Substituting for $I^{a,b,c}_{p(F)}$ in equation (6.2.3), the three-phase voltage at the faulted bus p is

$$E^{a,b,c}_{p(F)} = Z^{a,b,c}_F(Z^{a,b,c}_F + Z^{a,b,c}_{pp})^{-1}E^{a,b,c}_{p(0)} \qquad (6.2.6)$$

Similarly, the three-phase voltages at buses other than p can be obtained by substituting for $I^{a,b,c}_{p(F)}$ from equation (6.2.5). Then

$$E^{a,b,c}_{i(F)} = E^{a,b,c}_{i(0)} - Z^{a,b,c}_{ip}(Z^{a,b,c}_F + Z^{a,b,c}_{pp})^{-1}E^{a,b,c}_{p(0)} \qquad i \neq p \qquad (6.2.7)$$

When it is desirable to express the parameters of the fault circuit in the admittance form, the three-phase fault current at bus p is

$$I^{a,b,c}_{p(F)} = Y^{a,b,c}_F E^{a,b,c}_{p(F)} \qquad (6.2.8)$$

where $Y^{a,b,c}_F$ is the three-phase admittance matrix for the fault. Substituting $I^{a,b,c}_{p(F)}$ from equation (6.2.8), the pth equation of (6.2.2) becomes

$$E^{a,b,c}_{p(F)} = E^{a,b,c}_{p(0)} - Z^{a,b,c}_{pp}Y^{a,b,c}_F E^{a,b,c}_{p(F)} \qquad (6.2.9)$$

Solving equation (6.2.9) for $E^{a,b,c}_{p(F)}$ yields

$$E^{a,b,c}_{p(F)} = (U + Z^{a,b,c}_{pp}Y^{a,b,c}_F)^{-1}E^{a,b,c}_{p(0)} \qquad (6.2.10)$$

Substituting for $E^{a,b,c}_{p(F)}$ in equation (6.2.8), the three-phase current at the faulted bus p is

$$I^{a,b,c}_{p(F)} = Y^{a,b,c}_F(U + Z^{a,b,c}_{pp}Y^{a,b,c}_F)^{-1}E^{a,b,c}_{p(0)} \qquad (6.2.11)$$

Similarly, the three-phase voltages at buses other than p can be obtained by substituting for $I^{a,b,c}_{p(F)}$ from equation (6.2.11). Then

$$E^{a,b,c}_{i(F)} = E^{a,b,c}_{i(0)} - Z^{a,b,c}_{ip}Y^{a,b,c}_F(U + Z^{a,b,c}_{pp}Y^{a,b,c}_F)^{-1}E^{a,b,c}_{p(0)} \qquad i \neq p \qquad (6.2.12)$$

Fault currents flowing through the elements of the network can be calculated with the bus voltages obtained from equations (6.2.6) and (6.2.7) or from equations (6.2.10) and (6.2.12). These currents in terms of the voltages across the elements of the network are

$$\bar{i}^{a,b,c}_{(F)} = [y^{a,b,c}]\bar{v}^{a,b,c}_{(F)}$$

where the elements of the current vector are

$$i^{a,b,c}_{ij(F)} = \begin{bmatrix} i^a_{ij(F)} \\ i^b_{ij(F)} \\ i^c_{ij(F)} \end{bmatrix}$$

the elements of the voltage vector are

$$v_{ij(F)}^{a,b,c} = \begin{bmatrix} v_{ij(F)}^{a} \\ v_{ij(F)}^{b} \\ v_{ij(F)}^{c} \end{bmatrix}$$

and the elements of the primitive admittance matrix are

$$y_{ij,kl}^{a,b,c} = \begin{bmatrix} y_{ij,kl}^{aa} & y_{ij,kl}^{ab} & y_{ij,kl}^{ac} \\ y_{ij,kl}^{ba} & y_{ij,kl}^{bb} & y_{ij,kl}^{bc} \\ y_{ij,kl}^{ca} & y_{ij,kl}^{cb} & y_{ij,kl}^{cc} \end{bmatrix}$$

where $y_{ij,kl}^{bc}$ is the mutual admittance between phase b of network element i–j and phase c of network element k–l. The three-phase current in the network element i–j can be calculated from

$$i_{ij(F)}^{a,b,c} = \bar{y}_{ij,\rho\sigma}^{a,b,c}\bar{v}_{\rho\sigma(F)}^{a,b,c} \tag{6.2.13}$$

where $\rho\sigma$ refers to the element i–j as well as to elements mutually coupled to i–j. Since

$$\bar{v}_{\rho\sigma(F)}^{a,b,c} = \bar{E}_{\rho(F)}^{a,b,c} - \bar{E}_{\sigma(F)}^{a,b,c}$$

then equation (6.2.13) becomes

$$i_{ij(F)}^{a,b,c} = \bar{y}_{ij,\rho\sigma}^{a,b,c}(\bar{E}_{\rho(F)}^{a,b,c} - \bar{E}_{\sigma(F)}^{a,b,c}) \tag{6.2.14}$$

The formulas for fault currents and voltages derived in this section can be used for balanced and unbalanced three-phase short circuit studies.

6.3 Short circuit calculations for balanced three-phase network using Z_{BUS}

Transformation to symmetrical components

The formulas developed in the preceding section for calculation of fault currents and voltages can be simplified for a balanced three-phase network by using symmetrical components. The primitive impedance matrix for a stationary balanced three-phase element is

$$z_{pq}^{a,b,c} = \begin{bmatrix} z_{pq}^{s} & z_{pq}^{m} & z_{pq}^{m} \\ z_{pq}^{m} & z_{pq}^{s} & z_{pq}^{m} \\ z_{pq}^{m} & z_{pq}^{m} & z_{pq}^{s} \end{bmatrix}$$

This matrix can be diagonalized by the transformation $(T_s^*)^t z_{pq}^{a,b,c} T_s$ into

$$z_{pq}^{0,1,2} = \begin{bmatrix} z_{pq}^{(0)} & & \\ & z_{pq}^{(1)} & \\ & & z_{pq}^{(2)} \end{bmatrix}$$

where $z_{pq}^{(0)}$, $z_{pq}^{(1)}$, and $z_{pq}^{(2)}$ are the zero, positive, and negative sequence impedances, respectively. The positive and negative sequence impedances for a stationary balanced three-phase element are equal. In addition, it is generally accepted that positive and negative sequence impedances for rotating elements can be assumed equal for short circuit calculations.

In a similar manner, each $y_{ij,kl}^{a,b,c}$ in the primitive admittance matrix and each $Z_{ij}^{a,b,c}$ in the bus impedance matrix can be diagonalized by the transformation matrix T_s to obtain, respectively,

$$y_{ij,kl}^{0,1,2} = \begin{bmatrix} y_{ij,kl}^{(0)} & & \\ & y_{ij,kl}^{(1)} & \\ & & y_{ij,kl}^{(2)} \end{bmatrix} \quad \text{and} \quad Z_{ij}^{0,1,2} = \begin{bmatrix} Z_{ij}^{(0)} & & \\ & Z_{ij}^{(1)} & \\ & & Z_{ij}^{(2)} \end{bmatrix}$$

It is customary to assume that all bus voltages prior to the fault are equal in magnitude and phase angle. Assuming the magnitude of the line-to-ground voltage $E_{i(0)}$ equal to one per unit, then the ith bus voltage before the fault is

$$E_{i(0)}^{a,b,c} = \begin{bmatrix} 1 \\ a^2 \\ a \end{bmatrix}$$

Transforming into symmetrical components, that is

$$E_{i(0)}^{0,1,2} = (T_s^*)^t E_{i(0)}^{a,b,c}$$

then

$$E_{i(0)}^{0,1,2} = \begin{bmatrix} 0 \\ \sqrt{3} \\ 0 \end{bmatrix}$$

The fault impedance matrix $Z_F^{a,b,c}$ can be transformed by T_s into the matrix $Z_F^{0,1,2}$. The resulting matrix is diagonal if the fault is balanced. The fault impedance and admittance matrices in terms of three-phase and symmetrical components for various types of faults are given in Table 6.1.

Similarly, the equations for calculating fault currents and voltages can be written in terms of symmetrical components. The current at the faulted bus p is

$$I_{p(F)}^{0,1,2} = (Z_F^{0,1,2} + Z_{pp}^{0,1,2})^{-1} E_{p(0)}^{0,1,2} \tag{6.3.1}$$

or

$$I_{p(F)}^{0,1,2} = Y_F^{0,1,2}(U + Z_{pp}^{0,1,2} Y_F^{0,1,2})^{-1} E_{p(0)}^{0,1,2} \tag{6.3.2}$$

The voltage at the faulted bus p is

$$E_{p(F)}^{0,1,2} = Z_F^{0,1,2}(Z_F^{0,1,2} + Z_{pp}^{0,1,2})^{-1} E_{p(0)}^{0,1,2} \tag{6.3.3}$$

or

$$E_{p(F)}^{0,1,2} = (U + Z_{pp}^{0,1,2} Y_F^{0,1,2})^{-1} E_{p(0)}^{0,1,2} \tag{6.3.4}$$

The voltages at buses other than p are

$$E_{i(F)}^{0,1,2} = E_{i(0)}^{0,1,2} - Z_{ip}^{0,1,2}(Z_F^{0,1,2} + Z_{pp}^{0,1,2})^{-1} E_{p(0)}^{0,1,2} \tag{6.3.5}$$

or

$$E_{i(F)}^{0,1,2} = E_{i(0)}^{0,1,2} - Z_{ip}^{0,1,2} Y_F^{0,1,2}(U + Z_{pp}^{0,1,2} Y_F^{0,1,2})^{-1} E_{p(0)}^{0,1,2} \tag{6.3.6}$$

The fault current in the three-phase element i–j is

$$i_{ij(F)}^{0,1,2} = \bar{y}_{ij,\rho\sigma}^{0,1,2}(\bar{E}_{\rho(F)}^{0,1,2} - \bar{E}_{\sigma(F)}^{0,1,2}) \tag{6.3.7}$$

Three-phase-to-ground fault

Fault currents and voltages for a three-phase-to-ground fault can be obtained by substituting the corresponding fault impedance matrix, in

Table 6.1 *Fault impedance and admittance matrices*

Type of fault	*Three-phase components* $Z_F^{a,b,c}$	$Y_F^{a,b,c}$
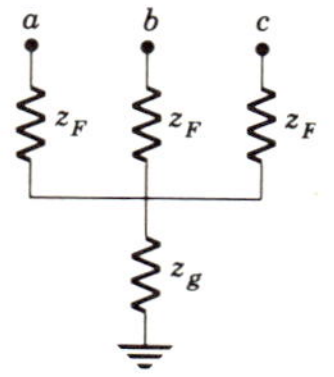 Three-phase-to-ground	$\begin{bmatrix} z_F + z_g & z_g & z_g \\ z_g & z_F + z_g & z_g \\ z_g & z_g & z_F + z_g \end{bmatrix}$	$\frac{1}{3}\begin{bmatrix} y_0 + 2y_F & y_0 - y_F & y_0 - y_F \\ y_0 - y_F & y_0 + 2y_F & y_0 - y_F \\ y_0 - y_F & y_0 - y_F & y_0 + 2y_F \end{bmatrix}$ where $y_0 = \dfrac{1}{z_F + 3z_g}$
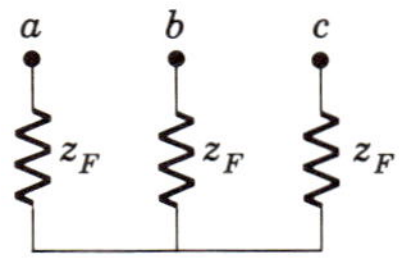 Three-phase	Not defined	$\frac{y_F}{3}\begin{bmatrix} 2 & -1 & -1 \\ -1 & 2 & -1 \\ -1 & -1 & 2 \end{bmatrix}$
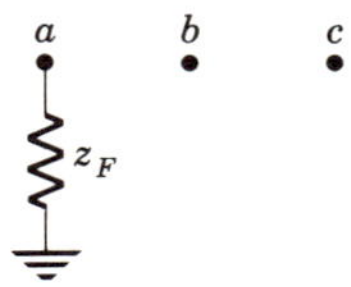 Line-to-ground	$\begin{bmatrix} z_F & 0 & 0 \\ 0 & \infty & 0 \\ 0 & 0 & \infty \end{bmatrix}$	$\begin{bmatrix} y_F & 0 & 0 \\ 0 & 0 & 0 \\ 0 & 0 & 0 \end{bmatrix}$
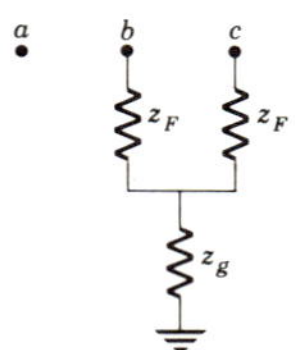 Line-to-line-to-ground	$\begin{bmatrix} \infty & 0 & 0 \\ 0 & z_F + z_g & z_g \\ 0 & z_g & z_F + z_g \end{bmatrix}$	$\begin{bmatrix} 0 & 0 & 0 \\ 0 & \dfrac{z_F + z_g}{z_F^2 + 2z_F z_g} & \dfrac{-z_g}{z_F^2 + 2z_F z_g} \\ 0 & \dfrac{-z_g}{z_F^2 + 2z_F z_g} & \dfrac{z_F + z_g}{z_F^2 + 2z_F z_g} \end{bmatrix}$
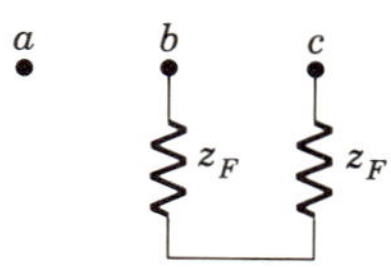 Line-to-line	Not defined	$\frac{y_F}{2}\begin{bmatrix} 0 & 0 & 0 \\ 0 & 1 & -1 \\ 0 & -1 & 1 \end{bmatrix}$

Symmetrical components	
$Z_F^{0,1,2}$	$Y_F^{0,1,2}$
$\begin{bmatrix} z_F + 3z_g & 0 & 0 \\ 0 & z_F & 0 \\ 0 & 0 & z_F \end{bmatrix}$	$\begin{bmatrix} y_0 & 0 & 0 \\ 0 & y_F & 0 \\ 0 & 0 & y_F \end{bmatrix}$ where $y_0 = \dfrac{1}{z_F + 3z_g}$
$\begin{bmatrix} \infty & 0 & 0 \\ 0 & z_F & 0 \\ 0 & 0 & z_F \end{bmatrix}$	$y_F \begin{bmatrix} 0 & 0 & 0 \\ 0 & 1 & 0 \\ 0 & 0 & 1 \end{bmatrix}$
Not defined	$\dfrac{y_F}{3} \begin{bmatrix} 1 & 1 & 1 \\ 1 & 1 & 1 \\ 1 & 1 & 1 \end{bmatrix}$
Not defined	$\dfrac{1}{3(z_F^2 + 2z_F z_g)} \begin{bmatrix} 2z_F & -z_F & -z_F \\ -z_F & 2z_F + 3z_g & -(z_F + 3z_g) \\ -z_F & -(z_F + 3z_g) & 2z_F + 3z_g \end{bmatrix}$
Not defined	$\dfrac{y_F}{2} \begin{bmatrix} 0 & 0 & 0 \\ 0 & 1 & -1 \\ 0 & -1 & 1 \end{bmatrix}$

terms of symmetrical components, into equations (6.3.1), (6.3.3), and (6.3.5). Both sides of the resulting equations can be premultiplied by T_s to obtain the corresponding formulas in terms of phase components.

The fault impedance matrix for a three-phase-to-ground fault is, from Table 6.1,

$$Z_F^{0,1,2} = \begin{bmatrix} z_F + 3z_g & & \\ & z_F & \\ & & z_F \end{bmatrix} \qquad (6.3.8)$$

The three-phase fault current and the bus voltages are obtained by substituting, from equation (6.3.8), for $Z_F^{0,1,2}$ in equations (6.3.1), (6.3.3), and (6.3.5). The current at the faulted bus p is

$$\begin{bmatrix} I_{p(F)}^{(0)} \\ I_{p(F)}^{(1)} \\ I_{p(F)}^{(2)} \end{bmatrix} = \left(\begin{bmatrix} z_F + 3z_g + Z_{pp}^{(0)} & & \\ & z_F + Z_{pp}^{(1)} & \\ & & z_F + Z_{pp}^{(1)} \end{bmatrix} \right)^{-1} \begin{bmatrix} 0 \\ \sqrt{3} \\ 0 \end{bmatrix}$$

which reduces to

$$\begin{bmatrix} I_{p(F)}^{(0)} \\ I_{p(F)}^{(1)} \\ I_{p(F)}^{(2)} \end{bmatrix} = \begin{bmatrix} 0 \\ \dfrac{\sqrt{3}}{z_F + Z_{pp}^{(1)}} \\ 0 \end{bmatrix} \qquad (6.3.9)$$

The phase components of the fault current at bus p can be obtained by premultiplying both sides of equation (6.3.9) by T_s. These currents are

$$\begin{bmatrix} I_{p(F)}^{a} \\ I_{p(F)}^{b} \\ I_{p(F)}^{c} \end{bmatrix} = \frac{1}{z_F + Z_{pp}^{(1)}} \begin{bmatrix} 1 \\ a^2 \\ a \end{bmatrix}$$

The voltage at the faulted bus p is

$$\begin{bmatrix} E_{p(F)}^{(0)} \\ E_{p(F)}^{(1)} \\ E_{p(F)}^{(2)} \end{bmatrix} = \begin{bmatrix} z_F + 3z_g & & \\ & z_F & \\ & & z_F \end{bmatrix} \begin{bmatrix} 0 \\ \dfrac{\sqrt{3}}{z_F + Z_{pp}^{(1)}} \\ 0 \end{bmatrix}$$

which reduces to

$$\begin{bmatrix} E_{p(F)}^{(0)} \\ E_{p(F)}^{(1)} \\ E_{p(F)}^{(2)} \end{bmatrix} = \begin{bmatrix} 0 \\ \dfrac{\sqrt{3}\, z_F}{z_F + Z_{pp}^{(1)}} \\ 0 \end{bmatrix}$$

The phase components of the fault voltage are

$$\begin{bmatrix} E_{p(F)}^{a} \\ E_{p(F)}^{b} \\ E_{p(F)}^{c} \end{bmatrix} = \frac{z_F}{z_F + Z_{pp}^{(1)}} \begin{bmatrix} 1 \\ a^2 \\ a \end{bmatrix}$$

The voltages at buses other than p are

$$\begin{bmatrix} E_{i(F)}^{(0)} \\ E_{i(F)}^{(1)} \\ E_{i(F)}^{(2)} \end{bmatrix} = \begin{bmatrix} 0 \\ \sqrt{3} \\ 0 \end{bmatrix} - \begin{bmatrix} Z_{ip}^{(0)} & & \\ & Z_{ip}^{(1)} & \\ & & Z_{ip}^{(2)} \end{bmatrix} \begin{bmatrix} 0 \\ \dfrac{\sqrt{3}}{z_F + Z_{pp}^{(1)}} \\ 0 \end{bmatrix}$$

which reduces to

$$\begin{bmatrix} E_{i(F)}^{(0)} \\ E_{i(F)}^{(1)} \\ E_{i(F)}^{(2)} \end{bmatrix} = \sqrt{3} \begin{bmatrix} 0 \\ 1 - \dfrac{Z_{ip}^{(1)}}{z_F + Z_{pp}^{(1)}} \\ 0 \end{bmatrix}$$

In phase components,

$$\begin{bmatrix} E_{i(F)}^{a} \\ E_{i(F)}^{b} \\ E_{i(F)}^{c} \end{bmatrix} = \left(1 - \frac{Z_{ip}^{(1)}}{z_F + Z_{pp}^{(1)}}\right) \begin{bmatrix} 1 \\ a^2 \\ a \end{bmatrix}$$

Table 6.2 Current and voltage formulas for three-phase-to-ground fault at bus p

Three-phase components	*Symmetrical components*
$I_{p(F)}^{a,b,c} = \dfrac{E_{p(0)}}{z_F + Z_{pp}^{(1)}} \begin{bmatrix} 1 \\ a^2 \\ a \end{bmatrix}$	$I_{p(F)}^{0,1,2} = \dfrac{\sqrt{3}\, E_{p(0)}}{z_F + Z_{pp}^{(1)}} \begin{bmatrix} 0 \\ 1 \\ 0 \end{bmatrix}$
$E_{p(F)}^{a,b,c} = \dfrac{z_F E_{p(0)}}{z_F + Z_{pp}^{(1)}} \begin{bmatrix} 1 \\ a^2 \\ a \end{bmatrix}$	$E_{p(F)}^{0,1,2} = \dfrac{\sqrt{3}\, z_F E_{p(0)}}{z_F + Z_{pp}^{(1)}} \begin{bmatrix} 0 \\ 1 \\ 0 \end{bmatrix}$
$E_{i(F)}^{a,b,c} = \left(E_{i(0)} - \dfrac{Z_{ip}^{(1)} E_{p(0)}}{z_F + Z_{pp}^{(1)}}\right) \begin{bmatrix} 1 \\ a^2 \\ a \end{bmatrix}$ $i \neq p$	$E_{i(F)}^{0,1,2} = \sqrt{3} \left(E_{i(0)} - \dfrac{Z_{ip}^{(1)} E_{p(0)}}{z_F + Z_{pp}^{(1)}}\right) \begin{bmatrix} 0 \\ 1 \\ 0 \end{bmatrix}$ $i \neq p$

The formulas derived in this section are summarized in Table 6.2. The line-to-ground voltage was assumed to be one per unit in the derivations. The formulas in Table 6.2 include the term for the line-to-ground voltage which can be set at any desired per unit value.

The currents in the network elements during the fault can be calculated from equation (6.3.7). Since the zero and negative sequence bus voltages are zero for a three-phase fault and there is no mutual coupling in the positive sequence network, that is, $y^{(1)}_{ij,\rho\sigma} = 0$ except when $\rho\sigma = ij$, then equation (6.3.7) reduces to

$$\begin{bmatrix} i^{(0)}_{ij(F)} \\ i^{(1)}_{ij(F)} \\ i^{(2)}_{ij(F)} \end{bmatrix} = \begin{bmatrix} 0 \\ y^{(1)}_{ij,ij}(E^{(1)}_{i(F)} - E^{(1)}_{j(F)}) \\ 0 \end{bmatrix}$$

In phase components,

$$\begin{bmatrix} i^{a}_{ij(F)} \\ i^{b}_{ij(F)} \\ i^{c}_{ij(F)} \end{bmatrix} = \frac{1}{\sqrt{3}} y^{(1)}_{ij,ij}(E^{(1)}_{i(F)} - E^{(1)}_{j(F)}) \begin{bmatrix} 1 \\ a^2 \\ a \end{bmatrix}$$

Line-to-ground fault

The fault admittance matrix for a line-to-ground fault in phase a is, from Table 6.1,

$$Y^{0,1,2}_{F} = \frac{y_F}{3} \begin{bmatrix} 1 & 1 & 1 \\ 1 & 1 & 1 \\ 1 & 1 & 1 \end{bmatrix} \tag{6.3.10}$$

The fault current and the bus voltages are obtained by substituting from equation (6.3.10) for $Y^{0,1,2}_{F}$ in equations (6.3.2), (6.3.4), and (6.3.6). The current at the faulted bus p is

$$\begin{bmatrix} I^{(0)}_{p(F)} \\ I^{(1)}_{p(F)} \\ I^{(2)}_{p(F)} \end{bmatrix} = \frac{y_F}{3} \begin{bmatrix} 1 & 1 & 1 \\ 1 & 1 & 1 \\ 1 & 1 & 1 \end{bmatrix} \left(\begin{bmatrix} 1 + Z^{(0)}_{pp}\frac{y_F}{3} & Z^{(0)}_{pp}\frac{y_F}{3} & Z^{(0)}_{pp}\frac{y_F}{3} \\ Z^{(1)}_{pp}\frac{y_F}{3} & 1 + Z^{(1)}_{pp}\frac{y_F}{3} & Z^{(1)}_{pp}\frac{y_F}{3} \\ Z^{(1)}_{pp}\frac{y_F}{3} & Z^{(1)}_{pp}\frac{y_F}{3} & 1 + Z^{(1)}_{pp}\frac{y_F}{3} \end{bmatrix} \right)^{-1} \begin{bmatrix} 0 \\ \sqrt{3} \\ 0 \end{bmatrix}$$

which reduces to

$$\begin{bmatrix} I^{(0)}_{p(F)} \\ I^{(1)}_{p(F)} \\ I^{(2)}_{p(F)} \end{bmatrix} = \frac{\sqrt{3}}{Z^{(0)}_{pp} + 2Z^{(1)}_{pp} + 3z_F} \begin{bmatrix} 1 \\ 1 \\ 1 \end{bmatrix} \tag{6.3.11}$$

The phase components of the fault current at bus p can be obtained by premultiplying both sides of equation (6.3.11) by T_s. These currents are

$$\begin{bmatrix} I^{a}_{p(F)} \\ I^{b}_{p(F)} \\ I^{c}_{p(F)} \end{bmatrix} = \begin{bmatrix} \dfrac{3}{Z^{(0)}_{pp} + 2Z^{(1)}_{pp} + 3z_F} \\ 0 \\ 0 \end{bmatrix}$$

The voltage at the faulted bus p is

$$\begin{bmatrix} E^{(0)}_{p(F)} \\ E^{(1)}_{p(F)} \\ E^{(2)}_{p(F)} \end{bmatrix} = \left(\begin{bmatrix} 1 + Z^{(0)}_{pp}\frac{y_F}{3} & Z^{(0)}_{pp}\frac{y_F}{3} & Z^{(0)}_{pp}\frac{y_F}{3} \\ Z^{(1)}_{pp}\frac{y_F}{3} & 1 + Z^{(1)}_{pp}\frac{y_F}{3} & Z^{(1)}_{pp}\frac{y_F}{3} \\ Z^{(1)}_{pp}\frac{y_F}{3} & Z^{(1)}_{pp}\frac{y_F}{3} & 1 + Z^{(1)}_{pp}\frac{y_F}{3} \end{bmatrix} \right)^{-1} \begin{bmatrix} 0 \\ \sqrt{3} \\ 0 \end{bmatrix}$$

which reduces to

$$\begin{bmatrix} E^{(0)}_{p(F)} \\ E^{(1)}_{p(F)} \\ E^{(2)}_{p(F)} \end{bmatrix} = \frac{\sqrt{3}}{Z^{(0)}_{pp} + 2Z^{(1)}_{pp} + 3z_F} \begin{bmatrix} -Z^{(0)}_{pp} \\ Z^{(0)}_{pp} + Z^{(1)}_{pp} + 3z_F \\ -Z^{(1)}_{pp} \end{bmatrix}$$

The phase components of the fault voltage are

$$\begin{bmatrix} E^{a}_{p(F)} \\ E^{b}_{p(F)} \\ E^{c}_{p(F)} \end{bmatrix} = \begin{bmatrix} \dfrac{3z_F}{Z^{(0)}_{pp} + 2Z^{(1)}_{pp} + 3z_F} \\ a^2 - \dfrac{Z^{(0)}_{pp} - Z^{(1)}_{pp}}{Z^{(0)}_{pp} + 2Z^{(1)}_{pp} + 3z_F} \\ a - \dfrac{Z^{(0)}_{pp} - Z^{(1)}_{pp}}{Z^{(0)}_{pp} + 2Z^{(1)}_{pp} + 3z_F} \end{bmatrix}$$

The voltages at buses other than p are

$$\begin{bmatrix} E^{(0)}_{i(F)} \\ E^{(1)}_{i(F)} \\ E^{(2)}_{i(F)} \end{bmatrix} = \begin{bmatrix} 0 \\ \sqrt{3} \\ 0 \end{bmatrix} - \begin{bmatrix} Z^{(0)}_{ip} & & \\ & Z^{(1)}_{ip} & \\ & & Z^{(1)}_{ip} \end{bmatrix} \frac{\sqrt{3}}{Z^{(0)}_{pp} + 2Z^{(1)}_{pp} + 3z_F} \begin{bmatrix} 1 \\ 1 \\ 1 \end{bmatrix}$$

which reduces to

$$\begin{bmatrix} E^{(0)}_{i(F)} \\ E^{(1)}_{i(F)} \\ E^{(2)}_{i(F)} \end{bmatrix} = \begin{bmatrix} 0 \\ \sqrt{3} \\ 0 \end{bmatrix} - \frac{\sqrt{3}}{Z^{(0)}_{pp} + 2Z^{(1)}_{pp} + 3z_F} \begin{bmatrix} Z^{(0)}_{ip} \\ Z^{(1)}_{ip} \\ Z^{(1)}_{ip} \end{bmatrix}$$

In phase components,

$$\begin{bmatrix} E^{a}_{i(F)} \\ E^{b}_{i(F)} \\ E^{c}_{i(F)} \end{bmatrix} = \begin{bmatrix} 1 \\ a^2 \\ a \end{bmatrix} - \frac{1}{Z^{(0)}_{pp} + 2Z^{(1)}_{pp} + 3z_F} \begin{bmatrix} Z^{(0)}_{ip} + 2Z^{(1)}_{ip} \\ Z^{(0)}_{ip} - Z^{(1)}_{ip} \\ Z^{(0)}_{ip} - Z^{(1)}_{ip} \end{bmatrix}$$

Table 6.3 Current and voltage formulas for line-to-ground fault (phase a) at bus p

Three-phase components	Symmetrical components
$I_{p(F)}^{a,b,c} = \dfrac{3E_{p(0)}}{Z_{pp}^{(0)} + 2Z_{pp}^{(1)} + 3z_F}\begin{bmatrix} 1 \\ 0 \\ 0 \end{bmatrix}$	$I_{p(F)}^{0,1,2} = \dfrac{\sqrt{3}\,E_{p(0)}}{Z_{pp}^{(0)} + 2Z_{pp}^{(1)} + 3z_F}\begin{bmatrix} 1 \\ 1 \\ 1 \end{bmatrix}$
$E_{p(F)}^{a,b,c} = E_{p(0)}\begin{bmatrix} \dfrac{3z_F}{Z_{pp}^{(0)} + 2Z_{pp}^{(1)} + 3z_F} \\ a^2 - \dfrac{Z_{pp}^{(0)} - Z_{pp}^{(1)}}{z_{pp}^{(0)} + 2Z_{pp}^{(1)} + 3z_F} \\ a - \dfrac{Z_{pp}^{(0)} - Z_{pp}^{(1)}}{Z_{pp}^{(0)} + 2Z_{pp}^{(1)} + 3z_F} \end{bmatrix}$	$E_{p(F)}^{0,1,2} = \dfrac{\sqrt{3}\,E_{p(0)}}{Z_{pp}^{(0)} + 2Z_{pp}^{(1)} + 3z_F}\begin{bmatrix} -Z_{pp}^{(0)} \\ Z_{pp}^{(0)} + Z_{pp}^{(1)} + 3z_F \\ -Z_{pp}^{(1)} \end{bmatrix}$
$E_{i(F)}^{a,b,c} = E_{i(0)}\begin{bmatrix} 1 \\ a^2 \\ a \end{bmatrix} - E_{p(0)}\begin{bmatrix} \dfrac{Z_{ip}^{(0)} + 2Z_{ip}^{(1)}}{Z_{pp}^{(0)} + 2Z_{pp}^{(1)} + 3z_F} \\ \dfrac{Z_{ip}^{(0)} - Z_{ip}^{(1)}}{Z_{pp}^{(0)} + 2Z_{pp}^{(1)} + 3z_F} \\ \dfrac{Z_{ip}^{(0)} - Z_{ip}^{(1)}}{Z_{pp}^{(3)} + 2Z_{pp}^{(1)} + 3z_F} \end{bmatrix}$ $i \neq p$	$E_{i(F)}^{0,1,2} = E_{i(0)}\begin{bmatrix} 0 \\ \sqrt{3} \\ 0 \end{bmatrix} - \dfrac{\sqrt{3}\,E_{p(0)}}{Z_{pp}^{(0)} + 2Z_{pp}^{(1)} + 3z_F}\begin{bmatrix} Z_{ip}^{(0)} \\ Z_{ip}^{(1)} \\ Z_{ip}^{(1)} \end{bmatrix}$ $i \neq p$

The formulas derived in this section are summarized in Table 6.3. The line-to-ground voltage was assumed to be one per unit in the derivations. The formulas in Table 6.3 include the term for the line-to-ground voltage which can be set at any desired per unit value.

The currents in the network elements during the fault can be calculated from equation (6.3.7).

6.4 Example of short circuit calculations using Z_{BUS}

The method of calculating short circuit currents and voltages will be illustrated for the sample system shown in Fig. 6.4*a*. The oriented connected graph of this system is shown in Fig. 6.4*b*. This sample system is identical to the one used in Sec. 5.9.

Problem

a. Using symmetrical components, calculate the following for a three-phase fault at bus 4:

i. Total fault current
ii. Bus voltages during fault
iii. Short circuit currents in lines connected to the faulted bus

b. Using symmetrical components, calculate the following for a line-to-ground fault at bus 4:

i. Total fault current
ii. Bus voltages during fault
iii. Short circuit currents in lines connected to the faulted bus.

c. Determine the maximum three-phase short circuit current that circuit breaker A must interrupt for a fault on the line side of the breaker.

Solution

a. The bus impedance matrix in terms of sequence quantities must be determined to calculate three-phase and line-to-ground fault currents using symmetrical components. Table 5.2 shows the three-phase impedances of the network elements. The zero, positive, and negative sequence impedances of the network elements can be obtained by means of the transformation matrix T_s, that is,

$$z_{pq}^{0,1,2} = (T_s^*)^t z_{pq}^{a,b,c} T_s$$

Assuming the impedance matrices of the generators are symmetric and using the average value -0.0225 for the off-diagonal elements, the sequence impedances are shown in Table 6.4.

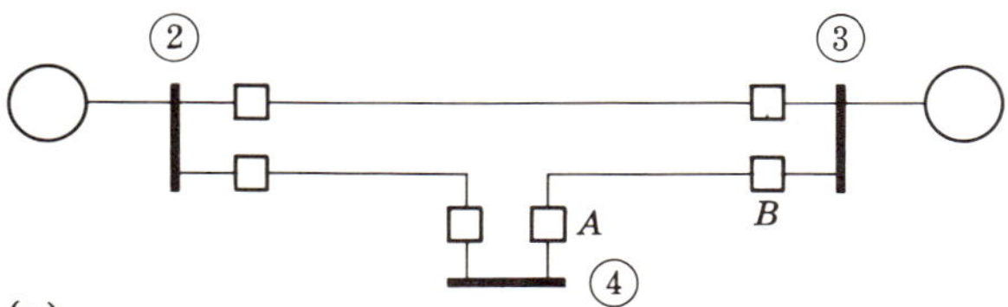

(*a*)

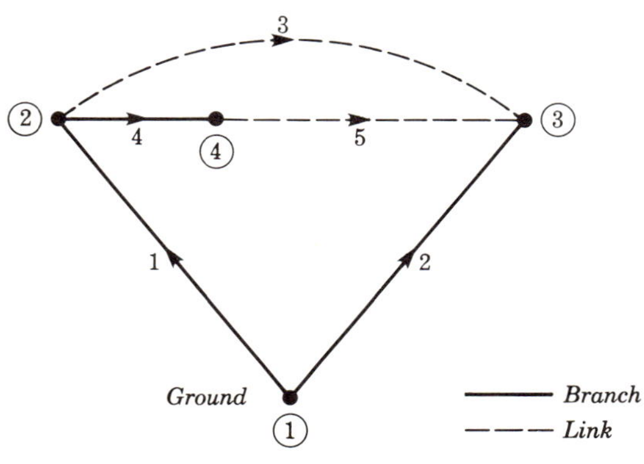

(*b*)

Fig. 6.4 Sample system for short circuit calculations. (a) Single line diagram of three-phase system; (b) oriented connected graph.

Table 6.4 Zero, positive, and negative sequence impedances for sample system

		Self	Mutual	
Element number	Bus code p–q	Impedance $z^{0,1,2}_{pq,pq}$	Bus code r–s	Impedance $z^{0,1,2}_{pq,rs}$
1	1–2	$\begin{bmatrix} 0.035 & & \\ & 0.1025 & \\ & & 0.1025 \end{bmatrix}$		
2	1–3	$\begin{bmatrix} 0.035 & & \\ & 0.1025 & \\ & & 0.1025 \end{bmatrix}$		
3	2–3	$\begin{bmatrix} 2.50 & & \\ & 1.00 & \\ & & 1.00 \end{bmatrix}$		
4	2–4	$\begin{bmatrix} 1.00 & & \\ & 0.40 & \\ & & 0.40 \end{bmatrix}$	2–3	$\begin{bmatrix} 0.60 & & \\ & & \\ & & \end{bmatrix}$
5	4–3	$\begin{bmatrix} 1.50 & & \\ & 0.60 & \\ & & 0.60 \end{bmatrix}$	2–3	$\begin{bmatrix} 0.90 & & \\ & & \\ & & \end{bmatrix}$

Since there is no coupling between the sequence impedances, the bus impedance matrix in terms of sequence quantities can be obtained by forming the positive, negative, and zero sequence bus impedance matrices independently. First, the positive sequence bus impedance matrix will be formed.

Step 1. Start with element 1, the branch from $p = 1$ to $q = 2$. The positive sequence bus impedance matrix for the partial network is

$Z_{BUS}^{(1)} =$

	②
②	0.1025

Step 2. Add element 2, the branch from $p = 1$ to $q = 3$. Then,

$Z_{BUS}^{(1)} =$

	②	③
②	0.1025	
③		0.1025

Step 3. Add element 4, the branch from $p = 2$ to $q = 4$. Thus,

$$Z_{24} = Z_{42} = Z_{22}$$
$$Z_{34} = Z_{43} = 0$$
$$Z_{44} = Z_{24} + z_{24,24}$$

and

$Z_{BUS}^{(1)} =$

	②	③	④
②	0.1025		0.1025
③		0.1025	
④	0.1025		0.5025

Step 4. Add element 5, the link from $p = 4$ to $q = 3$. The elements of the row and column corresponding to the fictitious node l are

$$Z_{l2} = Z_{2l} = Z_{42} - Z_{32}$$
$$Z_{l3} = Z_{3l} = Z_{43} - Z_{33}$$
$$Z_{l4} = Z_{4l} = Z_{44} - Z_{34}$$
$$Z_{ll} = Z_{4l} - Z_{3l} + z_{43,43}$$

and the augmented matrix is

	②	③	④	l
②	0.1025		0.1025	0.1025
③		0.1025		−0.1025
④	0.1025		0.5025	0.5025
l	0.1025	−0.1025	0.5025	1.2050

To eliminate the lth row and column the elements of the augmented matrix are modified as follows:

$$Z'_{22} = Z_{22} - Z_{2l}Z_{ll}^{-1}Z_{l2}$$
$$Z'_{33} = Z_{33} - Z_{3l}Z_{ll}^{-1}Z_{l3}$$
$$Z'_{44} = Z_{44} - Z_{4l}Z_{ll}^{-1}Z_{l4}$$
$$Z'_{23} = Z'_{32} = Z_{23} - Z_{2l}Z_{ll}^{-1}Z_{l3}$$
$$Z'_{24} = Z'_{42} = Z_{24} - Z_{2l}Z_{ll}^{-1}Z_{l4}$$
$$Z'_{34} = Z'_{43} = Z_{34} - Z_{3l}Z_{ll}^{-1}Z_{l4}$$

Thus,

$Z^{(1)}_{BUS}$ =

	②	③	④
②	0.0938	0.0087	0.0598
③	0.0087	0.0938	0.0427
④	0.0598	0.0427	0.2930

Step 5. Add element 3, the link from $p = 2$ to $q = 3$. As in the previous step,

$$Z_{l2} = Z_{2l} = Z_{22} - Z_{32}$$
$$Z_{l3} = Z_{3l} = Z_{23} - Z_{33}$$
$$Z_{l4} = Z_{4l} = Z_{24} - Z_{34}$$
$$Z_{ll} = Z_{2l} - Z_{3l} + z_{23,23}$$

	②	③	④	l
②	0.0938	0.0087	0.0598	0.0851
③	0.0087	0.0938	0.0427	−0.0851
④	0.0598	0.0427	0.2930	0.0171
l	0.0851	−0.0851	0.0171	1.1702

Eliminating the lth row and column, the final positive sequence bus impedance matrix is

$Z^{(1)}_{BUS}$ =

	②	③	④
②	0.0876	0.0149	0.0586
③	0.0149	0.0876	0.0439
④	0.0586	0.0439	0.2928

Since positive and negative primitive sequence impedances are equal, the positive and negative sequence bus impedance matrices are equal.

The procedure for forming the zero sequence bus impedance matrix is identical for the first four steps. The zero sequence bus impedance matrix of the partial network, before adding element 3, is

$Z_{BUS}^{(0)} =$

	②	③	④
②	0.0345	0.0005	0.0209
③	0.0005	0.0345	0.0141
④	0.0209	0.0141	0.6182

Step 5. Add element 3, the link from $p = 2$ to $q = 3$, which is coupled with the elements 4 and 5. The elements of the row and column corresponding to the fictitious node l are

$$Z_{l2} = Z_{22} - Z_{32} + \frac{y_{23,24}(Z_{22} - Z_{42}) + y_{23,43}(Z_{42} - Z_{32})}{y_{23,23}}$$

$$Z_{l3} = Z_{23} - Z_{33} + \frac{y_{23,24}(Z_{23} - Z_{43}) + y_{23,43}(Z_{43} - Z_{33})}{y_{23,23}}$$

$$Z_{l4} = Z_{24} - Z_{34} + \frac{y_{23,24}(Z_{24} - Z_{44}) + y_{23,43}(Z_{44} - Z_{34})}{y_{23,23}}$$

$$Z_{ll} = Z_{2l} - Z_{3l} + \frac{1 + y_{23,24}(Z_{2l} - Z_{4l}) + y_{23,43}(Z_{4l} - Z_{3l})}{y_{23,23}}$$

The zero sequence primitive impedance matrix is

	1–2	1–3	2–3	2–4	4–3
1–2	0.035				
1–3		0.035			
2–3			2.500	0.600	0.900
2–4			0.600	1.000	
4–3			0.900		1.500

The new element 3 is coupled only to elements 4 and 5 and it is sufficient to invert the submatrix containing the coupled elements.

$[z^{(0)}_{\rho\sigma,\rho\sigma}] =$

	2–3	2–4	4–3
2–3	2.500	0.600	0.900
2–4	0.600	1.000	
4–3	0.900		1.500

Thus,

$[y^{(0)}_{\rho\sigma,\rho\sigma}] =$

	2–3	2–4	4–3
2–3	0.625	−0.375	−0.375
2–4	−0.375	1.225	0.225
4–3	−0.375	0.225	0.892

and the augmented matrix is

	②	③	④	l
②	0.0345	0.0005	0.0209	0.0136
③	0.0005	0.0345	0.0141	−0.0136
④	0.0209	0.0141	0.6182	0.0027
l	0.0136	−0.0136	0.0027	1.6109

Eliminating the lth row and column, the final zero sequence bus impedance matrix is

$Z^{(0)}_{BUS} =$

	②	③	④
②	0.0344	0.0006	0.0209
③	0.0006	0.0344	0.0141
④	0.0209	0.0141	0.6182

Combining the elements of the three sequence impedance matrices, the bus impedance matrix is

$Z_{BUS}^{0,1,2} =$

		② 0	② 1	② 2	③ 0	③ 1	③ 2	④ 0	④ 1	④ 2
②	0	0.0344			0.0006			0.0209		
	1		0.0876			0.0149			0.0586	
	2			0.0876			0.0149			0.0586
③	0	0.0006			0.0344			0.0141		
	1		0.0149			0.0876			0.0439	
	2			0.0149			0.0876			0.0439
④	0	0.0209			0.0141			0.6182		
	1		0.0586			0.0439			0.2928	
	2			0.0586			0.0439			0.2928

Assuming the fault impedance is zero, the total fault current for a three-phase fault at bus 4 is

$$I_{4(F)}^{0,1,2} = \begin{bmatrix} 0 \\ \dfrac{\sqrt{3}}{Z_{44}^{(1)} + z_F} \\ 0 \end{bmatrix} = \begin{bmatrix} 0 \\ \dfrac{\sqrt{3}}{0.2928} \\ 0 \end{bmatrix} = \begin{bmatrix} 0 \\ 3.42\sqrt{3} \\ 0 \end{bmatrix}$$

The phase components of the fault current are

$$I_{4(F)}^{a,b,c} = T_s I_{4(F)}^{0,1,2} = 3.42 \begin{bmatrix} 1 \\ a^2 \\ a \end{bmatrix}$$

Bus voltages during fault are

$$E_{4(F)}^{0,1,2} = \begin{bmatrix} 0 \\ 0 \\ 0 \end{bmatrix}$$

$$E_{2(F)}^{0,1,2} = \begin{bmatrix} 0 \\ \sqrt{3} - \dfrac{Z_{24}^{(1)} \sqrt{3}}{Z_{44}^{(1)} + z_F} \\ 0 \end{bmatrix} = \begin{bmatrix} 0 \\ \sqrt{3} - \dfrac{0.0586}{0.2928} \sqrt{3} \\ 0 \end{bmatrix}$$

$$= \begin{bmatrix} 0 \\ 0.80 \sqrt{3} \\ 0 \end{bmatrix}$$

$$E_{3(F)}^{0,1,2} = \begin{bmatrix} 0 \\ \sqrt{3} - \dfrac{Z_{34}^{(1)} \sqrt{3}}{Z_{44}^{(1)} + z_F} \\ 0 \end{bmatrix} = \begin{bmatrix} 0 \\ \sqrt{3} - \dfrac{0.0439}{0.2928} \sqrt{3} \\ 0 \end{bmatrix}$$

$$= \begin{bmatrix} 0 \\ 0.85 \sqrt{3} \\ 0 \end{bmatrix}$$

The phase components of the voltages are

$$E_{4(F)}^{a,b,c} = T_s E_{4(F)}^{0,1,2} = \begin{bmatrix} 0 \\ 0 \\ 0 \end{bmatrix}$$

$$E_{2(F)}^{a,b,c} = T_s E_{2(F)}^{0,1,2} = 0.80 \begin{bmatrix} 1 \\ a^2 \\ a \end{bmatrix}$$

$$E_{3(F)}^{a,b,c} = T_s E_{3(F)}^{0,1,2} = 0.85 \begin{bmatrix} 1 \\ a^2 \\ a \end{bmatrix}$$

The short circuit currents in the lines connected to the faulted bus are

$$i_{43(F)}^{0,1,2} = \begin{bmatrix} 0 \\ y_{43,43}^{(1)}(E_{4(F)}^{(1)} - E_{3(F)}^{(1)}) \\ 0 \end{bmatrix} = \begin{bmatrix} 0 \\ \frac{1}{0.60}(0 - 0.85\sqrt{3}) \\ 0 \end{bmatrix}$$

$$= \begin{bmatrix} 0 \\ -1.42\sqrt{3} \\ 0 \end{bmatrix}$$

$$i^{0,1,2}_{24(F)} = \begin{bmatrix} 0 \\ y^{(1)}_{24,24}(E^{(1)}_{2(F)} - E^{(1)}_{4(F)}) \\ 0 \end{bmatrix} = \begin{bmatrix} 0 \\ \frac{1}{0.40}(0.80\sqrt{3} - 0) \\ 0 \end{bmatrix}$$

$$= \begin{bmatrix} 0 \\ 2.00\sqrt{3} \\ 0 \end{bmatrix}$$

The phase components of these currents are

$$i^{a,b,c}_{43(F)} = T_s i^{0,1,2}_{43(F)} = -1.42 \begin{bmatrix} 1 \\ a^2 \\ a \end{bmatrix}$$

$$i^{a,b,c}_{24(F)} = T_s i^{0,1,2}_{24(F)} = 2.00 \begin{bmatrix} 1 \\ a^2 \\ a \end{bmatrix}$$

b. The fault current for a line-to-ground fault at bus 4, assuming zero fault impedance, is

$$I_{4(F)}^{0,1,2} = \frac{\sqrt{3}}{Z_{44}^{(0)} + 2Z_{44}^{(1)} + 3z_F} \begin{bmatrix} 1 \\ 1 \\ 1 \end{bmatrix} = \frac{\sqrt{3}}{0.6182 + 0.5856} \begin{bmatrix} 1 \\ 1 \\ 1 \end{bmatrix}$$

$$= \begin{bmatrix} 0.83\sqrt{3} \\ 0.83\sqrt{3} \\ 0.83\sqrt{3} \end{bmatrix}$$

The phase components of the total fault current are

$$I_{4(F)}^{a,b,c} = T_s I_{4(F)}^{0,1,2} = \begin{bmatrix} 2.49 \\ 0 \\ 0 \end{bmatrix}$$

Bus voltages during the fault are

$$E_{4(F)}^{0,1,2} = \frac{\sqrt{3}}{Z_{44}^{(0)} + 2Z_{44}^{(1)} + 3z_F} \begin{bmatrix} -Z_{44}^{(0)} \\ Z_{44}^{(0)} + Z_{44}^{(1)} \\ -Z_{44}^{(1)} \end{bmatrix} = 0.83\sqrt{3} \begin{bmatrix} -0.6182 \\ 0.9110 \\ -0.2928 \end{bmatrix}$$

$$= \begin{bmatrix} -0.5131\sqrt{3} \\ 0.7561\sqrt{3} \\ -0.2430\sqrt{3} \end{bmatrix}$$

$$E_{2(F)}^{0,1,2} = \begin{bmatrix} 0 \\ \sqrt{3} \\ 0 \end{bmatrix} - \frac{\sqrt{3}}{Z_{44}^{(0)} + 2Z_{44}^{(1)} + 3z_F} \begin{bmatrix} Z_{42}^{(0)} \\ Z_{42}^{(1)} \\ Z_{42}^{(1)} \end{bmatrix}$$

$$= \begin{bmatrix} 0 \\ \sqrt{3} \\ 0 \end{bmatrix} - 0.83\sqrt{3} \begin{bmatrix} 0.0209 \\ 0.0586 \\ 0.0586 \end{bmatrix}$$

$$= \begin{bmatrix} -0.0173\sqrt{3} \\ 0.9514\sqrt{3} \\ -0.0486\sqrt{3} \end{bmatrix}$$

$$E_{3(F)}^{0,1,2} = \begin{bmatrix} 0 \\ \sqrt{3} \\ 0 \end{bmatrix} - \frac{\sqrt{3}}{Z_{44}^{(0)} + 2Z_{44}^{(1)} + 3z_F} \begin{bmatrix} Z_{43}^{(0)} \\ Z_{43}^{(1)} \\ Z_{43}^{(1)} \end{bmatrix}$$

$$= \begin{bmatrix} 0 \\ \sqrt{3} \\ 0 \end{bmatrix} - 0.83\sqrt{3} \begin{bmatrix} 0.0141 \\ 0.0439 \\ 0.0439 \end{bmatrix}$$

$$= \begin{bmatrix} -0.0117\sqrt{3} \\ 0.9636\sqrt{3} \\ -0.0364\sqrt{3} \end{bmatrix}$$

The phase components of the voltages are

$$E_{4(F)}^{a,b,c} = \begin{bmatrix} \dfrac{3z_F}{Z_{44}^{(0)} + 2Z_{44}^{(1)} + 3z_F} \\ a^2 - \dfrac{Z_{44}^{(0)} - Z_{44}^{(1)}}{Z_{44}^{(0)} + 2Z_{44}^{(1)} + 3z_F} \\ a - \dfrac{Z_{44}^{(0)} - Z_{44}^{(1)}}{Z_{44}^{(0)} + 2Z_{44}^{(1)} + 3z_F} \end{bmatrix} = \begin{bmatrix} 0 \\ a^2 - \dfrac{0.6182 - 0.2928}{0.6182 + 0.5856} \\ a - \dfrac{0.6182 - 0.2928}{0.6182 + 0.5856} \end{bmatrix}$$

$$= \begin{bmatrix} 0 \\ -0.77 - j0.866 \\ -0.77 + j0.866 \end{bmatrix}$$

$$E_{2(F)}^{a,b,c} = \begin{bmatrix} 1 \\ a^2 \\ a \end{bmatrix} - \begin{bmatrix} \dfrac{Z_{42}^{(0)} + 2Z_{42}^{(1)}}{Z_{44}^{(0)} + 2Z_{44}^{(1)} + 3z_F} \\ \dfrac{Z_{42}^{(0)} - Z_{42}^{(1)}}{Z_{44}^{(0)} + 2Z_{44}^{(1)} + 3z_F} \\ \dfrac{Z_{42}^{(0)} - Z_{42}^{(1)}}{Z_{44}^{(0)} + 2Z_{44}^{(1)} + 3z_F} \end{bmatrix}$$

$$= \begin{bmatrix} 1 \\ a^2 \\ a \end{bmatrix} - \begin{bmatrix} \dfrac{0.0209 + 0.1172}{0.6182 + 0.5856} \\ \dfrac{0.0209 - 0.0586}{0.6182 + 0.5856} \\ \dfrac{0.0209 - 0.0586}{0.6182 + 0.5856} \end{bmatrix} = \begin{bmatrix} 0.8853 + j0 \\ -0.4687 - j0.866 \\ -0.4687 + j0.866 \end{bmatrix}$$

$$E^{a,b,c}_{3(F)} = \begin{bmatrix} 1 \\ a^2 \\ a \end{bmatrix} - \begin{bmatrix} \dfrac{Z^{(0)}_{43} + 2Z^{(1)}_{43}}{Z^{(0)}_{44} + 2Z^{(1)}_{44} + 3z_F} \\ \dfrac{Z^{(0)}_{43} - Z^{(1)}_{43}}{Z^{(0)}_{44} + 2Z^{(1)}_{44} + 3z_F} \\ \dfrac{Z^{(0)}_{43} - Z^{(1)}_{43}}{Z^{(0)}_{44} + 2Z^{(1)}_{44} + 3z_F} \end{bmatrix} = \begin{bmatrix} 1 \\ a^2 \\ a \end{bmatrix} - \begin{bmatrix} \dfrac{0.0141 + 0.0878}{0.6182 + 0.5856} \\ \dfrac{0.0141 - 0.0439}{0.6182 + 0.5856} \\ \dfrac{0.0141 - 0.0439}{0.6182 + 0.5856} \end{bmatrix}$$

$$= \begin{bmatrix} 0.9154 + j0 \\ -0.4752 - j0.866 \\ -0.4752 + j0.866 \end{bmatrix}$$

Short circuit currents in the lines connected to the faulted bus are

$$i^{0,1,2}_{43(F)} = \begin{bmatrix} y^{(0)}_{43,43}(E^{(0)}_{4(F)} - E^{(0)}_{3(F)}) + y^{(0)}_{43,23}(E^{(0)}_{2(F)} - E^{(0)}_{3(F)}) + y^{(0)}_{43,24}(E^{(0)}_{2(F)} - E^{(0)}_{4(F)}) \\ y^{(1)}_{43,43}(E^{(1)}_{4(F)} - E^{(1)}_{3(F)}) \\ y^{(2)}_{43,43}(E^{(2)}_{4(F)} - E^{(2)}_{3(F)}) \end{bmatrix}$$

$$= \sqrt{3} \begin{bmatrix} 0.892(-0.5131 + 0.0117) + (-0.375)(-0.0173 + 0.0117) + 0.225(-0.0173 + 0.5131) \\ \dfrac{1}{0.6}(0.7561 - 0.9636) \\ \dfrac{1}{0.6}(-0.2430 + 0.0364) \end{bmatrix}$$

$$= \begin{bmatrix} -0.33\sqrt{3} \\ -0.34\sqrt{3} \\ -0.34\sqrt{3} \end{bmatrix}$$

$$i_{24(F)}^{0,1,2} = \begin{bmatrix} y_{24,24}^{(0)}(E_{2(F)}^{(0)} - E_{4(F)}^{(0)}) + y_{24,43}^{(0)}(E_{4(F)}^{(0)} - E_{3(F)}^{(0)}) \\ \quad + y_{24,23}^{(0)}(E_{2(F)}^{(0)} - E_{3(F)}^{(0)}) \\ y_{24,24}^{(1)}(E_{2(F)}^{(1)} - E_{4(F)}^{(1)}) \\ y_{24,24}^{(2)}(E_{2(F)}^{(2)} - E_{4(F)}^{(2)}) \end{bmatrix}$$

$$= \sqrt{3} \begin{bmatrix} 1.225(-0.0173 + 0.5131) + 0.225(-0.5131 + 0.0117) \\ \quad + (-0.375)(-0.0173 + 0.0117) \\ \frac{1}{0.4}(0.9514 - 0.7561) \\ \frac{1}{0.4}(-0.0486 + 0.2430) \end{bmatrix}$$

$$= \begin{bmatrix} 0.50\sqrt{3} \\ 0.49\sqrt{3} \\ 0.49\sqrt{3} \end{bmatrix}$$

The phase components of the currents in the lines connected to the fault bus are

$$i_{43(F)}^{a,b,c} = T_s i_{43(F)}^{0,1,2} = \begin{bmatrix} -1.02 \\ 0 \\ 0 \end{bmatrix} \quad \text{and} \quad i_{24(F)}^{a,b,c} = T_s i_{24(F)}^{0,1,2} = \begin{bmatrix} 1.47 \\ 0 \\ 0 \end{bmatrix}$$

c. The fault currents occurring for a fault on the line side of breaker A can be calculated by assuming the fault on bus 4, since both locations are electrically equivalent. When this type of fault occurs and breaker A opens before breaker B, the interrupted current will be the total fault current at bus 4 less the fault contribution flowing from bus 3 over line 5.

This current is

$$I^{a,b,c}_{4(F)} - i^{a,b,c}_{34(F)} = 3.42 \begin{bmatrix} 1 \\ a^2 \\ a \end{bmatrix} - 1.42 \begin{bmatrix} 1 \\ a^2 \\ a \end{bmatrix}$$

$$= 2.00 \begin{bmatrix} 1 \\ a^2 \\ a \end{bmatrix}$$

When breaker B opens first, breaker A must interrupt the total fault current which occurs when line 5 is open. The fault current can be calculated after modifying the bus impedance matrix to open line 5. To simulate the opening of line 5 a fictitious link, whose impedance is equal to the negative of the impedance of element 5, is added between buses 4 and 3. From Table 6.4, the impedance of this fictitious link in terms of the sequence quantities will be

$$z^{0,1,2}_{43,43(2)} = \begin{bmatrix} -1.5 & & \\ & -0.60 & \\ & & -0.60 \end{bmatrix}$$

The elements of the lth row and column are

$$Z_{l2} = Z_{2l} = Z_{42} - Z_{32}$$
$$Z_{l3} = Z_{3l} = Z_{43} - Z_{33}$$
$$Z_{l4} = Z_{4l} = Z_{44} - Z_{34}$$
$$Z_{ll} = Z_{4l} - Z_{3l} + z_{43,43(2)}$$

and the augmented positive sequence matrix is

	②	③	④	l
②	0.0876	0.0149	0.0586	0.0437
③	0.0149	0.0876	0.0439	−0.0437
④	0.0586	0.0439	0.2928	0.2489
l	0.0437	−0.0437	0.2489	−0.3074

The element Z'_{44} of the modified bus impedance matrix is

$$Z'_{44} = Z_{44} - Z_{4l}Z_{ll}^{-1}Z_{l3}$$

This new value is

$$Z_{44}^{(1)} = 0.4943$$

Then, the total fault current is

$$I_{4(F)}^{0,1,2} = \sqrt{3}\begin{bmatrix} 0 \\ \dfrac{1}{Z_{44}^{(1)} + z_F} \\ 0 \end{bmatrix} = \sqrt{3}\begin{bmatrix} 0 \\ \dfrac{1}{0.494} \\ 0 \end{bmatrix}$$

$$= \begin{bmatrix} 0 \\ 2.02\sqrt{3} \\ 0 \end{bmatrix}$$

The phase components of the total fault current are

$$I_{4(F)}^{a,b,c} = T_s I_{4(F)}^{0,1,2} = 2.02\begin{bmatrix} 1 \\ a^2 \\ a \end{bmatrix}$$

which are the maximum currents to be interrupted.

6.5 Short circuit calculations using Z_{LOOP}

Short circuit currents and voltages can be calculated using the loop impedance matrix for the simplified system given in Fig. 6.2 (Lantz, 1957). The loop currents of the simplified system are zero prior to the fault since all bus currents and off-nominal tap settings are neglected. It is necessary, therefore, to calculate the loop currents resulting from the fault in order to determine short circuit currents and voltages. The fault calculations can be performed using either three-phase quantities or symmetrical components. The method will be described using three-phase quantities.

The number of three-phase elements in the simplified system is equal to the number of network elements plus the number of machine equivalents. The number of nodes is equal to the number of buses n plus ground, that is $n + 1$. The number of links or basic loops, in the simplified system, is, then

$$l_n = (e + e_g) - (n + 1) + 1$$

or

$$l_n = e + e_g - n$$

where e is the number of three-phase network elements and e_g is the number of three-phase machine equivalents.

A fault at bus p is simulated by adding a link from the bus to ground. Using the representation of the system shown in Fig. 6.3, the voltages during the fault are

$$\bar{E}^{a,b,c}_{BUS(F)} = \bar{E}^{a,b,c}_{BUS(0)} + \overline{\Delta E}^{a,b,c}_{BUS} \tag{6.5.1}$$

where the vector $\overline{\Delta E}^{a,b,c}_{BUS}$ represents changes in bus voltages resulting from the faulted bus source voltage $E^{a,b,c}_{p(0)}$.

The performance equation of a network in the loop frame of reference is

$$\bar{E}^{a,b,c}_{LOOP} = Z^{a,b,c}_{LOOP}\bar{I}^{a,b,c}_{LOOP}$$

For the faulted system, shown in Fig. 6.3, the known loop voltage vector is

$$\bar{E}^{a,b,c}_{LOOP} = \begin{bmatrix} 0 \\ \cdots \\ 0 \\ E^{a,b,c}_{p(0)} \end{bmatrix}$$

The dimension of the loop impedance matrix, which includes the fault loop, is $3(l_n + 1) \times 3(l_n + 1)$. The unknown loop current vector due to the fault is

$$\bar{I}^{a,b,c}_{LOOP(F)} = \begin{bmatrix} I^{a,b,c}_{A(F)} \\ \cdots \\ I^{a,b,c}_{l_n(F)} \\ I^{a,b,c}_{L(F)} \end{bmatrix}$$

where $I^{a,b,c}_{L(F)}$ is the current associated with the fault loop. The loop currents can be calculated, then, from

$$\bar{I}^{a,b,c}_{LOOP(F)} = (Z^{a,b,c}_{LOOP})^{-1}\bar{E}^{a,b,c}_{LOOP}$$

The currents in all elements of the network during fault can be calculated from

$$\bar{i}^{a,b,c}_{(F)} = C\bar{I}^{a,b,c}_{LOOP(F)} \tag{6.5.2}$$

where C is the loop incidence matrix on a three-phase basis. The current vector can be partitioned as follows:

$$\bar{i}^{a,b,c}_{(F)} = \begin{bmatrix} i^{a,b,c}_{b(F)} \\ i^{a,b,c}_{l(F)} \end{bmatrix}$$

where $\bar{i}^{a,b,c}_{b(F)}$ = branch current vector

$\bar{i}^{a,b,c}_{l(F)}$ = link current vector

Then the vector of voltage changes is

$$\overline{\Delta E}^{a,b,c}_{BUS} = K^t[z^{a,b,c}_{bb}]\bar{i}^{a,b,c}_{b(F)}$$

where K = branch-path incidence matrix on a three-phase basis

$[z^{a,b,c}_{bb}]$ = primitive impedance matrix for branches

The bus voltages during the fault are obtained by adding the voltage changes to the voltages prior to the fault. Equation (6.5.1) becomes

$$\bar{E}^{a,b,c}_{BUS(F)} = \bar{E}^{a,b,c}_{BUS(0)} + K^t[z^{a,b,c}_{bb}]\bar{i}^{a,b,c}_{b(F)} \tag{6.5.3}$$

The current at the faulted bus is the same as the current in the auxiliary loop, that is, $I^{a,b,c}_{L(F)}$.

The method described can be employed to calculate faults at various locations in the system by adding links, one at a time, between the faulted bus and ground. This requires the formation and inversion of a loop impedance matrix for each different fault location. The necessary matrix operations required to provide short circuit data for a large number of locations, therefore, would be time-consuming.

An alternate method, in which links are added simultaneously between each bus and ground, requires the formation of a single loop impedance matrix and only one inversion of a submatrix (Byerly, Long, Baldwin, and King, 1958). In this method, the currents in the auxiliary loops are changed to simulate different fault locations. Phase currents are assumed for the auxiliary loop associated with a faulted bus p depending on the type of fault. Assuming one per unit phase current, the pth auxiliary loop current is

For a three-phase fault:

$$I_{Lp(F)}^{a,b,c} = \begin{bmatrix} 1 \\ a^2 \\ a \end{bmatrix}$$

For a line-to-ground fault on phase a:

$$I_{Lp(F)}^{a,b,c} = \begin{bmatrix} 1 \\ 0 \\ 0 \end{bmatrix}$$

For a line-to-line fault between phases b and c:

$$I_{Lp(F)}^{a,b,c} = \begin{bmatrix} 0 \\ 1 \\ -1 \end{bmatrix}$$

The currents in all other auxiliary loops are assumed to be zero.

The loop voltage and current vectors and the loop impedance matrix in the performance equation for the entire network, including auxiliary loops, can be partitioned as follows:

$$\begin{bmatrix} E_L^{a,b,c} \\ E_{L(F)}^{a,b,c} \end{bmatrix} = \begin{bmatrix} Z_L^{a,b,c} & Z_M^{a,b,c} \\ (Z_M^{a,b,c})^t & Z_A^{a,b,c} \end{bmatrix} \begin{bmatrix} I_L^{a,b,c} \\ I_{L(F)}^{a,b,c} \end{bmatrix} \tag{6.5.4}$$

In equation (6.5.4), the vectors $\bar{E}_L^{a,b,c}$ and $\bar{I}_L^{a,b,c}$ refer to the loops in the simplified system and $\bar{E}_{L(F)}^{a,b,c}$ and $\bar{I}_{L(F)}^{a,b,c}$ refer to the auxiliary loops.

The vector $\bar{I}_L^{a,b,c}$ can be calculated for a fault at bus p from equation (6.5.4) by assuming the auxiliary loop currents to be

$$\bar{I}_{L(F)}^{a,b,c} = \begin{bmatrix} 0 \\ \cdots \\ 0 \\ I_{Lp(F)}^{a,b,c} \\ 0 \\ \cdots \\ 0 \end{bmatrix} \tag{6.5.5}$$

where $I_{Lp(F)}^{a,b,c}$ is the assumed three-phase current vector of the pth auxiliary loop. From equation (6.5.4) it follows that

$$Z_L^{a,b,c}\bar{I}_L^{a,b,c} + Z_M^{a,b,c}\bar{I}_{L(F)}^{a,b,c} = \bar{E}_L^{a,b,c} \tag{6.5.6}$$

Since $\bar{E}_L^{a,b,c} = 0$, equation (6.5.6) becomes

$$Z_L^{a,b,c}\bar{I}_L^{a,b,c} + Z_M^{a,b,c}\bar{I}_{L(F)}^{a,b,c} = 0$$

Solving for the loop currents of the simplified system,

$$\bar{I}_L^{a,b,c} = -(Z_L^{a,b,c})^{-1}Z_M^{a,b,c}\bar{I}_{L(F)}^{a,b,c} \tag{6.5.7}$$

The auxiliary loop voltages, from equation (6.5.4), are

$$\bar{E}^{a,b,c}_{L(F)} = (Z^{a,b,c}_{M})^{t}\bar{I}^{a,b,c}_{L} + Z^{a,b,c}_{A}\bar{I}^{a,b,c}_{L(F)}$$

Substituting from equation (6.5.7) for $\bar{I}^{a,b,c}_{L}$, then

$$\bar{E}^{a,b,c}_{L(F)} = \{Z^{a,b,c}_{A} - (Z^{a,b,c}_{M})^{t}(Z^{a,b,c}_{L})^{-1}Z^{a,b,c}_{M}\}\bar{I}^{a,b,c}_{L(F)} \tag{6.5.8}$$

Equation (6.5.8) determines the auxiliary loop source voltages for the assumed auxiliary loop currents given by equation (6.5.5).

To determine actual fault current the voltage source in the pth auxiliary loop must equal $E^{a,b,c}_{p(0)}$, the pth bus voltage prior to the fault. The calculated source voltage of the pth auxiliary loop $E^{a,b,c}_{Lp(F)}$ is obtained from equation (6.5.8) using the assumed currents. The actual fault current at bus p is

For phase a:

$$I^{a}_{Lp(F)}(\text{actual}) = I^{a}_{Lp(F)}\ (\text{assumed})\ \frac{E^{a}_{p(0)}}{E^{a}_{Lp(F)}}$$

For phase b:

$$I^{b}_{Lp(F)}(\text{actual}) = I^{b}_{Lp(F)}\ (\text{assumed})\ \frac{E^{b}_{p(0)}}{E^{b}_{Lp(F)}}$$

and so forth.

The loop currents $\bar{I}^{a,b,c}_{L}$ of the simplified system can be obtained from equation (6.5.7) using the actual auxiliary loop currents. The branch currents can be calculated from equation (6.5.2) and the bus voltage, then, can be determined from equation (6.5.3).

In equation (6.5.8), the assumed auxiliary loop currents $\bar{I}^{a,b,c}_{L(F)}$ are flowing in the auxiliary links connecting network buses and ground and therefore are bus currents. The auxiliary loop voltages $\bar{E}^{a,b,c}_{L(F)}$ are the bus voltages resulting from the assumed currents. In equation (6.5.8), then,

$$Z^{a,b,c}_{A} - (Z^{a,b,c}_{M})^{t}(Z^{a,b,c}_{L})^{-1}Z^{a,b,c}_{M} = Z^{a,b,c}_{BUS}$$

In this method, therefore, the loop impedance matrix is used to determine the bus impedance matrix for the calculation of short circuits.

6.6 Example of short circuit calculations using Z_{LOOP}

Using the loop frame of reference, the method of calculating short circuit currents and voltages will be illustrated for a fault at bus 4 in the sample system shown in Fig. 6.4a.

Problem

Using symmetrical components, calculate the following for a three-phase fault at bus 4:

i. Total fault current
ii. Short circuit currents in all the lines of the network
iii. Bus voltages during the fault

Solution

The positive sequence loop impedance matrix for the system including the link representing the fault must be determined to calculate the three-phase fault conditions. The basic loops prior to fault are shown in the oriented connected graph in Fig. 6.5*a*. The basic loops of the graph including the fault link are shown in Fig. 6.5*b*. The basic loop incidence

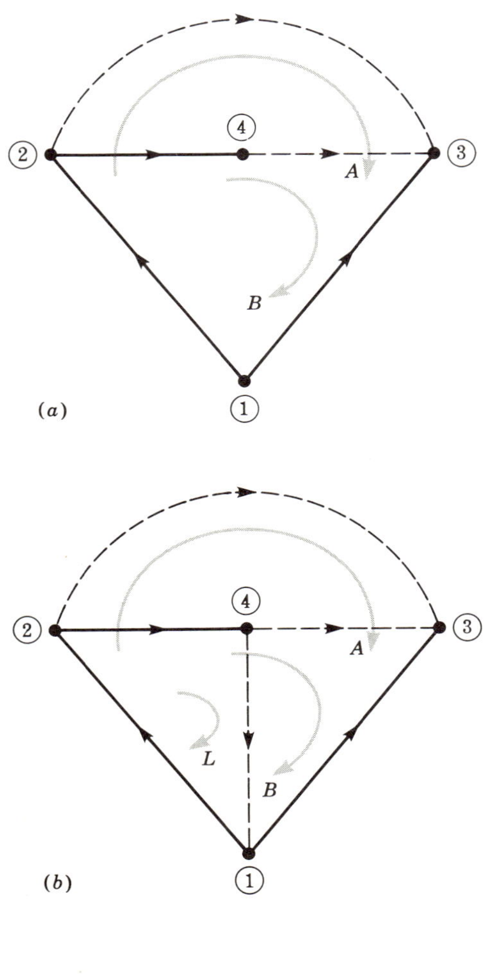

Fig. 6.5 Basic loops of oriented connected graph for sample power system. (a) Prior to fault; (b) during fault condition.

matrix for the fault condition is

$C =$

e \ l	A	B	L
1–2	1	1	1
1–3	−1	−1	
2–4		1	1
2–3	1		
4–3		1	
4–1			1

Using the sequence impedances given in Table 6.4 and letting the impedance of the fault link be zero, the positive sequence primitive impedance matrix is

$[z^{(1)}] =$

e \ e	1–2	1–3	2–4	2–3	4–3	4–1
1–2	0.1025					
1–3		0.1025				
2–4			0.4000			
2–3				1.0000		
4–3					0.6000	
4–1						0

The positive sequence loop impedance matrix is

$Z^{(1)}_{LOOP} = C^t[z^{(1)}]C =$

	A	B	L
A	1.2050	0.2050	0.1025
B	0.2050	1.2050	0.5025
L	0.1025	0.5025	0.5025

Assuming the line-to-ground voltages prior to the fault are equal to one per unit, the voltage source in the loop associated with the faulted bus is

$$E_{4(0)}^{a,b,c} = \begin{bmatrix} 1 \\ a^2 \\ a \end{bmatrix}$$

and the sequence voltages are

$$E_{4(0)}^{0,1,2} = (T_s^*)^t E_{4(0)}^{a,b,c} = \begin{bmatrix} 0 \\ \sqrt{3} \\ 0 \end{bmatrix}$$

The positive sequence loop currents are

$$\bar{I}_{LOOP}^{(1)} = (Z_{LOOP}^{(1)})^{-1} \bar{E}_{LOOP}^{(1)}$$

$$\begin{bmatrix} I_A^{(1)} \\ I_B^{(1)} \\ I_{L(F)}^{(1)} \end{bmatrix} = \begin{bmatrix} 0.855 & -0.125 & -0.050 \\ -0.125 & 1.442 & -1.416 \\ -0.050 & -1.416 & 3.416 \end{bmatrix} \begin{bmatrix} 0 \\ 0 \\ \sqrt{3} \end{bmatrix} = \begin{bmatrix} -0.05\sqrt{3} \\ -1.42\sqrt{3} \\ 3.42\sqrt{3} \end{bmatrix}$$

where $I_{L(F)}^{(1)}$ is the positive sequence current associated with the fault loop and equals the total positive sequence fault current. The sequence currents are

$$I_{L(F)}^{0,1,2} = \begin{bmatrix} 0 \\ 3.42\sqrt{3} \\ 0 \end{bmatrix}$$

Then, the phase components of the fault current are

$$I_{L(F)}^{a,b,c} = T_s I_{L(F)}^{0,1,2} = 3.42 \begin{bmatrix} 1 \\ a^2 \\ a \end{bmatrix}$$

The positive sequence currents in the elements of the network are

$$\bar{i}^{(1)} = C\bar{I}_{LOOP}^{(1)}$$

$$\begin{bmatrix} i_{12}^{(1)} \\ i_{13}^{(1)} \\ i_{24}^{(1)} \\ i_{23}^{(1)} \\ i_{43}^{(1)} \\ i_{41}^{(1)} \end{bmatrix} = \begin{array}{c} \\ 1\text{–}2 \\ 1\text{–}3 \\ 2\text{–}4 \\ 2\text{–}3 \\ 4\text{–}3 \\ 4\text{–}1 \end{array} \begin{array}{c} \begin{array}{ccc} A & B & L \end{array} \\ \begin{bmatrix} 1 & 1 & 1 \\ -1 & -1 & \\ & 1 & 1 \\ 1 & & \\ & 1 & \\ & & 1 \end{bmatrix} \end{array} \begin{bmatrix} -0.05\sqrt{3} \\ -1.42\sqrt{3} \\ 3.42\sqrt{3} \end{bmatrix} = \begin{bmatrix} 1.95\sqrt{3} \\ 1.47\sqrt{3} \\ 2.00\sqrt{3} \\ -0.05\sqrt{3} \\ -1.42\sqrt{3} \\ 3.42\sqrt{3} \end{bmatrix}$$

Then, the sequence and phase components are

$$i_{12}^{0,1,2} = \begin{bmatrix} 0 \\ 1.95\sqrt{3} \\ 0 \end{bmatrix} \qquad i_{12}^{a,b,c} = T_s i_{12}^{0,1,2} = 1.95 \begin{bmatrix} 1 \\ a^2 \\ a \end{bmatrix}$$

$$i_{13}^{0,1,2} = \begin{bmatrix} 0 \\ 1.47\sqrt{3} \\ 0 \end{bmatrix} \qquad i_{13}^{a,b,c} = T_s i_{13}^{0,1,2} = 1.47 \begin{bmatrix} 1 \\ a^2 \\ a \end{bmatrix}$$

$$i_{24}^{0,1,2} = \begin{bmatrix} 0 \\ 2.00\sqrt{3} \\ 0 \end{bmatrix} \qquad i_{24}^{a,b,c} = T_s i_{24}^{0,1,2} = 2.00 \begin{bmatrix} 1 \\ a^2 \\ a \end{bmatrix}$$

$$i_{23}^{0,1,2} = \begin{bmatrix} 0 \\ -0.05\sqrt{3} \\ 0 \end{bmatrix} \qquad i_{23}^{a,b,c} = T_s i_{23}^{0,1,2} = -0.05 \begin{bmatrix} 1 \\ a^2 \\ a \end{bmatrix}$$

$$i_{43}^{0,1,2} = \begin{bmatrix} 0 \\ -1.42\sqrt{3} \\ 0 \end{bmatrix} \qquad i_{43}^{a,b,c} = T_s i_{43}^{0,1,2} = -1.42 \begin{bmatrix} 1 \\ a^2 \\ a \end{bmatrix}$$

The branch-path incidence matrix is

$K =$

b \ path	②	③	④
1–2	−1		−1
1–3		−1	
2–4			−1

The positive sequence bus voltages due to the fault are

$$\overline{\Delta E}^{(1)}_{BUS} = K^t [z^{(1)}_{bb}] \bar{i}^{(1)}_b$$

$$\begin{array}{c} \\ \\ \end{array}
\begin{bmatrix} \Delta E_2 \\ \Delta E_3 \\ \Delta E_4 \end{bmatrix}
=
\begin{array}{c|ccc}
\text{path} \backslash b & 1\text{–}2 & 1\text{–}3 & 2\text{–}4 \\
\hline
② & -1 & & \\
③ & & -1 & \\
④ & -1 & & -1
\end{array}
\quad
\begin{array}{c|ccc}
 & 1\text{–}2 & 1\text{–}3 & 2\text{–}4 \\
\hline
1\text{–}2 & 0.1025 & & \\
1\text{–}3 & & 0.1025 & \\
2\text{–}4 & & & 0.4000
\end{array}
\begin{bmatrix} 1.95\sqrt{3} \\ 1.47\sqrt{3} \\ 2.00\sqrt{3} \end{bmatrix}
=
\begin{bmatrix} -0.20\sqrt{3} \\ -0.15\sqrt{3} \\ -1.00\sqrt{3} \end{bmatrix}$$

Since the positive sequence bus voltages prior to the fault are

$$E^{(1)}_{2(0)} = E^{(1)}_{3(0)} = E^{(1)}_{4(0)} = \sqrt{3}$$

the positive sequence bus voltages during the fault are

$$\bar{E}^{(1)}_{BUS(F)} = \bar{E}^{(1)}_{BUS(0)} + \overline{\Delta E}^{(1)}_{BUS}$$

$$\begin{bmatrix} E^{(1)}_{2(F)} \\ E^{(1)}_{3(F)} \\ E^{(1)}_{4(F)} \end{bmatrix} = \sqrt{3}\begin{bmatrix} 1 \\ 1 \\ 1 \end{bmatrix} + \sqrt{3}\begin{bmatrix} -0.20 \\ -0.15 \\ -1.00 \end{bmatrix} = \begin{bmatrix} 0.80\sqrt{3} \\ 0.85\sqrt{3} \\ 0 \end{bmatrix}$$

Then, the sequence and phase components of bus voltages during the fault are

$$E^{0,1,2}_{2(F)} = \begin{bmatrix} 0 \\ 0.80\sqrt{3} \\ 0 \end{bmatrix} \qquad E^{a,b,c}_{2(F)} = T_s E^{0,1,2}_{2(F)} = 0.80\begin{bmatrix} 1 \\ a^2 \\ a \end{bmatrix}$$

$$E^{0,1,2}_{3(F)} = \begin{bmatrix} 0 \\ 0.85\sqrt{3} \\ 0 \end{bmatrix} \qquad E^{a,b,c}_{3(F)} = T_s E^{0,1,2}_{3(F)} = 0.85\begin{bmatrix} 1 \\ a^2 \\ a \end{bmatrix}$$

$$E^{0,1,2}_{4(F)} = E^{a,b,c}_{4(F)} = \begin{bmatrix} 0 \\ 0 \\ 0 \end{bmatrix}$$

Using the alternative method described in Sec. 6.5, the loop impedance matrix is partitioned as follows:

$$Z^{0,1,2}_{LOOP} = \left[\begin{array}{c|c} Z^{0,1,2}_{L} & Z^{0,1,2}_{M} \\ \hline (Z^{0,1,2}_{M})^t & Z^{0,1,2}_{A} \end{array}\right]$$

For the sample system, then,

$$
Z_{LOOP}^{(1)} = \begin{array}{c|c|c|c} & A & B & L \\ \hline A & 1.2050 & 0.2050 & 0.1025 \\ \hline B & 0.2050 & 1.2050 & 0.5025 \\ \hline L & 0.1025 & 0.5025 & 0.5025 \end{array}
$$

Assuming the fault current equal to one per unit,

$$
I_{L4(F)}^{a,b,c} = \begin{bmatrix} 1 \\ a^2 \\ a \end{bmatrix} \quad \text{and} \quad I_{L4(F)}^{0,1,2} = (T_s^*)^t I_{L4(F)}^{a,b,c} = \begin{bmatrix} 0 \\ \sqrt{3} \\ 0 \end{bmatrix}
$$

The positive sequence voltage of the fault loop with the assumed value of fault current is

$$
E_{L(F)}^{(1)} = \{Z_A^{(1)} - (Z_M^{(1)})^t (Z_L^{(1)})^{-1} Z_M^{(1)}\} I_{L(F)}^{(1)}
$$

$$
= \left(0.5025 - \begin{bmatrix} 0.1025 & 0.5025 \end{bmatrix} \begin{bmatrix} 0.855 & -0.145 \\ -0.145 & 0.855 \end{bmatrix} \begin{bmatrix} 0.1025 \\ 0.5025 \end{bmatrix} \right) \sqrt{3}
$$

$$
= 0.293\sqrt{3}
$$

The actual positive sequence fault current is

$$
I_{L4(F)\,(\text{actual})}^{(1)} = I_{L4(F)\,(\text{assumed})}^{(1)} \frac{E_{4(0)}^{(1)}}{E_{L4(F)}^{(1)}}
$$

$$
= \sqrt{3}\left(\frac{\sqrt{3}}{0.293\sqrt{3}}\right) = 3.42\sqrt{3}
$$

Then, the sequence and phase components of the fault current are

$$
I_{L4(F)}^{0,1,2} = \sqrt{3} \begin{bmatrix} 0 \\ 3.42 \\ 0 \end{bmatrix} \qquad I_{L4(F)}^{a,b,c} = T_s I_{L4(F)}^{0,1,2} = 3.42 \begin{bmatrix} 1 \\ a^2 \\ a \end{bmatrix}
$$

The positive sequence loop currents are

$$\bar{I}_L^{(1)} = -(Z_L^{(1)})^{-1} Z_M^{(1)} I_{L(F)}^{(1)}$$

$$\begin{bmatrix} I_A^{(1)} \\ I_B^{(1)} \end{bmatrix} = - \begin{bmatrix} 0.855 & -0.145 \\ -0.145 & 0.855 \end{bmatrix} \begin{bmatrix} 0.1025 \\ 0.5025 \end{bmatrix} 3.42\sqrt{3} = \sqrt{3} \begin{bmatrix} -0.05 \\ -1.42 \end{bmatrix}$$

Using the loop currents, the currents in the elements and bus voltages can be calculated as previously shown.

6.7 Description of short circuit program

The majority of short circuit studies involve only the calculation of three-phase and line-to-ground faults. The American Electric Power Short Circuit Program, which is designed to calculate these faults, uses the positive and zero sequence bus impedance matrices as described in Sec. 6.3 and the simplified system representation presented in Sec. 6.2.

The input data describing the system is specified using power plant and substation names. Data for generators, synchronous condensers, and terminal points includes the station names and corresponding positive and zero sequence reactances. Lines are specified by two station names, one for each terminal, along with the line reactances. Each set of two lines which are mutually coupled requires a station name for each terminal and a mutual reactance. Transformers are specified by station names for each terminal, the number of windings and their connections, and the positive and zero sequence reactances. The input data may include also a study name, case number, and identifying remarks.

The program first assigns sequential bus numbers and then rearranges the network data to facilitate the formation of the positive and zero sequence bus impedance matrices. During this phase extensive data checks are performed. Next, the positive sequence bus impedance matrix is formed. This matrix is temporarily stored on an auxiliary storage device to provide space in memory for the next program segment. Then the zero sequence bus impedance matrix is formed and the positive sequence bus impedance matrix is retrieved for use in the fault calculations. Since these matrices are symmetrical only the diagonal elements and upper off-diagonal elements need to be formed and stored. The sequence of steps in the short circuit program is shown in Fig. 6.6.

Short circuits in megavolt-amperes (mva) are calculated for each bus and tabulated with the appropriate station name. The following results are obtained:

1. Total three-phase and line-to-ground bus fault currents
2. Three-phase and line-to-ground fault contributions in each line connected to the faulted bus
3. Zero sequence driving point reactance for the faulted bus
4. Total three-phase and line-to-ground bus fault currents and cor-

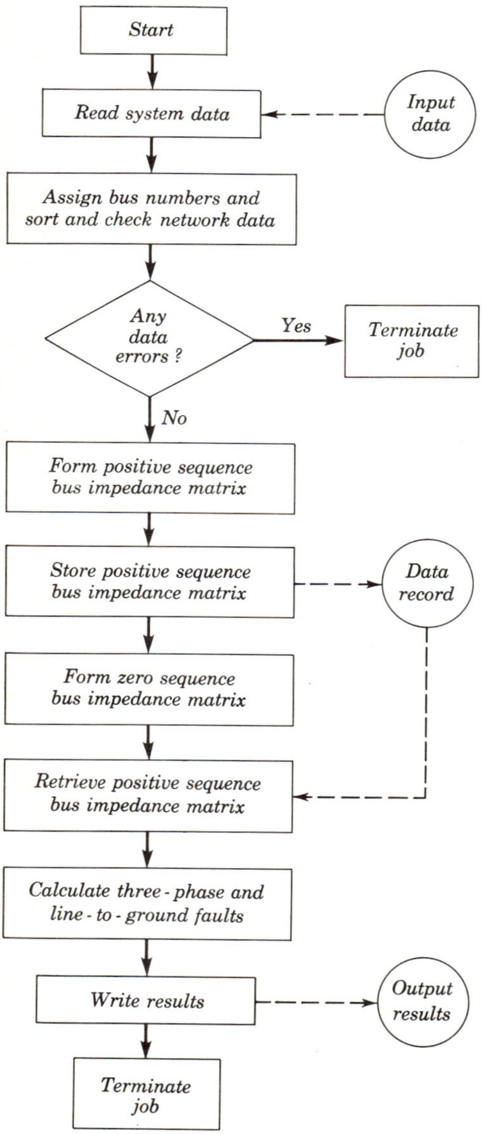

***Fig. 6.6** Simplified flow chart for the American Electric Power Short Circuit Program.*

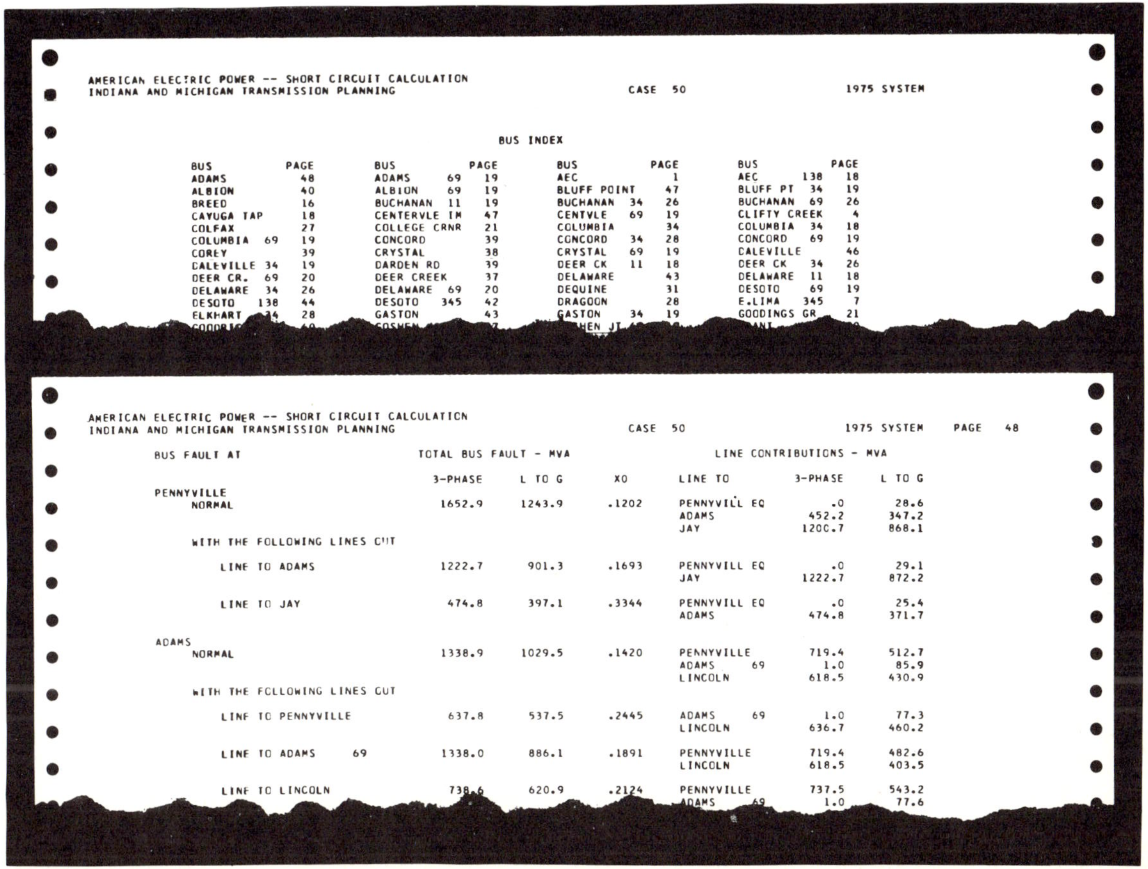

AMERICAN ELECTRIC POWER -- SHORT CIRCUIT CALCULATION
INDIANA AND MICHIGAN TRANSMISSION PLANNING   CASE 50   1975 SYSTEM

BUS INDEX

BUS	PAGE	BUS	PAGE	BUS	PAGE	BUS	PAGE
ADAMS	48	ADAMS 69	19	AEC	1	AEC 138	18
ALBION	40	ALBION 69	19	BLUFF POINT	47	BLUFF PT 34	19
BREED	16	BUCHANAN 11	19	BUCHANAN 34	26	BUCHANAN 69	26
CAYUGA TAP	18	CENTERVLE IM	47	CENTVLE 69	19	CLIFTY CREEK	4
COLFAX	27	COLLEGE CRNR	21	COLUMBIA	34	COLUMBIA 34	18
COLUMBIA 69	19	CONCORD	39	CONCORD 34	28	CONCORD 69	19
COREY	39	CRYSTAL	38	CRYSTAL 69	19	DALEVILLE	46
DALEVILLE 34	19	DARDEN RD	39	DEER CK 11	18	DEER CK 34	26
DEER CR. 69	20	DEER CREEK	37	DELAWARE	43	DELAWARE 11	18
DELAWARE 34	26	DELAWARE 69	20	DEQUINE	31	DESOTO 69	19
DESOTO 138	44	DESOTO 345	42	DRAGOON	28	E.LIMA 345	7
ELKHART 34	28	GASTON	43	GASTON 34	19	GOODINGS GR	21

AMERICAN ELECTRIC POWER -- SHORT CIRCUIT CALCULATION
INDIANA AND MICHIGAN TRANSMISSION PLANNING   CASE 50   1975 SYSTEM   PAGE 48

BUS FAULT AT	TOTAL BUS FAULT - MVA 3-PHASE	L TO G	XO	LINE CONTRIBUTIONS - MVA LINE TO	3-PHASE	L TO G
PENNYVILLE NORMAL	1652.9	1243.9	.1202	PENNYVILL EQ	.0	28.6
				ADAMS	452.2	347.2
				JAY	1200.7	868.1
WITH THE FOLLOWING LINES CUT						
LINE TO ADAMS	1222.7	901.3	.1693	PENNYVILL EQ	.0	29.1
				JAY	1222.7	872.2
LINE TO JAY	474.8	397.1	.3344	PENNYVILL EQ	.0	25.4
				ADAMS	474.8	371.7
ADAMS NORMAL	1338.9	1029.5	.1420	PENNYVILLE	719.4	512.7
				ADAMS 69	1.0	85.9
				LINCOLN	618.5	430.9
WITH THE FOLLOWING LINES CUT						
LINE TO PENNYVILLE	637.8	537.5	.2445	ADAMS 69	1.0	77.3
				LINCOLN	636.7	460.2
LINE TO ADAMS 69	1338.0	886.1	.1891	PENNYVILLE	719.4	482.6
				LINCOLN	618.5	403.5
LINE TO LINCOLN	738.6	620.9	.2124	PENNYVILLE	737.5	543.2
				ADAMS 69	1.0	77.6

Fig. 6.7 Sample output of the American Electric Power Short Circuit Program.

responding line contributions when lines connected to the faulted bus are opened one at a time

In order to locate quickly the short circuit results for individual stations, the printed output includes a title page with an index of stations listed alphabetically and their corresponding page numbers. A sample of this output is shown in Fig. 6.7.

In addition to these results which are obtained automatically for each bus, the following results can be obtained by special options:

1. Fault contributions in lines not connected directly to the faulted bus
2. Bus voltages during fault
3. Fault contributions following the switching of lines not connected to the faulted bus
4. Fault contributions following the switching of two or more lines connected to the faulted bus

Problems

6.1 The reactance data for the three-phase system shown in Fig. 6.8 is Generator:

$$x^{(1)} = x^{(2)} = 0.1$$
$$x^{(0)} = 0.04$$
$$x_g = 0.02$$

Transformer:

$$x^{(1)} = x^{(2)} = x^{(0)} = 0.1$$
$$x_g = 0.05$$

Form the positive and zero sequence bus impedance matrices and

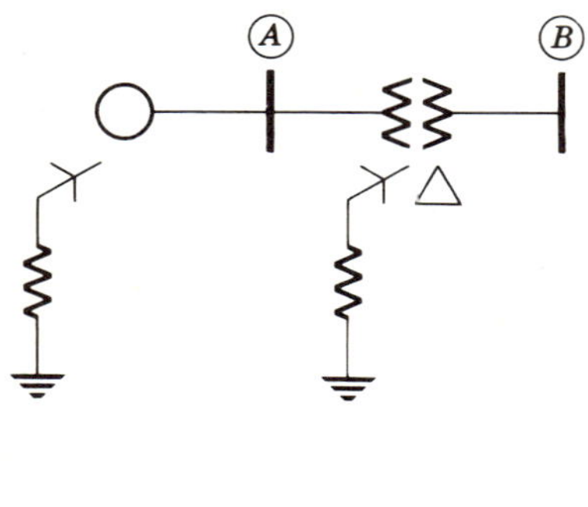

***Fig. 6.8** Sample system for Prob. 6.1.*

calculate the total fault current and the contributions from the generator and transformer for the following faults at bus A:

a. Three-phase-to-ground
b. Line-to-ground

6.2 For a fault at bus p let the admittance between each phase and neutral be y_a, y_b, and y_c, and let the admittance between neutral and ground be y_g.

a. Form the fault admittance matrix $Y_F^{a,b,c}$.
b. Verify that each fault admittance matrix $Y_F^{a,b,c}$ given in Table 6.1 is a special case for the condition that $y_a = y_b = y_c$.

6.3 Derive the equations for the total fault current in terms of symmetrical components and phase quantities for the following faults at bus p:

a. Three-phase (not grounded)
b. Line-to-line
c. Line-to-line-to-ground

6.4 From the equations for the fault currents draw the sequence networks and their connections to simulate the following faults:

a. Three-phase
b. Line-to-ground
c. Line-to-line
d. Line-to-line-to-ground

6.5 Using the sample system given in Prob. 5.3, compute the total fault current and bus voltages for the following faults at bus B:

a. Three-phase-to-ground
b. Line-to-ground
c. Line-to-line-to-ground

6.6 Using the same sample system given in Prob. 5.3, and assuming the transmission line B–C is balanced and its reactance is

$$x^{a,b,c} = \begin{bmatrix} 0.3 & 0.1 & 0.1 \\ 0.1 & 0.3 & 0.1 \\ 0.1 & 0.1 & 0.3 \end{bmatrix}$$

compute the total fault current and bus voltages for the following faults at bus B:

a. Three-phase-to-ground
b. Line-to-ground
c. Line-to-line-to-ground

Compare these results with those obtained from Prob. 6.5.

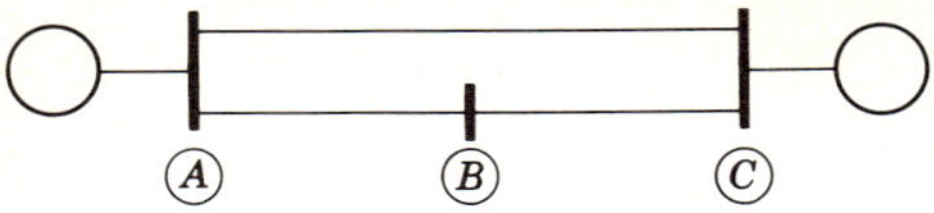

Fig. 6.9 Sample two-phase system for Prob. 6.7.

6.7 The reactance data for the two-phase system shown in Fig. 6.9 is Generators A and C:

$$x_{0A} = x_{0C} = \begin{bmatrix} 5 & 0 \\ 0 & 5 \end{bmatrix}$$

Transmission lines A–B and B–C:

$$x_{AB} = x_{BC} = \begin{bmatrix} 4 & \sqrt{2} \\ \sqrt{2} & 3 \end{bmatrix}$$

Transmission line A–C:

$$x_{AC} = \begin{bmatrix} 8 & 2\sqrt{2} \\ 2\sqrt{2} & 6 \end{bmatrix}$$

The bus voltages are

$$E_A = \begin{bmatrix} 1.1\underline{/30^\circ} \\ 1.1\underline{/120^\circ} \end{bmatrix}$$

$$E_C = \begin{bmatrix} 1.2\underline{/0^\circ} \\ 1.2\underline{/90^\circ} \end{bmatrix}$$

Compute the following:

a. The bus impedance matrix with ground as reference

b. The phase currents in each line

c. The ground current at bus B for a fault with the reactance

$$x_{0B} = \begin{array}{|c|c|} \hline 1 & 0 \\ \hline 0 & \infty \\ \hline \end{array}$$

d. The fault voltages at buses A and B

Bibliography

Brown, H. E., C. E. Person, L. K. Kirchmayer, and G. W. Stagg: Digital Calculation of Three-phase Short Circuits by Matrix Method, *Trans. AIEE*, vol. 79, pt. III, pp. 1277–1281, 1960.

Byerly, R. T., R. W. Long, C. J. Baldwin, Jr., and C. W. King: Digital Calculation of Power System Networks Under Faulted Conditions, *Trans. AIEE*, vol. 77, pt. III, pp. 1296–1307, 1958.

Coombe, L. W., and D. G. Lewis: Digital Calculation of Short-circuit Currents in Large Complex-impedance Networks, *Trans. AIEE*, vol. 75, pt. III, pp. 1394–1397, 1956.

El-Abiad, Ahmed H.: Digital Calculation of Line-to-ground Short Circuits by Matrix Method, *Trans. AIEE*, vol. 79, pt. III, pp. 323–332, 1960.

El-Abiad, A. H., Ruth Guidone, and G. W. Stagg: Calculation of Short Circuits Using a High-speed Digital Computer, *Trans. AIEE*, vol. 80, pt. III, pp. 702–707, 1961.

Lantz, Martin J.: Digital Short-circuit Solution of Power System Networks Including Mutual Impedance, *Trans. AIEE*, vol. 76, pt. III, pp. 1230–1235, 1957.

Rindt, L. J., and R. W. Long: New Short Circuit Program Provides Many Automatic Features, *IEEE Proc. Power Ind. Computer Appl. Conf., Phoenix, Ariz.*, pp. 167–188, 1963.

Siegel, J. C., and G. W. Bills: Nodal Representation of Large Complex-element Networks Including Mutual Reactances, *Trans. AIEE*, vol. 77, pt. III, pp. 1226–1229, 1958.

Taylor, G. E., G. K. Carter, and E. H. MacDonald: New Digital Computer Short-circuit Program for Relay Studies, *Trans. AIEE*, vol. 79, pt. III, pp. 1257–1264, 1960.

Toalston, Argil L.: Digital Solution of Short-circuit Currents for Networks Including Mutual Impedances, *Trans. AIEE*, vol. 78, pt. IIIB, pp. 1720–1725, 1959.

chapter 7
Solution of simultaneous algebraic equations

7.1 Introduction

The subject of numerical analysis pertains to that branch of mathematics which is a study of methods and their application for the numerical solution of mathematical problems. The mathematical relations of physical quantities must be established for a specific problem. It is then necessary to consider which of the available methods is most appropriate for obtaining a solution. Consideration must be given to both the speed with which the solution can be obtained and the resultant accuracy. The selection of a method is influenced also by the capability of the available computer.

Among the important and most frequently encountered problems in numerical analysis is the solution of sets of algebraic equations. A number of methods are available to solve sets of equations. All of these, however, fall into one of two general types: direct or iterative. A direct method, also referred to as an exact method, provides a solution in a definite number of arithmetic operations. The number of operations depends on the computational technique selected and the number of equations. Furthermore, if the coefficients of the equations form a symmetric matrix, the solution requires fewer arithmetic operations than for a problem of the same dimensions with a nonsymmetric matrix.

Except for the inevitable round-off error of intermediate or final results, a solution obtained by direct methods is exact. Failure to obtain an adequate solution for a consistent set of equations can be encountered by the loss of significant digits in the course of computation. For example, this could be a result of subtracting two numbers that are very near the same value. Thus, selection of a direct computational method and the number of significant digits retained throughout the process are important.

On the other hand, methods employing iterative techniques provide, in an orderly fashion, successive approximate solutions which may converge to results with acceptable accuracy. The rate of convergence, i.e., the required number of successive approximations or iterations, depends primarily on the coefficient matrix defining the physical system. Depending on the characteristics of this matrix, the successive approximate solutions may converge rapidly or slowly or even diverge. Thus, the formulation of a problem has a direct bearing on the time required for convergence. The iterative computational technique adopted also has an influence on the rate of convergence. An additional factor affecting the time of solution is the choice of the estimated initial values to start the computational process.

7.2 Direct methods for solution of linear algebraic equations

Simultaneous equations

A physical system in steady state can be represented by a set of simultaneous equations of the form

$$\begin{aligned} f_1(x_1, x_2, \ldots, x_n) &= y_1 \\ f_2(x_1, x_2, \ldots, x_n) &= y_2 \\ \cdots\cdots\cdots&\cdots\cdots \\ f_n(x_1, x_2, \ldots, x_n) &= y_n \end{aligned} \tag{7.2.1}$$

where f_i are functions relating the unknown variables x_j with the known parameters of the system. The system (7.2.1) is nonlinear if at least one of the functions f_i is nonlinear. If all f_i's are linear the system of equations is linear. A linear system can be represented in matrix form by

$$AX = Y$$

where A is a constant coefficient matrix.

Solution by determinants

Consider the following system of linear equations:

$$\begin{aligned} a_{11}x_1 + a_{12}x_2 + a_{13}x_3 + a_{14}x_4 &= y_1 \\ a_{21}x_1 + a_{22}x_2 + a_{23}x_3 + a_{24}x_4 &= y_2 \\ a_{31}x_1 + a_{32}x_2 + a_{33}x_3 + a_{34}x_4 &= y_3 \\ a_{41}x_1 + a_{42}x_2 + a_{43}x_3 + a_{44}x_4 &= y_4 \end{aligned}$$

The determinant of the coefficient matrix is

$$|A| = \begin{vmatrix} a_{11} & a_{12} & a_{13} & a_{14} \\ a_{21} & a_{22} & a_{23} & a_{24} \\ a_{31} & a_{32} & a_{33} & a_{34} \\ a_{41} & a_{42} & a_{43} & a_{44} \end{vmatrix}$$

Multiply the first equation by A_{11}, the cofactor of a_{11}, the second equation by A_{21}, the third by A_{31}, etc. The products are added to obtain

$$\begin{aligned} &x_1(a_{11}A_{11} + a_{21}A_{21} + a_{31}A_{31} + a_{41}A_{41}) \\ &+ x_2(a_{12}A_{11} + a_{22}A_{21} + a_{32}A_{31} + a_{42}A_{41}) \\ &+ x_3(a_{13}A_{11} + a_{23}A_{21} + a_{33}A_{31} + a_{43}A_{41}) \\ &+ x_4(a_{14}A_{11} + a_{24}A_{21} + a_{34}A_{31} + a_{44}A_{41}) \\ &= y_1A_{11} + y_2A_{21} + y_3A_{31} + y_4A_{41} \end{aligned} \tag{7.2.2}$$

From the relationships of determinant and cofactors the term

$$x_1(a_{11}A_{11} + a_{21}A_{21} + a_{31}A_{31} + a_{41}A_{41}) = x_1|A|$$

and the remaining terms are zero, i.e.,

$$x_j(a_{1j}A_{11} + a_{2j}A_{21} + a_{3j}A_{31} + a_{4j}A_{41}) = 0 \qquad j \neq 1$$

The equation (7.2.2) becomes

$$x_1|A| = y_1A_{11} + y_2A_{21} + y_3A_{31} + y_4A_{41} \tag{7.2.3}$$

The right-hand side of equation (7.2.3) is the determinant of the matrix obtained by replacing the first column of A with the known parameters y_1, y_2, y_3, and y_4. Designating this determinant by $|A_1|$,

$$|A_1| = \begin{vmatrix} y_1 & a_{12} & a_{13} & a_{14} \\ y_2 & a_{22} & a_{23} & a_{24} \\ y_3 & a_{32} & a_{33} & a_{34} \\ y_4 & a_{42} & a_{43} & a_{44} \end{vmatrix}$$

and solving equation (7.2.3) for x_1,

$$x_1 = \frac{|A_1|}{|A|}$$

if $|A| \neq 0$. Similarly,

$$x_2 = \frac{|A_2|}{|A|}$$

$$x_3 = \frac{|A_3|}{|A|}$$

$$x_4 = \frac{|A_4|}{|A|}$$

where $|A_2|$, $|A_3|$, and $|A_4|$ are the determinants obtained by replacing respectively the second, third, and fourth columns of A with the known parameters. This method of solution is known as Cramer's rule.

Gauss elimination method

The successive elimination of unknowns is the simplest and most practical direct method of solving a system of linear equations. Many variations of the original scheme attributed to Gauss have been proposed for organizing the computation to minimize the number of arithmetic steps, to reduce round-off error, and to minimize the chance of error in hand calculations. A detailed description of the elimination method applied to a system of four equations will be used to illustrate the method.

The given set of linear equations is

$$\begin{aligned} a_{11}x_1 + a_{12}x_2 + a_{13}x_3 + a_{14}x_4 &= y_1 \\ a_{21}x_1 + a_{22}x_2 + a_{23}x_3 + a_{24}x_4 &= y_2 \\ a_{31}x_1 + a_{32}x_2 + a_{33}x_3 + a_{34}x_4 &= y_3 \\ a_{41}x_1 + a_{42}x_2 + a_{43}x_3 + a_{44}x_4 &= y_4 \end{aligned} \tag{7.2.4}$$

As an initial step, divide the first equation by its leading coefficient $a_{11} \neq 0$ to obtain

$$x_1 + \frac{a_{12}}{a_{11}} x_2 + \frac{a_{13}}{a_{11}} x_3 + \frac{a_{14}}{a_{11}} x_4 = \frac{y_1}{a_{11}}$$

Using b_{1j} and g_1 to designate the resulting coefficients and constant term, respectively, the system of equations (7.2.4) becomes

$$\begin{aligned} x_1 + b_{12}x_2 + b_{13}x_3 + b_{14}x_4 &= g_1 \\ a_{21}x_1 + a_{22}x_2 + a_{23}x_3 + a_{24}x_4 &= y_2 \\ a_{31}x_1 + a_{32}x_2 + a_{33}x_3 + a_{34}x_4 &= y_3 \\ a_{41}x_1 + a_{42}x_2 + a_{43}x_3 + a_{44}x_4 &= y_4 \end{aligned} \tag{7.2.5}$$

Next, transform the set of equations to a new set in which the leading coefficients of the second, third, and fourth equations are zero. This is accomplished by multiplying the first equation of the system (7.2.5) by the leading coefficient of the second equation and then subtracting the resulting equation from the second equation to obtain

$$\begin{aligned} (a_{21} - a_{21})x_1 + (a_{22} - a_{21}b_{12})x_2 + (a_{23} - a_{21}b_{13})x_3 + (a_{24} - a_{21}b_{14})x_4 \\ = y_2 - a_{21}g_1 \end{aligned}$$

Designate these intermediate coefficients by $a_{2j}^{(1)}$ and $y_2^{(1)}$ to obtain

$$a_{22}^{(1)}x_2 + a_{23}^{(1)}x_3 + a_{24}^{(1)}x_4 = y_2^{(1)}$$

in which x_1 has been eliminated. Following the same procedure for subsequent equations, the third equation becomes

$$(a_{31} - a_{31})x_1 + (a_{32} - a_{31}b_{12})x_2 + (a_{33} - a_{31}b_{13})x_3 + (a_{34} - a_{31}b_{14})x_4 = y_3 - a_{31}g_1$$

or, using the new notation,

$$a_{32}^{(1)}x_2 + a_{33}^{(1)}x_3 + a_{34}^{(1)}x_4 = y_3^{(1)}$$

and the fourth equation is

$$(a_{41} - a_{41})x_1 + (a_{42} - a_{41}b_{12})x_2 + (a_{43} - a_{41}b_{13})x_3 + (a_{44} - a_{41}b_{14})x_4 = y_4 - a_{41}g_1$$

or

$$a_{42}^{(1)}x_2 + a_{43}^{(1)}x_3 + a_{44}^{(1)}x_4 = y_4^{(1)}$$

The final transformed equations after eliminating x_1 from the second, third, and fourth equations of (7.2.5) are

$$\begin{aligned} x_1 + b_{12}x_2 + b_{13}x_3 + b_{14}x_4 &= g_1 \\ a_{22}^{(1)}x_2 + a_{23}^{(1)}x_3 + a_{24}^{(1)}x_4 &= y_2^{(1)} \\ a_{32}^{(1)}x_2 + a_{33}^{(1)}x_3 + a_{34}^{(1)}x_4 &= y_3^{(1)} \\ a_{42}^{(1)}x_2 + a_{43}^{(1)}x_3 + a_{44}^{(1)}x_4 &= y_4^{(1)} \end{aligned} \tag{7.2.6}$$

This procedure is repeated for the second, third, and fourth equations to eliminate x_2 from the third and fourth equations. Dividing the second equation of (7.2.6) by its leading coefficient,

$$x_2 + \frac{a_{23}^{(1)}}{a_{22}^{(1)}}x_3 + \frac{a_{24}^{(1)}}{a_{22}^{(1)}}x_4 = \frac{y_2^{(1)}}{a_{22}^{(1)}}$$

where $a_{22}^{(1)} \neq 0$. Denoting these new coefficients by b_{2j} and g_2, respectively, the system of equations (7.2.6) becomes

$$\begin{aligned} x_1 + b_{12}x_2 + b_{13}x_3 + b_{14}x_4 &= g_1 \\ x_2 + b_{23}x_3 + b_{24}x_4 &= g_2 \\ a_{32}^{(1)}x_2 + a_{33}^{(1)}x_3 + a_{34}^{(1)}x_4 &= y_3^{(1)} \\ a_{42}^{(1)}x_2 + a_{43}^{(1)}x_3 + a_{44}^{(1)}x_4 &= y_4^{(1)} \end{aligned} \tag{7.2.7}$$

Multiply the second equation in (7.2.7) by the leading coefficient of the third equation and subtract the resulting equation from the third equation to obtain

$$(a_{32}^{(1)} - a_{32}^{(1)})x_2 + (a_{33}^{(1)} - a_{32}^{(1)}b_{23})x_3 + (a_{34}^{(1)} - a_{32}^{(1)}b_{24})x_4 = y_3^{(1)} - a_{32}^{(1)}g_2$$

Denote these new intermediate coefficients by $a_{3j}^{(2)}$ and $y_3^{(2)}$ to obtain

$$a_{33}^{(2)}x_3 + a_{34}^{(2)}x_4 = y_3^{(2)}$$

with x_2 eliminated. Similarly for the fourth equation,

$$a_{43}^{(2)}x_3 + a_{44}^{(2)}x_4 = y_4^{(2)}$$

The transformed equations after eliminating x_2 from the third and fourth equations of (7.2.7) are

$$\begin{aligned} x_1 + b_{12}x_2 + b_{13}x_3 + b_{14}x_4 &= g_1 \\ x_2 + b_{23}x_3 + b_{24}x_4 &= g_2 \\ a_{33}^{(2)}x_3 + a_{34}^{(2)}x_4 &= y_3^{(2)} \\ a_{43}^{(2)}x_3 + a_{44}^{(2)}x_4 &= y_4^{(2)} \end{aligned} \tag{7.2.8}$$

Continuing, the third equation in (7.2.8) becomes

$$x_3 + b_{34}x_4 = g_3$$

and the fourth is

$$a_{44}^{(3)}x_4 = y_4^{(3)}$$

The transformed equations of (7.2.8) are

$$\begin{aligned} x_1 + b_{12}x_2 + b_{13}x_3 + b_{14}x_4 &= g_1 \\ x_2 + b_{23}x_3 + b_{24}x_4 &= g_2 \\ x_3 + b_{34}x_4 &= g_3 \\ a_{44}^{(3)}x_4 &= y_4^{(3)} \end{aligned}$$

Divide the fourth equation by $a_{44}^{(3)} \neq 0$ to obtain the final transformed equations

$$\begin{aligned} x_1 + b_{12}x_2 + b_{13}x_3 + b_{14}x_4 &= g_1 \\ x_2 + b_{23}x_3 + b_{24}x_4 &= g_2 \\ x_3 + b_{34}x_4 &= g_3 \\ x_4 &= g_4 \end{aligned} \tag{7.2.9}$$

This is a triangular set of equations, the solution of which is the solution of the original set of equations (7.2.4).

The value of x_4 is obtained directly and is substituted into the third equation of the triangular set (7.2.9) to obtain a solution for x_3. Both x_4 and x_3 are substituted in the second equation to obtain x_2. All three values are substituted in the first equation to obtain x_1. The process of obtaining the triangular system of equations is referred to as the *forward course* and the process for obtaining the solution is called *back substitution.*

The general equations in the Gauss elimination method for transforming the coefficients and parameters a_{ij} and y_i of equations (7.2.4) into b_{ij} and g_i of equations (7.2.9) are, at the kth step in the process,

$$a_{ij}^{(k)} = a_{ij}^{(k-1)} - \frac{a_{ik}^{(k-1)}a_{kj}^{(k-1)}}{a_{kk}^{(k-1)}} \qquad i, j \geq k + 1$$

$$y_i^{(k)} = y_i^{(k-1)} - \frac{a_{ik}^{(k-1)}y_k^{(k-1)}}{a_{kk}^{(k-1)}} \qquad i \geq k + 1$$

$$b_{kj} = \frac{a_{kj}^{(k-1)}}{a_{kk}^{(k-1)}} \qquad j \geq k + 1$$

$$g_k = \frac{y_k^{(k-1)}}{a_{kk}^{(k-1)}}$$

The final solution is then

$$x_i = g_i - \sum_{j=i+1}^{n} b_{ij}x_j \qquad i = n, n - 1, n - 2, \ldots , 2, 1$$

If the leading coefficient is zero at any step in the forward course, then an alternate variable must be selected for elimination. If the leading coefficient is zero at the last step, there is no unique solution for the set of equations.

Crout method or compact form

The forward course of the elimination method may be performed more directly by a modification of the Gauss elimination method accredited to Crout. This method eliminates the need for explicitly determining and recording all coefficients of the modified equations in each step of the process. In applying this technique it is more convenient to work directly with the augmented matrix:

$$\hat{A} = \begin{bmatrix} a_{11} & a_{12} & a_{13} & a_{14} & y_1 \\ a_{21} & a_{22} & a_{23} & a_{24} & y_2 \\ a_{31} & a_{32} & a_{33} & a_{34} & y_3 \\ a_{41} & a_{42} & a_{43} & a_{44} & y_4 \end{bmatrix}$$

Uniformity in notation is gained by redesignating the constant terms y_1, y_2, y_3, and y_4 as a_{15}, a_{25}, a_{35}, and a_{45}, respectively, in the augmented matrix. Thus,

$$\hat{A} = \begin{bmatrix} a_{11} & a_{12} & a_{13} & a_{14} & a_{15} \\ a_{21} & a_{22} & a_{23} & a_{24} & a_{25} \\ a_{31} & a_{32} & a_{33} & a_{34} & a_{35} \\ a_{41} & a_{42} & a_{43} & a_{44} & a_{45} \end{bmatrix}$$

where the elements on the principal diagonal are those elements whose row and column indices are equal.

The basis of the Crout method is to determine the elements for an auxiliary matrix

$$\hat{B} = \begin{bmatrix} e_{11} & f_{12} & f_{13} & f_{14} & f_{15} \\ e_{21} & e_{22} & f_{23} & f_{24} & f_{25} \\ e_{31} & e_{32} & e_{33} & f_{34} & f_{35} \\ e_{41} & e_{42} & e_{43} & e_{44} & f_{45} \end{bmatrix}$$

of the same dimension as $\hat{A}$, in which the elements above the diagonal will be the same as the coefficients of the triangular set of equations obtained in the Gauss elimination method. These elements are designated by f_{ij} for $i < j$ in $\hat{B}$. The elements occupying the positions on and below the diagonal, designated by e_{ij} for $i \geq j$, are the only necessary additional intermediate values that must be calculated and recorded in this computational process. In the Gauss elimination method, these positions would contain ones on the diagonal and zeros below after the completion of the forward course. The recording of the intermediate values makes it possible to calculate all elements of $\hat{B}$, in an orderly process, from the elements of $\hat{A}$ and from the previously calculated elements of $\hat{B}$.

The elements of the first column of $\hat{B}$ are identical to the elements of the first column of $\hat{A}$. That is,

$$e_{11} = a_{11}$$
$$e_{21} = a_{21}$$
$$e_{31} = a_{31}$$
$$e_{41} = a_{41}$$

The remaining elements of the first row of $\hat{B}$ are obtained by dividing the corresponding elements of $\hat{A}$ by the diagonal element a_{11}:

$$f_{12} = \frac{a_{12}}{a_{11}}$$

$$f_{13} = \frac{a_{13}}{a_{11}}$$

$$f_{14} = \frac{a_{14}}{a_{11}}$$

$$f_{15} = \frac{a_{15}}{a_{11}}$$

The order of the computation then proceeds as follows: determine the remaining elements of the second column, the second row, the third column, third row, etc. The remaining elements of the second column are obtained from the equations

$$e_{22} = a_{22} - e_{21}f_{12}$$
$$e_{32} = a_{32} - e_{31}f_{12}$$
$$e_{42} = a_{42} - e_{41}f_{12}$$

The remaining elements of the second row to the right of the diagonal are determined from the equations

$$f_{23} = \frac{a_{23} - e_{21}f_{13}}{e_{22}}$$
$$f_{24} = \frac{a_{24} - e_{21}f_{14}}{e_{22}} \tag{7.2.10}$$
$$f_{25} = \frac{a_{25} - e_{21}f_{15}}{e_{22}}$$

Each of these equations results from a combination of steps. For example, in the first equation of (7.2.10) the numerator is the intermediate value

$$e_{23} = a_{23} - e_{21}f_{13} \tag{7.2.11}$$

and f_{23} was obtained from

$$f_{23} = \frac{e_{23}}{e_{22}} \tag{7.2.12}$$

Combining equations (7.2.11) and (7.2.12) yields the formula for f_{23} shown in equation (7.2.10). The need to determine explicitly the element e_{23} is eliminated by using equation (7.2.10).

Continuing, the remaining elements of the third column are determined by

$$e_{33} = a_{33} - e_{31}f_{13} - e_{32}f_{23}$$
$$e_{43} = a_{43} - e_{41}f_{13} - e_{42}f_{23}$$

and the remaining elements of the third row by

$$f_{34} = \frac{a_{34} - e_{31}f_{14} - e_{32}f_{24}}{e_{33}}$$
$$f_{35} = \frac{a_{35} - e_{31}f_{15} - e_{32}f_{25}}{e_{33}}$$

Finally,

$$e_{44} = a_{44} - e_{41}f_{14} - e_{42}f_{24} - e_{43}f_{34}$$

and

$$f_{45} = \frac{a_{45} - e_{41}f_{15} - e_{42}f_{25} - e_{43}f_{35}}{e_{44}}$$

Crout's method is shown pictorially in the following diagram.

e_{11}	f_{12}	f_{13}	f_{14}	f_{15}		f_{17}		f_{19}	
e_{21}	e_{22}	f_{23}	f_{24}	f_{25}		f_{27}		f_{29}	
e_{31}	e_{32}	e_{33}	f_{34}	f_{35}		f_{37}		f_{39}	
e_{41}	e_{42}	e_{43}	a_{44}	a_{45}	a_{46}	a_{47}	a_{48}	a_{49}	Step 4
			a_{54}	a_{55}				a_{59}	
e_{61}	e_{62}	e_{63}	a_{64}						
			a_{74}						
e_{81}	e_{82}	e_{83}	a_{84}	a_{85}				a_{89}	
			Step 4						

Assume that the first three columns and three rows have been calculated and the next step is to calculate e_{44} and the remaining terms in the fourth column, e_{54}, e_{64}, e_{74}, and e_{84}. The following operations are performed to obtain e_{44}.

$$e_{44} = a_{44} - \begin{bmatrix} e_{41} & e_{42} & e_{43} \end{bmatrix} \begin{bmatrix} f_{14} \\ f_{24} \\ f_{34} \end{bmatrix}$$

The value e_{44} immediately replaces a_{44}. This minimizes the storage requirement for a computer solution. Similar operations are performed to obtain the remaining column values. For example,

$$e_{64} = a_{64} - \begin{bmatrix} e_{61} & e_{62} & e_{63} \end{bmatrix} \begin{bmatrix} f_{14} \\ f_{24} \\ f_{34} \end{bmatrix}$$

The next step is to calculate f_{45} and the remaining terms in the fourth row.

$$f_{45} = \frac{a_{45} - \begin{bmatrix} e_{41} & e_{42} & e_{43} \end{bmatrix} \begin{bmatrix} f_{15} \\ f_{25} \\ f_{35} \end{bmatrix}}{e_{44}}$$

The value f_{45} replaces a_{45}.

The general equations for the Crout method are

$$e_{ij} = a_{ij} - \sum_{k=1}^{j-1} e_{ik} f_{kj} \qquad i \geq j$$

and

$$f_{ij} = \frac{1}{e_{ii}} \left(a_{ij} - \sum_{k=1}^{i-1} e_{ik} f_{kj} \right) \qquad i < j$$

The final values of x_1, x_2, x_3, etc., are obtained by back substitution after all elements of $\hat{B}$ have been determined. These are

$$x_i = f_{i,n+1} - \sum_{k=i+1}^{n} f_{ik} x_k \qquad i = n, n-1, \ldots, 2, 1$$

Gauss-Jordan method

A modification of the Gauss elimination method, known as the Gauss-Jordan method, eliminates the need for back substitution to obtain the values of the unknowns. This method provides a uniform procedure and is adaptable to computer calculation.

The first step of eliminating x_1 is performed as before. The second step is modified to the extent that after x_2 is eliminated from succeeding equations, x_2 is eliminated also from the previous equation. Similarly, after elimination of x_3 in step 3, x_3 is eliminated from the preceding equations. Finally, in the fourth step of the elimination process, for equations (7.2.4) it is necessary only to eliminate x_4 from the preceding equations. At this stage each of the equations has been reduced to an equation involving a single unknown, x_i. The solution then is obtained directly. The Gauss-Jordan method results in a diagonal set of equations while the Gauss elimination method results in a triangular set of equations.

At any stage of the elimination process in both the Gauss elimination and Gauss-Jordan methods a loss of significant digits may occur in the calculated coefficients. To overcome this difficulty the order of elimination of the variables can be changed. A method that improves the accuracy is *pivotal condensation.* The procedure is as follows: select the largest element (pivot) of the coefficient matrix and eliminate the corresponding x_j. Repeat this process for the reduced system. For the final solution, the order of the values x_j will depend on the order of selection for the pivotal elements.

Evaluation of determinants

The Gauss elimination method may be applied to the evaluation of determinants. Consider the determinant

$$|A| = \begin{vmatrix} a_{11} & a_{12} & a_{13} \\ a_{21} & a_{22} & a_{23} \\ a_{31} & a_{32} & a_{33} \end{vmatrix}$$

First, remove the factor a_{11} from the first row to obtain

$$|A| = a_{11}\begin{vmatrix} 1 & b_{12} & b_{13} \\ a_{21} & a_{22} & a_{23} \\ a_{31} & a_{32} & a_{33} \end{vmatrix}$$

where $b_{12} = a_{12}/a_{11}$ and $b_{13} = a_{13}/a_{11}$. Next, multiply the elements of the first row of the determinant by the leading element a_{21} of the second row and subtract the resultant products from the elements of the second row. Repeating this operation for subsequent rows, the determinant has the form

$$|A| = a_{11}\begin{vmatrix} 1 & b_{12} & b_{13} \\ 0 & a_{22}^{(1)} & a_{23}^{(1)} \\ 0 & a_{32}^{(1)} & a_{33}^{(1)} \end{vmatrix}$$

That is,

$$|A| = a_{11}\begin{vmatrix} a_{22}^{(1)} & a_{23}^{(1)} \\ a_{32}^{(1)} & a_{33}^{(1)} \end{vmatrix}$$

Repeating the process,

$$|A| = a_{11}a_{22}^{(1)}\begin{vmatrix} 1 & b_{23} \\ 0 & a_{33}^{(2)} \end{vmatrix}$$

which is equivalent to

$$|A| = a_{11}a_{22}^{(1)}|a_{33}^{(2)}|$$

The value of the determinant is then

$$|A| = a_{11}a_{22}^{(1)}a_{33}^{(2)}$$

Except for the last multiplications to obtain the value of the determinant, this process is identical to that performed in the forward course of the elimination method. In accordance with Cramer's rule the solution of a linear system is found by

$$x_i = \frac{|A_i|}{|A|} \qquad i = 1, 2, \ldots, n$$

and requires the evaluation of $n + 1$ determinants of order n. Thus the computation required for the Gauss elimination method to obtain a complete solution only slightly exceeds that required to evaluate a single determinant. This shows the inefficiency of the use of Cramer's rule for the solution of linear sets of equations.

Solution of multiple sets of equations and matrix inversion

The Gauss elimination method may be applied to the simultaneous solution of several sets of linear equations for which only the known constants differ. This is accomplished by adjoining all the constant vectors $y_i^{(1)}$, $y_i^{(2)}, \ldots, y_i^{(n)}$ to the coefficient matrix as follows:

$$\begin{bmatrix} a_{11} & a_{12} & \cdots & a_{1n} & y_1^{(1)} & y_1^{(2)} & \cdots & y_1^{(n)} \\ \cdots & \cdots & \cdots & \cdots & \cdots & \cdots & \cdots & \cdots \\ a_{n1} & a_{n2} & \cdots & a_{nn} & y_n^{(1)} & y_n^{(2)} & \cdots & y_n^{(n)} \end{bmatrix}$$

The elimination process is then applied to the entire array. Each back substitution must be performed separately for the Gauss elimination method, but for the Gauss-Jordan method all solutions are obtained directly.

If many constant vectors are given, it may be advantageous to obtain the inverse of the coefficient matrix and then multiply this inverse by each of the constant vectors in turn to obtain each solution. Given the set of linear equations,

$$\begin{aligned} a_{11}x_1 + a_{12}x_2 + a_{13}x_3 + a_{14}x_4 &= y_1^{(1)} \\ a_{21}x_1 + a_{22}x_2 + a_{23}x_3 + a_{24}x_4 &= y_2^{(1)} \\ a_{31}x_1 + a_{32}x_2 + a_{33}x_3 + a_{34}x_4 &= y_3^{(1)} \\ a_{41}x_1 + a_{42}x_2 + a_{43}x_3 + a_{44}x_4 &= y_4^{(1)} \end{aligned} \tag{7.2.13}$$

the solution is, from equation (7.2.3),

$$\begin{aligned} x_1 &= b_{11}y_1^{(1)} + b_{12}y_2^{(1)} + b_{13}y_3^{(1)} + b_{14}y_4^{(1)} \\ x_2 &= b_{21}y_1^{(1)} + b_{22}y_2^{(1)} + b_{23}y_3^{(1)} + b_{24}y_4^{(1)} \\ x_3 &= b_{31}y_1^{(1)} + b_{32}y_2^{(1)} + b_{33}y_3^{(1)} + b_{34}y_4^{(1)} \\ x_4 &= b_{41}y_1^{(1)} + b_{42}y_2^{(1)} + b_{43}y_3^{(1)} + b_{44}y_4^{(1)} \end{aligned} \tag{7.2.14}$$

where

$$b_{ij} = \frac{A_{ji}}{|A|}$$

and A_{ji} is the cofactor of a_{ji}. The equations (7.2.14) are written in matrix notation as

$$X^{(1)} = A^{-1}Y^{(1)}$$

where

$$A^{-1} = \begin{bmatrix} b_{11} & b_{12} & b_{13} & b_{14} \\ b_{21} & b_{22} & b_{23} & b_{24} \\ b_{31} & b_{32} & b_{33} & b_{34} \\ b_{41} & b_{42} & b_{43} & b_{44} \end{bmatrix}$$

is the inverse of the coefficient matrix A of equations (7.2.13) and $X^{(1)}$ is the solution vector corresponding to the constant vector $Y^{(1)}$. Solutions for the sets with the different constants $y_i^{(2)}, y_i^{(3)}, \ldots, y_i^{(n)}$ are obtained in a similar fashion.

$$\begin{aligned} X^{(2)} &= A^{-1}Y^{(2)} \\ X^{(3)} &= A^{-1}Y^{(3)} \\ &\cdots\cdots\cdots \\ X^{(n)} &= A^{-1}Y^{(n)} \end{aligned}$$

The inverse matrix A^{-1} can be obtained in the following manner. The elements of the first column of A^{-1}, that is, b_{11}, b_{21}, b_{31}, and b_{41}, are equal to the corresponding values obtained for x_1, x_2, x_3, and x_4 when the system of equations (7.2.13) is solved by letting $y_1^{(1)} = 1.0$ and $y_2^{(1)} = y_3^{(1)} = y_4^{(1)} = 0$. This is readily shown by substituting for the values of $y_i^{(1)}$ in equations (7.2.14). Similarly, the second column of A^{-1} is obtained from the solution with $y_2^{(2)} = 1.0$ and $y_1^{(2)} = y_3^{(2)} = y_4^{(2)} = 0$, the third column when $y_3^{(3)} = 1.0$ and $y_1^{(3)} = y_2^{(3)} = y_4^{(3)} = 0$, etc.

The scheme for solving several sets of equations with the same coefficient matrix but with different constant terms is employed to determine the elements of A^{-1}, by using the augmented array

$$\begin{bmatrix} a_{11} & a_{12} & a_{13} & a_{14} & 1 & 0 & 0 & 0 \\ a_{21} & a_{22} & a_{23} & a_{24} & 0 & 1 & 0 & 0 \\ a_{31} & a_{32} & a_{33} & a_{34} & 0 & 0 & 1 & 0 \\ a_{41} & a_{42} & a_{43} & a_{44} & 0 & 0 & 0 & 1 \end{bmatrix}$$

where

$$U = \begin{bmatrix} 1 & 0 & 0 & 0 \\ 0 & 1 & 0 & 0 \\ 0 & 0 & 1 & 0 \\ 0 & 0 & 0 & 1 \end{bmatrix}$$

Using the Gauss-Jordan method, the final array will be

$$\begin{bmatrix} 1 & 0 & 0 & 0 & b_{11} & b_{12} & b_{13} & b_{14} \\ 0 & 1 & 0 & 0 & b_{21} & b_{22} & b_{23} & b_{24} \\ 0 & 0 & 1 & 0 & b_{31} & b_{32} & b_{33} & b_{34} \\ 0 & 0 & 0 & 1 & b_{41} & b_{42} & b_{43} & b_{44} \end{bmatrix}$$

where the unit matrix replaces the coefficient matrix and the b_{ij} are the elements of A^{-1}.

7.3 Example of solution of linear equations by direct methods

The application of direct methods for the solution of linear algebraic equations will be illustrated by calculating the short circuit currents for the network shown in Fig. 7.1.

Problem

For a fault on bus 3

a. Calculate the short circuit currents by the Gauss elimination method.
b. Calculate the short circuit currents and obtain the inverse of the coefficient matrix by the Gauss-Jordan method.

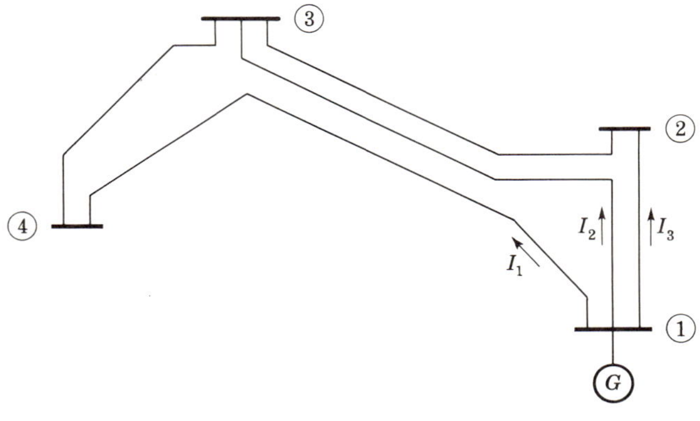

Fig. 7.1 Sample system for solution of simultaneous linear equations.

Solution

The data for the network is given in Table 7.1. The impedance of the generator is 0.01 and the voltage behind the generator is assumed equal to one per unit. The loop equations of the network are

$$1.0 = 0.01(I_1 + I_2 + I_3) + (0.3380 + 0.2790)I_1 + 0.1830I_2$$
$$1.0 = 0.01(I_1 + I_2 + I_3) + 0.4740I_2 + (0.0251 + 0.1360)I_3 + 0.1830I_1$$
$$1.0 = 0.01(I_1 + I_2 + I_3) + (0.5000 + 0.1860)I_3 + (0.0251 + 0.1360)I_2$$

Combining terms

$$0.6270I_1 + 0.1930I_2 + 0.0100I_3 = 1.0$$
$$0.1930I_1 + 0.4840I_2 + 0.1711I_3 = 1.0$$
$$0.0100I_1 + 0.1711I_2 + 0.6960I_3 = 1.0$$

a. The forward course in the Gauss elimination method for the solution of the linear equations is shown in Table 7.2. The original coefficients of the matrix and constant terms are given in part *a*. Included also is the control sum obtained by adding the coefficients and constant term of each row. If the same operations are performed on this sum as on the coefficients and constant term, the control sum will equal at each stage the sum of the elements of the row. This provides a check on the arithmetic operations of the process.

The process is initiated by dividing all elements in the first row by 0.6270, the leading coefficient. The resulting elements are given in the first row of part *b* of Table 7.2. The elements of this new row are multiplied then by 0.1930, the leading coefficient of the second row. The resulting terms are subtracted from the elements of the second row to obtain a new second row as shown in part *b*. Next, the elements of the first row are multiplied by 0.0100 and the resultant terms are subtracted from the elements of the third row. This procedure is repeated for the second and third rows by first dividing the elements of the second row by

Table 7.1 Impedance data for sample system

Self		*Mutual*	
Bus code p–q	*Impedance* $z_{pq,pq}$	*Bus code* r–s	*Impedance* $z_{pq,rs}$
1–2	0.5000	1–3	0.0251
1–3	0.4740	2–3	0.1360
1–4	0.3380	1–3	0.1830
2–3	0.1860		
3–4	0.2790		

Table 7.2 Forward course of Gauss elimination method

I_1	I_2	I_3		*Check sum*
0.6270	0.1930	0.0100	1.0	1.8300
0.1930	0.4840	0.1711	1.0	1.8481
0.0100	0.1711	0.6960	1.0	1.8771
		(*a*)		
1.0	0.307815	0.015949	1.594896	2.918660
0	0.424592	0.168022	0.692185	1.284799
0	0.168022	0.695841	0.984051	1.847913
		(*b*)		
1.0	0.307815	0.015949	1.594896	2.918660
0	1.0	0.395726	1.630236	3.025961
0	0	0.629350	0.710135	1.339485
		(*c*)		
1.0	0.307815	0.015949	1.594896	2.918660
0	1.0	0.395726	1.630236	3.025961
0	0	1.0	1.128363	2.128363
		(*d*)		

0.424592. This process is continued until the equations are transformed into the triangular set of equations shown in part *d* of Table 7.2. This completes the forward course. Then, by back substitution

$$I_3 = 1.128363$$
$$I_2 = 1.630236 - 0.395726 I_3 = 1.183713$$
$$I_1 = 1.594896 - 0.307815 I_2 - 0.015949 I_3 = 1.212535$$

b. The Gauss-Jordan method for obtaining the solution of the equations and the inverse of the coefficient matrix is shown in Table 7.3. Part *a* of this table includes the original coefficient matrix, the unit matrix, constant terms, and control sum. The elimination process is continued until the original matrix is transformed into the unit matrix. The unit matrix is replaced by the coefficients of the inverse matrix as shown in part *d*. The solution values are obtained directly and replace the

Table 7.3 Solution and matrix inversion by Gauss-Jordan method

I_1	I_2	I_3					Check sum
0.6270	0.1930	0.0100	1.0	0	0	1.0	2.8300
0.1930	0.4840	0.1711	0	1.0	0	1.0	2.8481
0.0100	0.1711	0.6960	0	0	1.0	1.0	2.8771
			(*a*)				
1.0	0.307815	0.015949	1.594896	0	0	1.594896	4.513557
0	0.424592	0.168022	−0.307815	1.0	0	0.692185	1.976983
0	0.168022	0.695841	−0.015949	0	1.0	0.984051	2.831964
			(*b*)				
1.0	0	−0.105861	1.818052	−0.724967	0	1.093085	3.080310
0	1.0	0.395726	−0.724967	2.355202	0	1.630236	4.656195
0	0	0.629350	0.105861	−0.395726	1.0	0.710135	2.049621
			(*c*)				
1.0	0	0	1.835859	−0.791531	0.168207	1.212535	3.425070
0	1.0	0	−0.791531	2.604029	−0.628785	1.183713	3.367423
0	0	1.0	0.168207	−0.628785	1.588941	1.128363	3.256727
			(*d*)				

constant terms. The solution is

$$I_1 = 1.212535$$
$$I_2 = 1.183713$$
$$I_3 = 1.128363$$

7.4 Iterative methods for solution of linear algebraic equations

Iterative solution

Given the system of linear equations,

$$\begin{aligned} a_{11}x_1 + a_{12}x_2 + a_{13}x_3 + a_{14}x_4 &= y_1 \\ a_{21}x_1 + a_{22}x_2 + a_{23}x_3 + a_{24}x_4 &= y_2 \\ a_{31}x_1 + a_{32}x_2 + a_{33}x_3 + a_{34}x_4 &= y_3 \\ a_{41}x_1 + a_{42}x_2 + a_{43}x_3 + a_{44}x_4 &= y_4 \end{aligned} \tag{7.4.1}$$

From the first equation,

$$a_{11}x_1 = y_1 - a_{12}x_2 - a_{13}x_3 - a_{14}x_4$$

Dividing by the coefficient a_{11}, x_1 can be expressed as a function of x_2, x_3, and x_4.

$$x_1 = \frac{1}{a_{11}}(y_1 - a_{12}x_2 - a_{13}x_3 - a_{14}x_4)$$

In a similar manner the second, third, and fourth equations can be rewritten so that the set of equations (7.4.1) is in the form

$$\begin{aligned} x_1 &= \frac{1}{a_{11}}(y_1 - a_{12}x_2 - a_{13}x_3 - a_{14}x_4) \\ x_2 &= \frac{1}{a_{22}}(y_2 - a_{21}x_1 - a_{23}x_3 - a_{24}x_4) \\ x_3 &= \frac{1}{a_{33}}(y_3 - a_{31}x_1 - a_{32}x_2 - a_{34}x_4) \\ x_4 &= \frac{1}{a_{44}}(y_4 - a_{41}x_1 - a_{42}x_2 - a_{43}x_3) \end{aligned} \tag{7.4.2}$$

An arbitrary set of initial values can be selected for x_1, x_2, x_3, and x_4, which then may be substituted in the right-hand sides of equations (7.4.2) to calculate new x values. For convenience the initial values can be chosen as $x_1^{(0)} = y_1/a_{11}$, $x_2^{(0)} = y_2/a_{22}$, $x_3^{(0)} = y_3/a_{33}$, $x_4^{(0)} = y_4/a_{44}$. If the results match the initial values within the specified tolerance, a solution has been obtained. If the selected and calculated values differ, a new selection must be tried. The calculated values obtained in the first trial can be used directly as estimates of the unknown x's for the second trial. The process then becomes automatic. Finally, when a selected set of values results in the same calculated values, within the specified tolerance, a solution has been obtained.

This repetitive process for obtaining a solution of a set of equations is known as an *iterative method.* It is applicable to those systems of equations for which the diagonal elements of the coefficient matrix are large in comparison to the off-diagonal elements. In general, for these systems, each successive step of the iterative process results in calculated values of the unknowns that are closer to being equal to the final solution. Then the iteration process is said to converge. For other systems of equations the iterative process may result in calculated values that oscillate or diverge from a solution.

Gauss and Gauss-Seidel iterative methods

Explicitly, the iterative process described requires the selection of initial values

$$x_1^{(0)}, x_2^{(0)}, \ldots, x_n^{(0)}$$

to use in the formulas

$$x_1^{k+1} = \frac{y_1}{a_{11}} - \frac{a_{12}}{a_{11}} x_2{}^k - \frac{a_{13}}{a_{11}} x_3{}^k \cdots - \frac{a_{1n}}{a_{11}} x_n{}^k$$

$$x_2^{k+1} = \frac{y_2}{a_{22}} - \frac{a_{21}}{a_{22}} x_1{}^k - \frac{a_{23}}{a_{22}} x_3{}^k \cdots - \frac{a_{2n}}{a_{22}} x_n{}^k$$

$$\cdots\cdots\cdots\cdots\cdots\cdots\cdots\cdots$$

$$x_n^{k+1} = \frac{y_n}{a_{nn}} - \frac{a_{n1}}{a_{nn}} x_1{}^k - \frac{a_{n2}}{a_{nn}} x_2{}^k \cdots - \frac{a_{n,n-1}}{a_{nn}} x_{n-1}^k$$

in order to calculate a second estimate

$$x_1^{(1)}, x_2^{(1)}, \ldots, x_n^{(1)}$$

The superscript k refers to the iteration count. These calculated values are used in the formulas to obtain a third estimate

$$x_1^{(2)}, x_2^{(2)}, \ldots, x_n^{(2)}$$

and so forth. This technique is referred to as the Gauss iterative method in which the new estimates are substituted only after all equations have been processed.

An alternative method is to make an immediate substitution in subsequent equations for each new value of x as it is obtained. Thus, in the solution of x_i the values used would be

$$x_1^{k+1}, x_2^{k+1}, \ldots, x_{i-1}^{k+1}, x_{i+1}^k, \ldots, x_n{}^k$$

This iteration scheme, called Gauss-Seidel, has the following formulas:

$$x_1^{k+1} = \frac{y_1}{a_{11}} - \frac{a_{12}}{a_{11}} x_2{}^k \cdots\cdots\cdots\cdots - \frac{a_{1n}}{a_{11}} x_n{}^k$$

$$x_2^{k+1} = \frac{y_2}{a_{22}} - \frac{a_{21}}{a_{22}} x_1^{k+1} \cdots\cdots\cdots\cdots - \frac{a_{2n}}{a_{22}} x_n{}^k$$

$$\cdots\cdots\cdots\cdots\cdots\cdots\cdots\cdots$$

$$x_i^{k+1} = \frac{y_i}{a_{ii}} - \frac{a_{i1}}{a_{ii}} x_1^{k+1} \cdots - \frac{a_{i,i-1}}{a_{ii}} x_{i-1}^{k+1} - \frac{a_{i,i+1}}{a_{ii}} x_{i+1}^k \cdots - \frac{a_{in}}{a_{ii}} x_n{}^k$$

$$\cdots\cdots\cdots\cdots\cdots\cdots\cdots\cdots$$

$$x_n^{k+1} = \frac{y_n}{a_{nn}} - \frac{a_{n1}}{a_{nn}} x_1^{k+1} \cdots\cdots\cdots\cdots - \frac{a_{n,n-1}}{a_{nn}} x_{n-1}^{k+1} \tag{7.4.3}$$

Successive approximation and correction techniques

The iterative methods thus far described have been applied for determining new values for the variables. This technique can be referred to as a method of *successive approximation.*

A variation to this is a procedure called the method of *successive correction* in which corrections to the initial selected values for the variables are calculated. For this scheme each variable x_i is the sum of its initial value $x_i^{(0)}$ and the corrections obtained at each step of the process, i.e.,

$$x_i^k = x_i^{(0)} + (x_i^{(1)} - x_i^{(0)}) + (x_i^{(2)} - x_i^{(1)}) \cdots + (x_i^k - x_i^{k-1})$$

Let the difference in the values of x_i obtained in two successive iterations be

$$\alpha_i^{k+1} = x_i^{k+1} - x_i^k$$

Replacing x_i^{k+1} and x_i^k with their equivalent expressions from equations (7.4.3), the formula for calculating successive corrections is

$$\alpha_i^{k+1} = \frac{1}{a_{ii}}(y_i - a_{i1}x_1^{k+1} \cdots - a_{i,i-1}x_{i-1}^{k+1} - a_{i,i+1}x_{i+1}^{k} \cdots - a_{in}x_n^k)$$
$$- \frac{1}{a_{ii}}(y_i - a_{i1}x_1^{k} \cdots - a_{i,i-1}x_{i-1}^{k} - a_{i,i+1}x_{i+1}^{k-1} \cdots - a_{in}x_n^{k-1})$$

Combining terms,

$$\alpha_i^{k+1} = \frac{1}{a_{ii}}\{(y_i - y_i) - a_{i1}(x_1^{k+1} - x_1^k) \cdots - a_{i,i-1}(x_{i-1}^{k+1} - x_{i-1}^k)$$
$$- a_{i,i+1}(x_{i+1}^{k} - x_{i+1}^{k-1}) \cdots - a_{in}(x_n^k - x_n^{k-1})\}$$

The resultant formula in terms of α is

$$\alpha_i^{k+1} = \frac{1}{a_{ii}}(-a_{i1}\alpha_1^{k+1} - a_{i2}\alpha_2^{k+1} \cdots - a_{i,i-1}\alpha_{i-1}^{k+1}$$
$$- a_{i,i+1}\alpha_{i+1}^{k} \cdots - a_{in}\alpha_n^k) \quad (7.4.4)$$

where

$$\alpha_n^{k+1} = \frac{1}{a_{nn}}(-a_{n1}\alpha_1^{k+1} - a_{n2}\alpha_2^{k+1} \cdots - a_{n,n-1}\alpha_{n-1}^{k+1})$$

The iterative process is initiated by selecting a set of initial values for the variables x_1, x_2, x_3, and x_4. These values are substituted in the formulas (7.4.3) to obtain new values for the variables $x_1^{(1)}$, $x_2^{(1)}$, $x_3^{(1)}$, and $x_4^{(1)}$ and the first correction terms $\alpha_1^{(1)}$, $\alpha_2^{(1)}$, $\alpha_3^{(1)}$, and $\alpha_4^{(1)}$ are determined. Equation (7.4.4) is used to calculate successive corrections $\alpha_1^{(2)}$, $\alpha_2^{(2)}$, etc. Each newly calculated α_i^{k+1} is used in successive equations. The process is continued until the values of all correction terms are less than a specified tolerance. Final values of the variables x_i are obtained from the equation

$$x_i = x_i^{(0)} + \sum_k \alpha_i^k$$

Relaxation method

The methods of Gauss and Gauss-Seidel are used to solve linear algebraic equations by successive approximations or corrections. These methods treat the equations in the order they are specified. The method of relaxation makes possible the application of a variety of schemes that alter the order.

Consider the system of equations

$$\begin{aligned}
x_1 + b_{12}x_2 + b_{13}x_3 + b_{14}x_4 - z_1 &= 0\\
b_{21}x_1 + x_2 + b_{23}x_3 + b_{24}x_4 - z_2 &= 0\\
b_{31}x_1 + b_{32}x_2 + x_3 + b_{34}x_4 - z_3 &= 0\\
b_{41}x_1 + b_{42}x_2 + b_{43}x_3 + x_4 - z_4 &= 0
\end{aligned}$$

where $b_{ij} = \dfrac{a_{ij}}{a_{ii}}$

$$z_i = \frac{y_i}{a_{ii}}$$

As in the Gauss and Gauss-Seidel iterative methods an initial set of values is selected for x_1, x_2, x_3, and x_4. Designating these initial values as $x_i^{(0)}$, the values $R_i^{(0)}$ obtained as a result of the initial substitution are:

$$\begin{aligned}
x_1^{(0)} + b_{12}x_2^{(0)} + b_{13}x_3^{(0)} + b_{14}x_4^{(0)} - z_1 &= R_1^{(0)}\\
b_{21}x_1^{(0)} + x_2^{(0)} + b_{23}x_3^{(0)} + b_{24}x_4^{(0)} - z_2 &= R_2^{(0)}\\
b_{31}x_1^{(0)} + b_{32}x_2^{(0)} + x_3^{(0)} + b_{34}x_4^{(0)} - z_3 &= R_3^{(0)}\\
b_{41}x_1^{(0)} + b_{42}x_2^{(0)} + b_{43}x_3^{(0)} + x_4^{(0)} - z_4 &= R_4^{(0)}
\end{aligned}$$

The relaxation procedure consists of estimating new values for the variables until all R_i's, called *residuals*, become negligible. The usual procedure is to select the largest residual R_i^k resulting from the kth iteration and calculate the change in x_i^k required to reduce R_i^k to zero. This change is

$$\Delta x_i^k = -R_i^k$$

and the new estimate for the variable is

$$x_i^{k+1} = x_i^k + \Delta x_i^k$$

Substituting x_i^{k+1} causes each of the other residuals to change. They are recalculated from

$$R_j^{k+1} = R_j^k + b_{ji}\,\Delta x_i^k \qquad j \neq i$$

The largest residual resulting from these calculations is then selected and the process is repeated. Thus, the order in which the variables are reestimated depends solely on the magnitude of the residuals. The solution is obtained when all R_i's are reduced to within a specified tolerance.

A variety of schemes can be employed for reducing the residuals. The method just described reduces the largest residual to zero for an iteration. An alternative would be to reduce several or all residuals by a fixed amount, say δ. This is referred to as *block iteration*. The selection of δ can be such as to result in a residual of opposite sign, i.e., make a change in x_i larger than required for making R_i equal to zero. This is called *overrelaxation*, whereas *underrelaxation* refers to the selection of a new x_i which is not sufficient to reduce R_i to zero. These schemes may be employed interchangeably at each step of the process.

7.5 Example of solution of linear equations by iterative methods

The iterative solution of linear equations will be illustrated using the same problem that was solved in Sec. 7.3.

Problem

Calculate the short circuit currents I_1, I_2, and I_3 for a fault on bus 3 of the network shown in Fig. 7.1 using the following methods:

a. Gauss iteration
b. Gauss-Seidel iteration
c. Relaxation

Solution

Rewrite the loop equations

$$\begin{aligned}
0.6270I_1 + 0.1930I_2 + 0.0100I_3 &= 1.0\\
0.1930I_1 + 0.4840I_2 + 0.1711I_3 &= 1.0\\
0.0100I_1 + 0.1711I_2 + 0.6960I_3 &= 1.0
\end{aligned}$$

for iterative solution as follows:

$$\begin{aligned}
I_1 &= 1.594896 - 0.307815I_2 - 0.015949I_3\\
I_2 &= 2.066116 - 0.398760I_1 - 0.353512I_3\\
I_3 &= 1.436782 - 0.014368I_1 - 0.245833I_2
\end{aligned}$$

Table 7.4 Solution by Gauss iterative method

Iteration count	I_1	I_2	I_3
0	1.0	1.0	1.0
1	1.271132	1.313844	1.176581
2	1.171710	1.143303	1.095532
3	1.225497	1.211601	1.138885
4	1.203783	1.174827	1.121322
5	1.215383	1.189694	1.130675
6	1.210657	1.181763	1.126853
7	1.213160	1.184998	1.128871
8	1.212132	1.183287	1.128039
9	1.212672	1.183991	1.128475
10	1.212448	1.183622	1.128294
11	1.212564	1.183775	1.128389
12	1.212515	1.183695	1.128349

a. Select for an initial estimated solution $I_1^{(0)} = I_2^{(0)} = I_3^{(0)} = 1.0$. Substitute these values in the equations to obtain a new estimate, $I_1^{(1)}$, $I_2^{(1)}$, and $I_3^{(1)}$, as shown in Table 7.4. Repeat the process until the changes in all variables are equal or less than 0.0001. The solution values using the Gauss iterative method are obtained in the twelfth iteration.

b. The results obtained by the Gauss-Seidel iterative method are shown in Table 7.5. The initial values $I_1^{(0)} = I_2^{(0)} = I_3^{(0)} = 1.0$ are substituted in the first equation to obtain $I_1^{(1)} = 1.271132$. Next, $I_1^{(1)} = 1.271132$ and $I_2^{(0)} = I_3^{(0)} = 1.0$ are substituted in the second equation to obtain $I_2^{(1)} = 1.205727$. This process is repeated until the changes in all variables are equal or less than 0.0001. The solution values are obtained in the sixth iteration.

c. The loop equations are rewritten for the relaxation method as follows:

$$
\begin{aligned}
I_1 + 0.307815I_2 + 0.015949I_3 - 1.594896 &= R_1 \\
0.398760I_1 + I_2 + 0.353512I_3 - 2.066116 &= R_2 \\
0.014368I_1 + 0.245833I_2 + I_3 - 1.436782 &= R_3
\end{aligned}
$$

Substitute the initial values $I_1^{(0)} = I_2^{(0)} = I_3^{(0)} = 1.0$ in these equations to calculate the residuals R_i as shown in Table 7.6. Recompute I_2 to reduce the maximum residual $R_2^{(0)} = -0.313844$ to zero as follows:

$$
\begin{aligned}
\Delta I_2^{(0)} &= -R_2^{(0)} = 0.313844 \\
I_2^{(1)} &= I_2^{(0)} + \Delta I_2^{(0)} = 1.313844
\end{aligned}
$$

Table 7.5 Solution by Gauss-Seidel iterative method

Iteration count	I_1	I_2	I_3
0	1.0	1.0	1.0
1	1.271132	1.205727	1.122111
2	1.205858	1.188588	1.127262
3	1.211052	1.184696	1.128145
4	1.212236	1.183912	1.128320
5	1.212474	1.183755	1.128355
6	1.212522	1.183724	1.128362

Table 7.6 Solution by relaxation method

Iteration count	I_1	I_2	I_3	R_1	R_2	R_3
0	1.0	1.0	1.0	−0.271132	−0.313844	−0.176581
1	1.0	1.313844	1.0	−0.174526	0	−0.099428
2	1.174526	1.313844	1.0	0	0.069594	−0.096920
3	1.174526	1.313844	1.096920	0.001546	0.103856	0
4	1.174526	1.209988	1.096920	−0.030423	0	−0.025531
5	1.204949	1.209988	1.096920	0	0.012131	−0.025094
6	1.204949	1.209988	1.122014	0.000400	0.021002	0
7	1.204949	1.188986	1.122014	−0.006064	0	−0.005163
8	1.211013	1.188986	1.122014	0	0.002419	−0.005076
9	1.211013	1.188986	1.127090	0.000081	0.004214	0
10	1.211013	1.184772	1.127090	−0.001216	0	−0.001036
11	1.212229	1.184772	1.127090	0	0.000484	−0.001019
12	1.212229	1.184772	1.128109	0.000016	0.000844	0
13	1.212229	1.183928	1.128109	−0.000244	0	−0.000207
14	1.212473	1.183928	1.128109	0	0.000098	−0.000203
15	1.212473	1.183928	1.128312	0.000003	0.000170	0
16	1.212473	1.183758	1.128312	−0.000050	0	−0.000042

Next, recompute the residuals,

$$
\begin{aligned}
R_1^{(1)} &= R_1^{(0)} + b_{12}\Delta I_2^{(0)} \\
&= -0.271132 + 0.307815(0.313844) \\
&= -0.271132 + 0.096606 = -0.174526 \\
R_3^{(1)} &= R_3^{(0)} + b_{32}\Delta I_2^{(0)} \\
&= -0.176581 + 0.245833(0.313844) \\
&= -0.176581 + 0.077153 = -0.099428
\end{aligned}
$$

Repeat the process for the new maximum residual. Continue until all residuals are equal or less than 0.0001. The solution values are obtained in the sixteenth iteration.

7.6 Methods for solution of nonlinear algebraic equations

Iterative solution

In the Gauss elimination method the system of equations

$$\begin{aligned} f_1(x_1, x_2, \ldots, x_n) &= y_1 \\ f_2(x_1, x_2, \ldots, x_n) &= y_2 \\ \cdots\cdots\cdots\cdots \\ f_n(x_1, x_2, \ldots, x_n) &= y_n \end{aligned} \tag{7.6.1}$$

is reduced to the system

$$\begin{aligned} q_1(x_2, x_3, \ldots, x_n) &= r_1 \\ q_2(x_2, x_3, \ldots, x_n) &= r_2 \\ \cdots\cdots\cdots\cdots \\ q_{n-1}(x_2, x_3, \ldots, x_n) &= r_{n-1} \end{aligned}$$

by the elimination of the variable x_1. This process is continued a sufficient number of times until one equation with one unknown is obtained that can be solved readily for that unknown. The remaining unknowns can be obtained by back substitution in the intermediate relations developed in the elimination process. This method is not always applicable for the solution of nonlinear equations since one or more of the unknowns cannot always be eliminated from the system of equations (7.6.1).

In the iterative methods the system of equations (7.6.1) is written in the form

$$\begin{aligned} x_1 &= y_1 + \phi_1(x_2, x_3, \ldots, x_n) \\ x_2 &= y_2 + \phi_2(x_1, x_3, \ldots, x_n) \\ \cdots\cdots\cdots\cdots \\ x_n &= y_n + \phi_n(x_1, x_2, \ldots, x_{n-1}) \end{aligned}$$

and the computation is initiated by selecting an approximate solution

$$x_1^{(0)}, x_2^{(0)}, \ldots, x_n^{(0)}$$

which is used to obtain a new approximation

$$x_1^{(1)}, x_2^{(1)}, \ldots, x_n^{(1)}$$

This in turn is used to obtain a third approximation, etc. The process is continued until all changes in the x_i's in succeeding iterations are within a specified tolerance. The iterative method is applicable to the solution of both linear and nonlinear systems of equations.

Newton-Raphson method

Given a set of nonlinear equations,

$$
\begin{aligned}
f_1(x_1, x_2, \ldots, x_n) &= y_1 \\
f_2(x_1, x_2, \ldots, x_n) &= y_2 \\
\cdots\cdots\cdots&\cdots\cdots \\
f_n(x_1, x_2, \ldots, x_n) &= y_n
\end{aligned}
\tag{7.6.2}
$$

and the initial estimate for the solution vector

$$x_1^{(0)}, x_2^{(0)}, \ldots, x_n^{(0)}$$

Assume $\Delta x_1, \Delta x_2, \ldots, \Delta x_n$ are the corrections required for $x_1^{(0)}, x_2^{(0)}, \ldots, x_n^{(0)}$ respectively, so that the equations (7.6.2) are solved, i.e.,

$$
\begin{aligned}
f_1(x_1^{(0)} + \Delta x_1, x_2^{(0)} + \Delta x_2, \ldots, x_n^{(0)} + \Delta x_n) &= y_1 \\
f_2(x_1^{(0)} + \Delta x_1, x_2^{(0)} + \Delta x_2, \ldots, x_n^{(0)} + \Delta x_n) &= y_2 \\
\cdots\cdots\cdots\cdots\cdots\cdots\cdots&\cdots\cdots \\
f_n(x_1^{(0)} + \Delta x_1, x_2^{(0)} + \Delta x_2, \ldots, x_n^{(0)} + \Delta x_n) &= y_n
\end{aligned}
\tag{7.6.3}
$$

Each equation of the set (7.6.3) can be expanded by Taylor's theorem for a function of two or more variables. For example, the following is obtained for the first equation:

$$
\begin{aligned}
f_1(x_1^{(0)} + \Delta x_1, x_2^{(0)} + \Delta x_2, \ldots, x_n^{(0)} + \Delta x_n) &= f_1(x_1^{(0)}, x_2^{(0)}, \ldots, x_n^{(0)}) \\
&\quad + \Delta x_1 \left.\frac{\partial f_1}{\partial x_1}\right|_0 + \Delta x_2 \left.\frac{\partial f_1}{\partial x_2}\right|_0 \cdots + \Delta x_n \left.\frac{\partial f_1}{\partial x_n}\right|_0 + \Phi_1
\end{aligned}
$$

where Φ_1 is a function of higher powers of $\Delta x_1, \Delta x_2, \ldots, \Delta x_n$ and second, third, etc., derivatives of the function f_1. If the initial estimate for x_i's is near the solution value, the Δx_i's will be relatively small and all terms of higher powers can be neglected. The linear set of equations resulting is as follows:

$$
\begin{aligned}
f_1(x_1^{(0)}, x_2^{(0)}, \ldots, x_n^{(0)}) + \Delta x_1 \left.\frac{\partial f_1}{\partial x_1}\right|_0 + \Delta x_2 \left.\frac{\partial f_1}{\partial x_2}\right|_0 \cdots + \Delta x_n \left.\frac{\partial f_1}{\partial x_n}\right|_0 &= y_1 \\
f_2(x_1^{(0)}, x_2^{(0)}, \ldots, x_n^{(0)}) + \Delta x_1 \left.\frac{\partial f_2}{\partial x_1}\right|_0 + \Delta x_2 \left.\frac{\partial f_2}{\partial x_2}\right|_0 \cdots + \Delta x_n \left.\frac{\partial f_2}{\partial x_n}\right|_0 &= y_2 \\
\cdots\cdots\cdots\cdots\cdots\cdots\cdots\cdots\cdots\cdots\cdots\cdots\cdots\cdots&\cdots\cdots \\
f_n(x_1^{(0)}, x_2^{(0)}, \ldots, x_n^{(0)}) + \Delta x_1 \left.\frac{\partial f_n}{\partial x_1}\right|_0 + \Delta x_2 \left.\frac{\partial f_n}{\partial x_2}\right|_0 \cdots + \Delta x_n \left.\frac{\partial f_n}{\partial x_n}\right|_0 &= y_n
\end{aligned}
\tag{7.6.4}
$$

In matrix form, equations (7.6.4) are

$$\begin{bmatrix} y_1 - f_1(x_1^{(0)}, x_2^{(0)}, \ldots, x_n^{(0)}) \\ y_2 - f_2(x_1^{(0)}, x_2^{(0)}, \ldots, x_n^{(0)}) \\ \cdots\cdots\cdots\cdots \\ y_n - f_n(x_1^{(0)}, x_2^{(0)}, \ldots, x_n^{(0)}) \end{bmatrix} = \begin{bmatrix} \left.\frac{\partial f_1}{\partial x_1}\right|_0 & \left.\frac{\partial f_1}{\partial x_2}\right|_0 & \cdots & \left.\frac{\partial f_1}{\partial x_n}\right|_0 \\ \left.\frac{\partial f_2}{\partial x_1}\right|_0 & \left.\frac{\partial f_2}{\partial x_2}\right|_0 & \cdots & \left.\frac{\partial f_2}{\partial x_n}\right|_0 \\ \cdots & \cdots & \cdots & \cdots \\ \left.\frac{\partial f_n}{\partial x_1}\right|_0 & \left.\frac{\partial f_n}{\partial x_2}\right|_0 & \cdots & \left.\frac{\partial f_n}{\partial x_n}\right|_0 \end{bmatrix} \begin{bmatrix} \Delta x_1 \\ \Delta x_2 \\ \cdots \\ \Delta x_n \end{bmatrix}$$

or

$$D = JC$$

where J is the Jacobian for the functions f_i and C is the change vector Δx_i. The elements of the matrices D and J are evaluated by substituting the current values of x_i's. Hence a solution for the Δx_i can be obtained by the application of any method for the solution of a system of linear equations. The new values for x_i's are calculated from

$$x_i^{(1)} = x_i^{(0)} + \Delta x_i$$

The process is repeated until two successive values for each x_i differ only by a specified tolerance. In this process the elements of J can be reevaluated each iteration, or only every kth iteration provided the Δx_i are changing slowly.

7.7 Example of solution of nonlinear equations

Problem

Solve the nonlinear equations

$$y^2 - 4x - 4 = 0$$
$$2y - x - 2 = 0$$

using the Newton-Raphson method.

Solution

The curves of the two equations are shown in Fig. 7.2 and the point of intersection gives the solution. Selecting the point $x^{(0)} = -1$ and

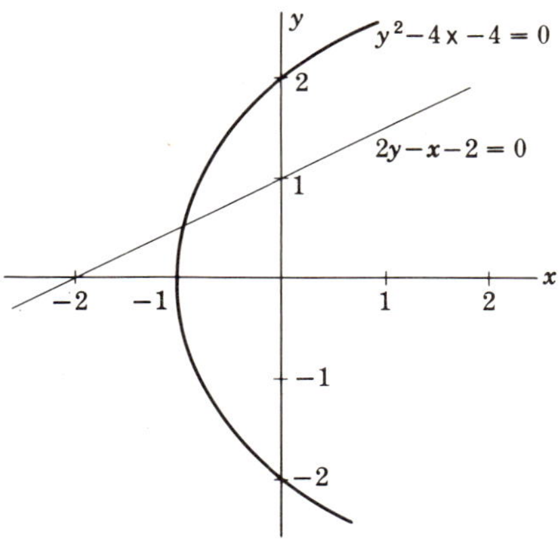

***Fig. 7.2** Graph of nonlinear equations.*

$y^{(0)} = 1$ as the initial approximation and substituting,

$$f(x^{(0)},y^{(0)}) = y^2 - 4x - 4 = 1 + 4 - 4 = 1$$
$$g(x^{(0)},y^{(0)}) = 2y - x - 2 = 2 + 1 - 2 = 1$$
$$\frac{\partial f}{\partial x} = -4$$
$$\frac{\partial f}{\partial y} = 2y = 2$$
$$\frac{\partial g}{\partial x} = -1$$
$$\frac{\partial g}{\partial y} = 2$$

Substituting in

$$f(x^{(0)},y^{(0)}) + \Delta x\,\frac{\partial f}{\partial x} + \Delta y\,\frac{\partial f}{\partial y} = 0$$
$$g(x^{(0)},y^{(0)}) + \Delta x\,\frac{\partial g}{\partial x} + \Delta y\,\frac{\partial g}{\partial y} = 0$$

the following linear simultaneous equations are obtained:

$$1 - 4\Delta x + 2\Delta y = 0$$
$$1 - \Delta x + 2\Delta y = 0$$

Solving,

$$\Delta x = 0$$
$$\Delta y = -0.5$$

Thus

$$x^{(1)} = x^{(0)} + \Delta x = -1$$
$$y^{(1)} = y^{(0)} + \Delta y = 0.5$$

Repeating the process with the new estimates,

$$f(x^{(1)},y^{(1)}) = 0.25 + 4 - 4 = 0.25$$
$$g(x^{(1)},y^{(1)}) = 1 + 1 - 2 = 0$$
$$\frac{\partial f}{\partial x} = -4$$
$$\frac{\partial f}{\partial y} = 1$$
$$\frac{\partial g}{\partial x} = -1$$
$$\frac{\partial g}{\partial y} = 2$$

and the linear equations are

$$0.25 - 4\Delta x + \Delta y = 0$$
$$-\Delta x + 2\Delta y = 0$$

Solving,

$$\Delta x = 0.07143$$
$$\Delta y = 0.03571$$

and

$$x^{(2)} = x^{(1)} + \Delta x = -1.0 + 0.07143 = -0.92857$$
$$y^{(2)} = y^{(1)} + \Delta y = 0.5 + 0.03571 = 0.53571$$

Substituting $x^{(2)}$ and $y^{(2)}$ in the original equations,

$$f(x^{(2)},y^{(2)}) = 0.28699 + 3.71428 - 4 = 0.00127$$
$$g(x^{(2)},y^{(2)}) = 1.07142 + 0.92857 - 2 = -0.00001$$

The values of x and y are close to the solution values. Form the linear equations using the same coefficients $\partial f/\partial x$, $\partial f/\partial y$, $\partial g/\partial x$, and $\partial g/\partial y$ to obtain

$$0.00127 - 4\Delta x + \Delta y = 0$$
$$-0.00001 - \Delta x + 2\Delta y = 0$$

Solving,

$$\Delta x = 0.00035$$
$$\Delta y = 0.00018$$

Since the changes are within the tolerance of 0.0005, the final solution is:

$$x^{(3)} = x^{(2)} + \Delta x = -0.92857 + 0.00035 = -0.92822$$
$$y^{(3)} = y^{(2)} + \Delta y = 0.53571 + 0.00018 = 0.53589$$

Substituting $x^{(3)}$ and $y^{(3)}$ in the original equation for the final check,

$$f(x^{(3)},y^{(3)}) = 0.28718 + 3.71288 - 4 = 0.00006$$
$$g(x^{(3)},y^{(3)}) = 1.07178 + 0.92822 - 2 = 0.00000$$

The equations have a second solution that can be obtained by choosing the initial values from the graph in Fig. 7.2 near the coordinates for a second point of intersection.

7.8 Comparison of methods

The best method to apply for the solution of any set of equations representing a physical system depends on the characteristics of the system.

The direct methods, in general, require many arithmetic calculations, but the number of operations required can be determined in advance. This facilitates the evaluation of the efficiency of these methods. However, the round-off errors are accumulated at each stage, and for a large system the error may increase so that the solution is invalid.

The iterative methods are most successful when each diagonal element in the coefficient matrix is large in magnitude relative to the other elements in its row. Round-off errors in these methods tend to be corrected at successive stages of the process.

The advantage associated with the relaxation method is that the residual of largest magnitude is known at each stage and the modification of the corresponding x is performed to reduce this residual to zero. A preassigned cyclic order is not required, therefore, as in other iterative methods. The decision process to take advantage of this technique requires more computer logic and machine time for each iteration.

The Newton-Raphson method is applicable when the truncation error obtained by neglecting the functions Φ_i is not significant, that is, if all Δx_i are small in magnitude and the initial values selected for x_i are reasonably close to an actual solution.

Problems

7.1 Show that the equations

$$3x_1 + ax_2 + x_3 = 5$$
$$x_1 - 2x_2 - 2ax_3 = -3$$
$$4x_1 + x_2 + ax_3 = 6$$

have

a. A unique solution when $a \neq \pm 1$

b. No solution when $a = -1$

c. An infinite number of solutions when $a = 1$

7.2 Solve by Cramer's rule without introducing round-off error the following set of equations:

$$\begin{aligned} x + y + z &= 1.75 \\ 0.5x + 2y + 3z &= 2.25 \\ 2x + 3y + 4z &= 4.50 \end{aligned}$$

7.3 Solve the following system of equations by Crout's method:

$$\begin{aligned} 4x + y + 2z &= 1 \\ 2x - 2y + z &= 3 \\ x + 2y - 0.5z &= -2 \end{aligned}$$

7.4 Solve the following system of equations by the Gauss elimination method:

$$\begin{aligned} 9.3x_1 + 3.0x_2 - 2.1x_3 + x_4 &= 9.3 \\ 3.0x_1 + 6.0x_2 + 1.2x_3 + 0.5x_4 &= 9.0 \\ -2.1x_1 + 1.2x_2 + 8.4x_3 + 0.4x_4 &= 4.2 \\ x_1 + 0.5x_2 + 0.4x_3 + 5.0x_4 &= 5.5 \end{aligned}$$

7.5 Starting with the initial values,

$$\begin{aligned} x_1^{(0)} &= 1.0 \\ x_2^{(0)} &= 1.5 \\ x_3^{(0)} &= 0.5 \\ x_4^{(0)} &= 1.1 \end{aligned}$$

solve the system of equations given in Prob. 7.4 by the Gauss-Seidel iterative method.

7.6 Solve the system of equations given in Prob. 7.4 by matrix inversion.

7.7 Starting with initial values equal to zero, solve the following system of equations by the relaxation method:

$$\begin{aligned} 6x_1 - 2x_2 + 3x_3 + 93 &= R_1 \\ 2x_1 + 5x_2 - x_3 + 49 &= R_2 \\ 2x_1 - x_2 - 10x_3 + 185 &= R_3 \end{aligned}$$

7.8 Solve by the Newton-Raphson method the following system of equations:

$$\begin{aligned} x^2 + xy + z &= 1.20 \\ y^2 + yz + x &= 1.76 \\ x + 2z &= 1.50 \end{aligned}$$

using the initial values

$$x^{(0)} = 0$$
$$y^{(0)} = 1$$
$$z^{(0)} = 0.75$$

7.9 Solve the following system of equations and obtain the inverse of the coefficient matrix by the Gauss-Jordan method:

$$\begin{aligned} x_1 + 2x_2 - 2x_3 + x_4 &= 4 \\ 2x_1 + 5x_2 - 2x_3 + 3x_4 &= 7 \\ -2x_1 - 2x_2 + 5x_3 + 3x_4 &= -1 \\ x_1 + 3x_2 + 3x_3 + 2x_4 &= 0 \end{aligned}$$

Bibliography

Crandall, Stephen H.: "Engineering Analysis—A Survey of Numerical Procedures," McGraw-Hill Book Company, New York, 1956.

Faddeev, D. K., and V. N. Faddeeva: "Computational Methods of Linear Algebra," W. H. Freeman and Company, San Francisco, 1963.

Faddeeva, V. N.: "Computational Methods of Linear Algebra," Dover Publications, Inc., New York, 1959.

Hildebrand, F. B.: "Introduction to Numerical Analysis," McGraw-Hill Book Company, New York, 1956.

Kunz, Kaiser S.: "Numerical Analysis," McGraw-Hill Book Company, New York, 1957.

McCormick, John M., and Mario G. Salvadori: "Numerical Methods in FORTRAN," Prentice-Hall, Inc., Englewood Cliffs, N.J., 1964.

McCracken, Daniel D., and William S. Dorn: "Numerical Methods and FORTRAN Programming," John Wiley & Sons, Inc., New York, 1964.

Milne, William Edmund: "Numerical Calculus," Princeton University Press, Princeton, N.J., 1949.

Salvadori, Mario G., and Melvin L. Baron: "Numerical Methods in Engineering," Prentice-Hall, Inc., Englewood Cliffs, N.J., 1952.

Scarborough, James B.: "Numerical Mathematical Analysis," The Johns Hopkins Press, Baltimore, 1950.

chapter 8
Load flow studies

8.1 Introduction

Load flow calculations provide power flows and voltages for a specified power system subject to the regulating capability of generators, condensers, and tap changing under load transformers as well as specified net interchange between individual operating systems. This information is essential for the continuous evaluation of the current performance of a power system and for analyzing the effectiveness of alternative plans for system expansion to meet increased load demand. These analyses require the calculation of numerous load flows for both normal and emergency operating conditions.

The load flow problem consists of the calculation of power flows and voltages of a network for specified terminal or bus conditions. A single-phase representation is adequate since power systems are usually balanced. Associated with each bus are four quantities: the real and reactive power, the voltage magnitude, and the phase angle. Three types of buses are represented in the load flow calculation and at a bus, two of the four quantities are specified. It is necessary to select one bus, called the slack bus, to provide the additional real and reactive power to supply the transmission losses, since these are unknown until the final solution is obtained. At this bus the voltage magnitude and phase angle are specified. The remaining buses of the system are designated either as voltage controlled buses or load buses. The real power and voltage magnitude are specified at a voltage controlled bus. The real and reactive powers are specified at a load bus.

Network connections are described by using code numbers assigned to each bus. These numbers specify the terminals of transmission lines and transformers. Code numbers are used also to identify the types of buses, the location of static capacitors, shunt reactors, and those network elements in which off-nominal turns ratios of transformers are to be represented.

The two primary considerations in the development of an effective engineering computer program are: (1) the formulation of a mathematical description of the problem; and (2) the application of a numerical method for a solution. The analysis of the problem must also consider the interrelation between these two factors.

The mathematical formulation of the load flow problem results in a system of algebraic nonlinear equations. These equations can be established by using either the bus or loop frame of reference. The coefficients of the equations depend on the selection of the independent variables, i.e., voltages or currents. Thus, either the admittance or impedance network matrices can be used.

Early approaches to the digital solution of load flows employed the loop frame of reference in admittance form. The loop admittance matrix was obtained by a matrix inversion. These methods did not have widespread application because of the tedious data preparation required to specify the network loops. Furthermore, the required matrix inversion was time-consuming and had to be repeated for each subsequent case involving network changes. Later approaches used the bus frame of reference in the admittance form to describe the system. This method gained widespread application because of the simplicity of data preparation and the ease with which the bus admittance matrix could be formed and modified for network changes in subsequent cases. Also, combinations of voltages and currents have been used as the independent variables. This formulation uses a hybrid matrix consisting of impedance, admittance, current-ratio, and voltage-ratio elements. The ability to formulate efficiently the network matrices has led to the use of the bus frame of reference in the impedance form. However, the majority of load flow programs for large power system studies still employ methods using the bus admittance matrix. This approach remains the most economical from the point of view of computer time and memory requirements.

The solution of the algebraic equations describing the power system are based on an iterative technique because of their nonlinearity. The solution must satisfy Kirchhoff's laws, i.e., the algebraic sum of all flows at a bus must equal zero, and the algebraic sum of all voltages in a loop must equal zero. One or the other of these laws is used as a test for convergence of the solution in the iterative computational method. Other constraints placed on the solution are: the capability limits of reactive power sources; the tap setting range of tap changing under load transformers; and the specified power interchange between interconnected systems.

8.2 Power system equations

Network performance equations

The equation describing the performance of the network of a power system using the bus frame of reference in impedance form is

$$\bar{E}_{BUS} = Z_{BUS}\bar{I}_{BUS} \tag{8.2.1}$$

or in admittance form is

$$\bar{I}_{BUS} = Y_{BUS}\bar{E}_{BUS} \tag{8.2.2}$$

The bus impedance and admittance matrices can be formed for the network including the ground bus. The elements of the matrices, then, will include the effects of shunt elements to ground such as static capacitors and reactors, line charging, and shunt elements of transformer equivalents. When the ground bus is included and selected as the reference node, the bus voltages in the network performance equations (8.2.1) and (8.2.2) are measured with respect to ground.

If the ground bus is not included in the network, the elements of the bus impedance and admittance matrices will not include the effects of shunt elements and one of the buses of the network must be selected as the reference node. In this case, the effects of shunt elements are treated as current sources at the buses of the network and the bus voltages in the performance equations (8.2.1) and (8.2.2) are measured with respect to the selected reference bus.

Using the loop frame of reference, the network performance equation in impedance form is

$$\bar{E}_{LOOP} = Z_{LOOP}\bar{I}_{LOOP}$$

or in admittance form is

$$\bar{I}_{LOOP} = Y_{LOOP}\bar{E}_{LOOP}$$

When the loop impedance and admittance matrices are formed for the network not including shunt elements, the dimension of the matrices is $l \times l$, where l is the number of links or basic loops calculated from

$$l = e - n + 1$$

e is the number of elements, excluding the shunt connections, and n is the number of nodes. In this case, the effects of shunt elements are treated as current sources at the buses of the network.

If the shunt elements e_s are included in forming the loop matrices, the number of elements of the network is increased by e_s. The total number of elements is, then, $e + e_s$ and the number of nodes is increased to $n + 1$. Consequently, the number of loops and the dimension of the loop matrices are increased by $e_s - 1$.

The different forms of network equations are summarized in Table 8.1.

Table 8.1 ***Network equations***

Frame of reference	Parameter form: Impedance	Parameter form: Admittance
Bus	$\bar{E}_{BUS} = Z_{BUS}\bar{I}_{BUS}$	$\bar{I}_{BUS} = Y_{BUS}\bar{E}_{BUS}$
Loop	$\bar{E}_{LOOP} = Z_{LOOP}\bar{I}_{LOOP}$	$\bar{I}_{LOOP} = Y_{LOOP}\bar{E}_{LOOP}$

Bus loading equations

The real and reactive power at any bus p is

$$P_p - jQ_p = E_p^* I_p$$

and the current is

$$I_p = \frac{P_p - jQ_p}{E_p^*} \tag{8.2.3}$$

where I_p is positive when flowing into the system.

In the formulation of the network equation, if the shunt elements to ground are included in the parameter matrix, then equation (8.2.3) is the total current at the bus. On the other hand, if the shunt elements are not included in the parameter matrix, the total current at bus p is

$$I_p = \frac{P_p - jQ_p}{E_p^*} - y_p E_p$$

where y_p is the total shunt admittance at the bus and $y_p E_p$ is the shunt current flowing from bus p to ground.

Line flow equations

After the iterative solution of bus voltages is completed, line flows can be calculated. The current at bus p in the line connecting bus p to q is

$$i_{pq} = (E_p - E_q)y_{pq} + E_p \frac{y'_{pq}}{2}$$

where y_{pq} = line admittance

y'_{pq} = total line charging admittance

$E_p \frac{y'_{pq}}{2}$ = current contribution at bus p due to line charging

The power flow, real and reactive, is

$$P_{pq} - jQ_{pq} = E_p^* i_{pq}$$

or

$$P_{pq} - jQ_{pq} = E_p^*(E_p - E_q)y_{pq} + E_p^* E_p \frac{y'_{pq}}{2} \tag{8.2.4}$$

where at bus p the real power flow from bus p to q is P_{pq} and the reactive is Q_{pq}. Similarly, at bus q the power flow from q to p is

$$P_{qp} - jQ_{qp} = E_q^*(E_q - E_p)y_{pq} + E_q^* E_q \frac{y'_{pq}}{2} \tag{8.2.5}$$

The power loss in line p–q is the algebraic sum of the power flows determined from equations (8.2.4) and (8.2.5).

8.3 Solution techniques

Gauss iterative method using Y_{BUS}

The solution of the load flow problem is initiated by assuming voltages for all buses except the slack bus, where the voltage is specified and remains fixed. Then, currents are calculated for all buses except the slack bus s from the bus loading equation

$$I_p = \frac{P_p - jQ_p}{E_p^*} \qquad p = 1, 2, \ldots, n \quad p \neq s \tag{8.3.1}$$

where n is the number of buses in the network. The performance of the network can be obtained from the equation

$$\bar{I}_{BUS} = Y_{BUS} \bar{E}_{BUS} \tag{8.3.2}$$

Selecting the ground as the reference bus, a set of $n - 1$ simultaneous equations can be written in the form

$$E_p = \frac{1}{Y_{pp}} \left(I_p - \sum_{\substack{q=1 \\ q \neq p}}^{n} Y_{pq} E_q \right) \qquad p = 1, 2, \ldots, n \quad p \neq s \tag{8.3.3}$$

The bus currents calculated from equation (8.3.1), the slack bus voltage, and the estimated bus voltages are substituted into equation (8.3.3) to obtain a new set of bus voltages. These new voltages are used in equation (8.3.1) to recalculate bus currents for a subsequent solution of equation (8.3.3). The process is continued until changes in all bus voltages are negligible. After the voltage solution has been obtained, the power at the slack bus and line flows can be calculated.

The network equation (8.3.3) and the bus loading equation (8.3.1) can be combined to obtain

$$E_p = \frac{1}{Y_{pp}}\left(\frac{P_p - jQ_p}{E_p^*} - \sum_{\substack{q=1\\q\neq p}}^{n} Y_{pq}E_q\right) \qquad \begin{matrix} p = 1, 2, \ldots, n \\ p \neq s \end{matrix} \tag{8.3.4}$$

which involves only bus voltages as variables. Formulating the load flow problem in this manner results in a set of nonlinear equations that can be solved by an iterative method.

A significant reduction in the computing time for a solution will be obtained by performing as many arithmetic operations as possible before initiating the iterative calculation. Letting

$$\frac{1}{Y_{pp}} = L_p$$

equation (8.3.4) can be written

$$E_p = \frac{(P_p - jQ_p)L_p}{E_p^*} - \sum_{\substack{q=1\\q\neq p}}^{n} Y_{pq}L_pE_q \qquad \begin{matrix} p = 1, 2, \ldots, n \\ p \neq s \end{matrix} \tag{8.3.5}$$

Letting

$$(P_p - jQ_p)L_p = KL_p$$

and

$$Y_{pq}L_p = YL_{pq}$$

then, the bus voltage equation (8.3.5) becomes

$$E_p = \frac{KL_p}{E_p^*} - \sum_{\substack{q=1\\q\neq p}}^{n} YL_{pq}E_q \qquad \begin{matrix} p = 1, 2, \ldots, n \\ p \neq s \end{matrix} \tag{8.3.6}$$

The normal procedure for a load flow study is to assume a balanced system and to use a single-phase representation equivalent to the positive sequence network. Since there is no mutual coupling, the bus admittance matrix can be formed by inspection and many of its elements will be zero. Selecting bus 2 as the slack bus in the system shown in Fig. 8.1, the for-

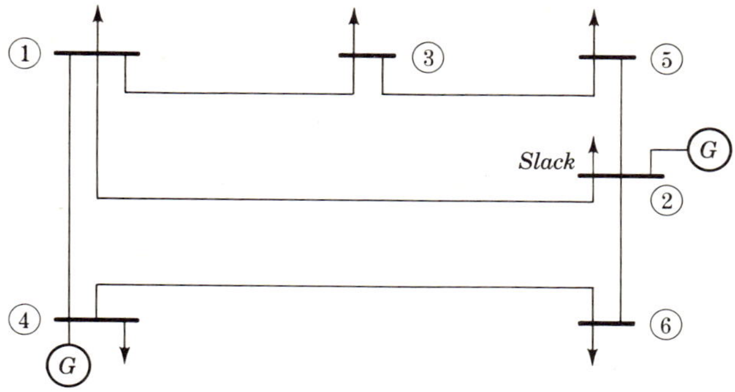

$Y_{BUS} =$

Bus p \ Bus q	1	2	3	4	5	6
1	Y_{11}	Y_{12}	Y_{13}	Y_{14}		
2	Y_{21}	Y_{22}			Y_{25}	Y_{26}
3	Y_{31}		Y_{33}		Y_{35}	
4	Y_{41}			Y_{44}		Y_{46}
5		Y_{52}	Y_{53}		Y_{55}	
6		Y_{62}		Y_{64}		Y_{66}

Fig. 8.1 Single line diagram and bus admittance matrix of a power system.

mulas for the Gauss iterative solution are

$$E_1^{k+1} = \frac{KL_1}{(E_1^k)^*} - YL_{12}E_2 - YL_{13}E_3^k - YL_{14}E_4^k$$

$$E_2 = \text{specified fixed value}$$

$$E_3^{k+1} = \frac{KL_3}{(E_3^k)^*} - YL_{31}E_1^k - YL_{35}E_5^k$$

$$E_4^{k+1} = \frac{KL_4}{(E_4^k)^*} - YL_{41}E_1^k - YL_{46}E_6^k$$

$$E_5^{k+1} = \frac{KL_5}{(E_5^k)^*} - YL_{52}E_2 - YL_{53}E_3^k$$

$$E_6^{k+1} = \frac{KL_6}{(E_6^k)^*} - YL_{62}E_2 - YL_{64}E_4^k$$

where the superscript k refers to the iteration count. The sequence of

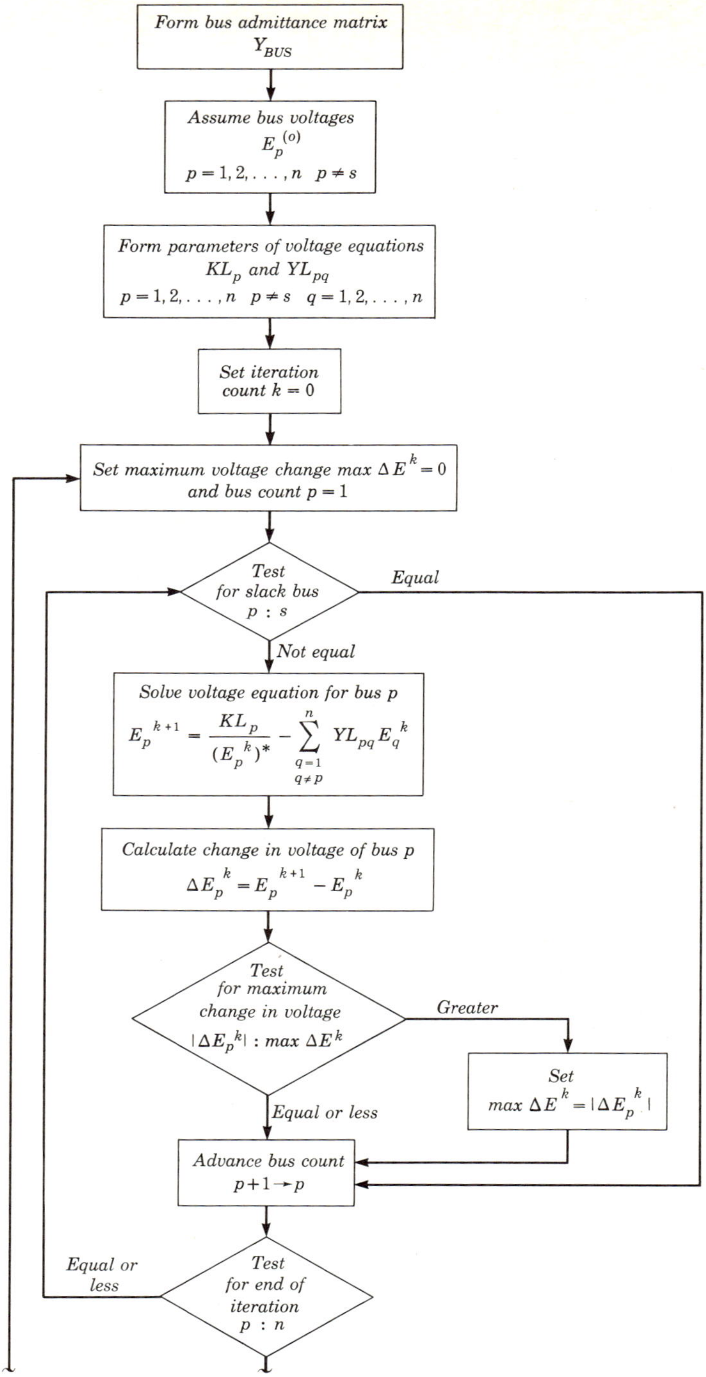

Fig. 8.2 Load flow solution by the Gauss iterative method using Y_{BUS}.

steps for the load flow solution by the Gauss iterative method is shown in Fig. 8.2.

Gauss-Seidel iterative method using Y_{BUS}

The bus voltage equation (8.3.6) also can be solved by the Gauss-Seidel iterative method (Glimn and Stagg, 1957). In this method the new calculated voltage E_p^{k+1} immediately replaces $E_p{}^k$ and is used in the solution of the subsequent equations. For the system shown in Fig. 8.1, the formulas for this method are

$$E_1^{k+1} = \frac{KL_1}{(E_1{}^k)^*} - YL_{12}E_2 - YL_{13}E_3{}^k - YL_{14}E_4{}^k$$

$$E_2 = \text{specified fixed value}$$

$$E_3^{k+1} = \frac{KL_3}{(E_3{}^k)^*} - YL_{31}E_1^{k+1} - YL_{35}E_5{}^k$$

$$E_4^{k+1} = \frac{KL_4}{(E_4{}^k)^*} - YL_{41}E_1^{k+1} - YL_{46}E_6{}^k$$

$$E_5^{k+1} = \frac{KL_5}{(E_5{}^k)^*} - YL_{52}E_2 - YL_{53}E_3^{k+1}$$

$$E_6^{k+1} = \frac{KL_6}{(E_6{}^k)^*} - YL_{62}E_2 - YL_{64}E_4^{k+1}$$

The sequence of steps for the load flow solution by the Gauss-Seidel iterative method is shown in Fig. 8.3.

Relaxation method using Y_{BUS}

The equations for bus currents are used for the solution of the load flow problem by the relaxation method (Jordan, 1957). From the network

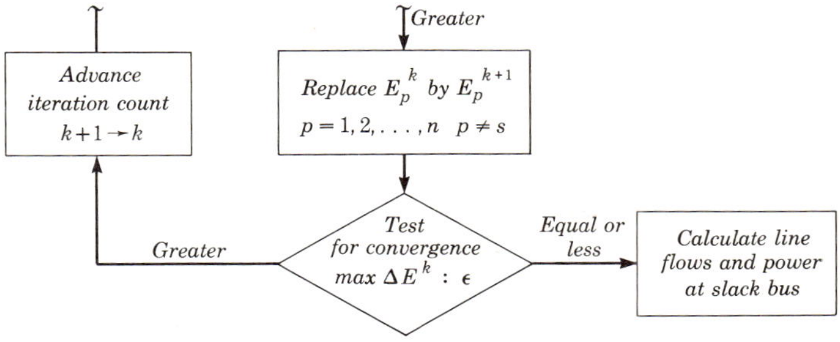

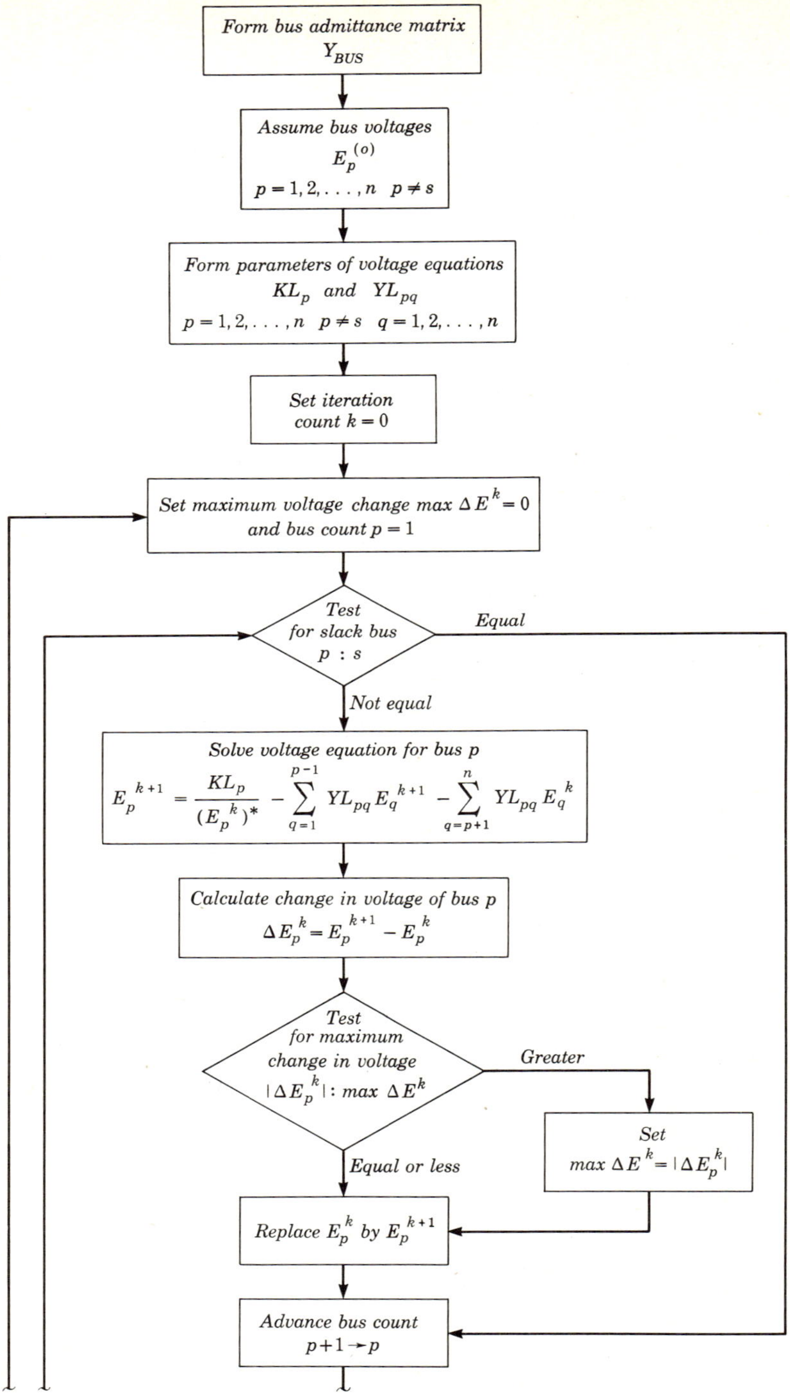

Fig. 8.3 Load flow solution by the Gauss-Seidel iterative method using Y_{BUS}.

performance equation (8.3.2), the current at bus p is

$$I_p = Y_{p1}E_1 + Y_{p2}E_2 \cdots + Y_{pp}E_p \cdots + Y_{pn}E_n$$

This equation can be rewritten as

$$Y_{p1}E_1 + Y_{p2}E_2 \cdots + Y_{pp}E_p \cdots + Y_{pn}E_n - I_p = R_p$$

where R_p is a residual and represents the error in current at bus p resulting from the assumed voltage solution. For the system shown in Fig. 8.1, the formulas for the relaxation method are

$$\begin{aligned} Y_{11}E_1^k + Y_{12}E_2 + Y_{13}E_3^k + Y_{14}E_4^k - I_1^k &= R_1^k \\ Y_{31}E_1^k + Y_{33}E_3^k + Y_{35}E_5^k - I_3^k &= R_3^k \\ Y_{41}E_1^k + Y_{44}E_4^k + Y_{46}E_6^k - I_4^k &= R_4^k \\ Y_{52}E_2 + Y_{53}E_3^k + Y_{55}E_5^k - I_5^k &= R_5^k \\ Y_{62}E_2 + Y_{64}E_4^k + Y_{66}E_6^k - I_6^k &= R_6^k \end{aligned} \tag{8.3.7}$$

where the equation for the slack bus has been excluded since E_2 is specified and remains fixed.

With the set of bus voltages

$$E_1^{(0)}, E_2, E_3^{(0)}, E_4^{(0)}, E_5^{(0)}, E_6^{(0)}$$

bus currents are calculated from equation (8.3.1) and then bus residuals are calculated from equations (8.3.7). A voltage correction is obtained for that bus at which the residual R_p is a maximum. If the current at bus p remained constant, the residual R_p would be reduced to zero by the voltage correction

$$\Delta E_p^k = -\frac{R_p^k}{Y_{pp}}$$

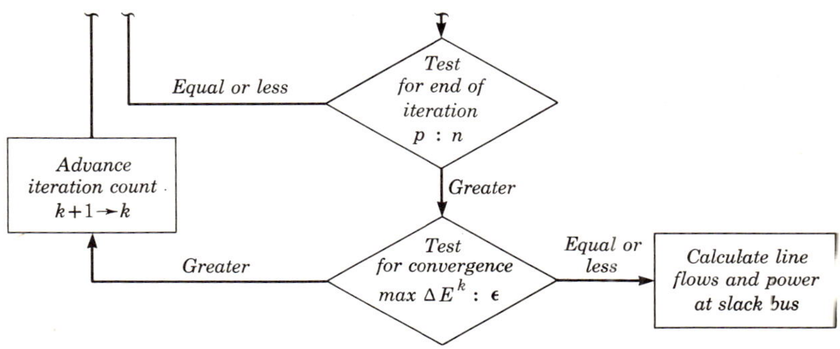

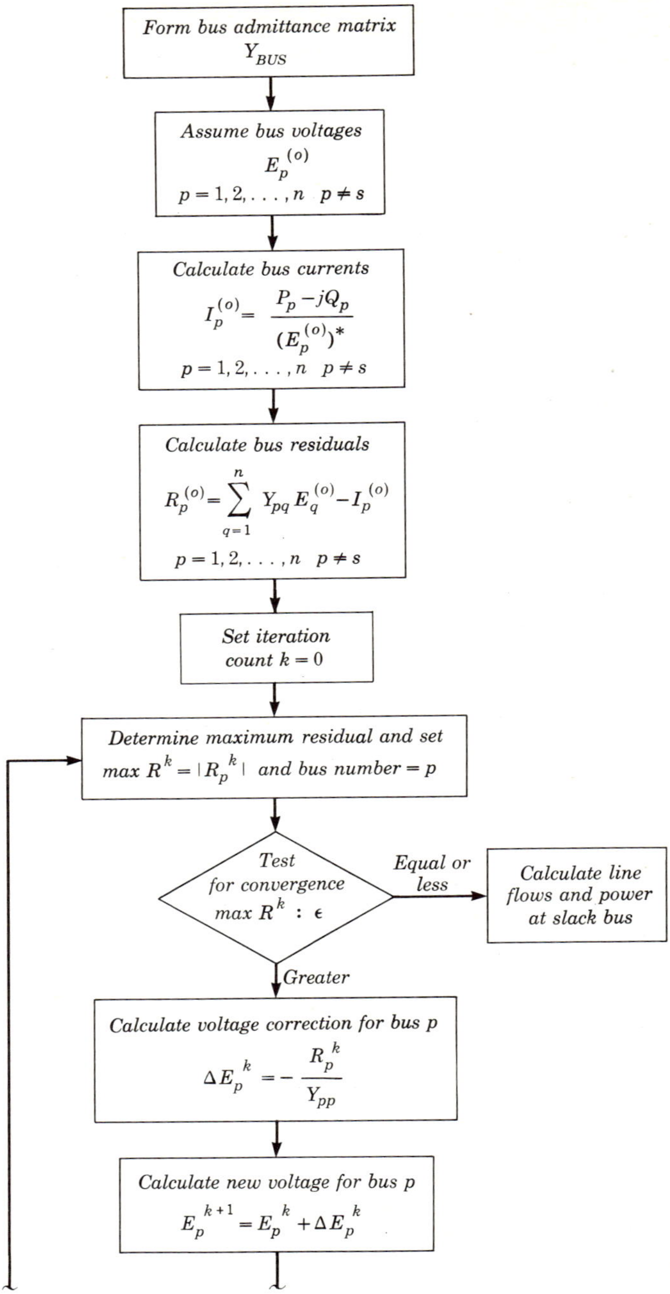

Fig. 8.4 Load flow solution by the relaxation method using Y_{BUS}.

An improved estimate of voltage for bus p is then

$$E_p^{k+1} = E_p^k + \Delta E_p^k$$

and the new current is

$$I_p^{k+1} = \frac{P_p - jQ_p}{(E_p^{k+1})^*}$$

As a result of the change in the current, the actual residual at bus p is

$$R_p^{k+1} = I_p^k - I_p^{k+1}$$

Using the voltage E_p^{k+1}, the new residuals for buses other than p and the slack bus are calculated from

$$R_q^{k+1} = R_q^k + Y_{qp}\Delta E_p^k \qquad q = 1, 2, \ldots, n \quad q \neq p, q \neq s$$

This process is repeated, each time correcting the voltage corresponding to the largest residual, until all residuals are less than or equal to a specified tolerance. The sequence of steps for the load flow solution by the relaxation method is shown in Fig. 8.4.

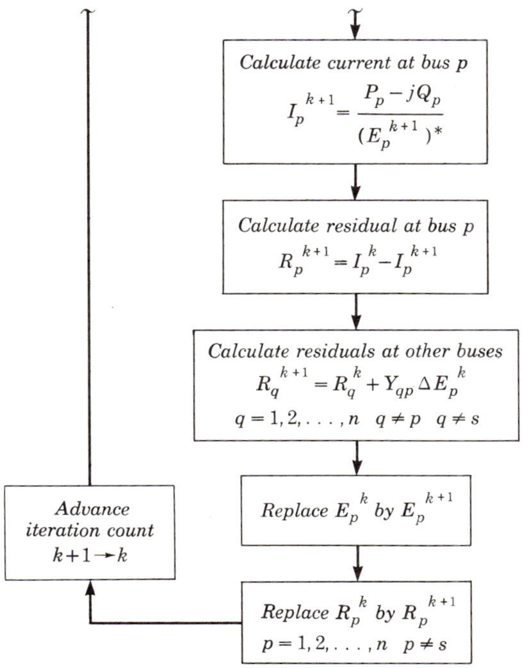

Newton-Raphson method using Y_{BUS}

The load flow problem can be solved by the Newton-Raphson method using a set of nonlinear equations to express the specified real and reactive powers in terms of bus voltages (Van Ness and Griffin, 1961). The power at bus p is

$$P_p - jQ_p = E_p^* I_p \tag{8.3.8}$$

Substituting from the network performance equation (8.3.2) for I_p in (8.3.8),

$$P_p - jQ_p = E_p^* \sum_{q=1}^{n} Y_{pq} E_q \tag{8.3.9}$$

Since $E_p = e_p + jf_p$ and $Y_{pq} = G_{pq} - jB_{pq}$, equation (8.3.9) becomes

$$P_p - jQ_p = (e_p - jf_p) \sum_{q=1}^{n} (G_{pq} - jB_{pq})(e_q + jf_q)$$

Separating the real and imaginary parts,

$$\begin{aligned} P_p &= \sum_{q=1}^{n} \{e_p(e_q G_{pq} + f_q B_{pq}) + f_p(f_q G_{pq} - e_q B_{pq})\} \\ Q_p &= \sum_{q=1}^{n} \{f_p(e_q G_{pq} + f_q B_{pq}) - e_p(f_q G_{pq} - e_q B_{pq})\} \end{aligned} \tag{8.3.10}$$

This formulation results in a set of nonlinear simultaneous equations, two for each bus of the system. The real and reactive powers P_p and Q_p are known and the real and imaginary components of voltage e_p and f_p are unknown for all buses except the slack bus, where the voltage is specified and remains fixed. Thus there are $2(n - 1)$ equations to be solved for a load flow solution.

The Newton-Raphson method requires that a set of linear equations be formed expressing the relationship between the changes in real and reactive powers and the components of bus voltages as follows:

$$\begin{bmatrix} \Delta P_1 \\ \cdots \\ \Delta P_{n-1} \\ \hline \Delta Q_1 \\ \cdots \\ \Delta Q_{n-1} \end{bmatrix} = \left[\begin{array}{ccc|ccc} \dfrac{\partial P_1}{\partial e_1} & \cdots & \dfrac{\partial P_1}{\partial e_{n-1}} & \dfrac{\partial P_1}{\partial f_1} & \cdots & \dfrac{\partial P_1}{\partial f_{n-1}} \\ \cdots & \cdots & \cdots & \cdots & \cdots & \cdots \\ \dfrac{\partial P_{n-1}}{\partial e_1} & \cdots & \dfrac{\partial P_{n-1}}{\partial e_{n-1}} & \dfrac{\partial P_{n-1}}{\partial f_1} & \cdots & \dfrac{\partial P_{n-1}}{\partial f_{n-1}} \\ \hline \dfrac{\partial Q_1}{\partial e_1} & \cdots & \dfrac{\partial Q_1}{\partial e_{n-1}} & \dfrac{\partial Q_1}{\partial f_1} & \cdots & \dfrac{\partial Q_1}{\partial f_{n-1}} \\ \cdots & \cdots & \cdots & \cdots & \cdots & \cdots \\ \dfrac{\partial Q_{n-1}}{\partial e_1} & \cdots & \dfrac{\partial Q_{n-1}}{\partial e_{n-1}} & \dfrac{\partial Q_{n-1}}{\partial f_1} & \cdots & \dfrac{\partial Q_{n-1}}{\partial f_{n-1}} \end{array}\right] \begin{bmatrix} \Delta e_1 \\ \cdots \\ \Delta e_{n-1} \\ \hline \Delta f_1 \\ \cdots \\ \Delta f_{n-1} \end{bmatrix} \tag{8.3.11}$$

where the coefficient matrix is the Jacobian and the nth bus is the slack. In matrix form, equation (8.3.11) is

$$\begin{bmatrix} \Delta P \\ \Delta Q \end{bmatrix} = \begin{bmatrix} J_1 & J_2 \\ J_3 & J_4 \end{bmatrix} \begin{bmatrix} \Delta e \\ \Delta f \end{bmatrix}$$

Equations for determining the elements of the Jacobian can be derived from the bus power equations. The real power from equation (8.3.10) is

$$P_p = e_p(e_pG_{pp} + f_pB_{pp}) + f_p(f_pG_{pp} - e_pB_{pp}) + \sum_{\substack{q=1 \\ q \neq p}}^{n} \{e_p(e_qG_{pq} + f_qB_{pq}) + f_p(f_qG_{pq} - e_qB_{pq})\} \tag{8.3.12}$$

$$p = 1, 2, \ldots, n-1$$

Differentiating, the off-diagonal elements of J_1 are

$$\frac{\partial P_p}{\partial e_q} = e_pG_{pq} - f_pB_{pq} \qquad q \neq p$$

and the diagonal elements of J_1 are

$$\frac{\partial P_p}{\partial e_p} = 2e_pG_{pp} + f_pB_{pp} - f_pB_{pp} + \sum_{\substack{q=1 \\ q \neq p}}^{n} (e_qG_{pq} + f_qB_{pq}) \tag{8.3.13}$$

However, the equation for the current at bus p is

$$I_p = c_p + jd_p = (G_{pp} - jB_{pp})(e_p + jf_p) + \sum_{\substack{q=1 \\ q \neq p}}^{n} (G_{pq} - jB_{pq})(e_q + jf_q)$$

which can be separated into the real and imaginary parts

$$c_p = e_pG_{pp} + f_pB_{pp} + \sum_{\substack{q=1 \\ q \neq p}}^{n} (e_qG_{pq} + f_qB_{pq})$$

$$d_p = f_pG_{pp} - e_pB_{pp} + \sum_{\substack{q=1 \\ q \neq p}}^{n} (f_qG_{pq} - e_qB_{pq}) \tag{8.3.14}$$

$$p = 1, 2, \ldots, n-1$$

Therefore, the expression for the diagonal elements of J_1 can be simplified by substituting the real component of current c_p in equation (8.3.13) to obtain

$$\frac{\partial P_p}{\partial e_p} = e_p G_{pp} - f_p B_{pp} + c_p$$

From equation (8.3.12), the off-diagonal elements of J_2 are

$$\frac{\partial P_p}{\partial f_q} = e_p B_{pq} + f_p G_{pq} \qquad q \neq p$$

and the diagonal elements of J_2 are

$$\frac{\partial P_p}{\partial f_p} = e_p B_{pp} + 2f_p G_{pp} - e_p B_{pp} + \sum_{\substack{q=1 \\ q \neq p}}^{n} (f_q G_{pq} - e_q B_{pq}) \tag{8.3.15}$$

The imaginary component of current from equation (8.3.14) is substituted in (8.3.15) to obtain

$$\frac{\partial P_p}{\partial f_p} = e_p B_{pp} + f_p G_{pp} + d_p$$

The reactive power from equation (8.3.10) is

$$Q_p = f_p(e_p G_{pp} + f_p B_{pp}) - e_p(f_p G_{pp} - e_p B_{pp}) + \sum_{\substack{q=1 \\ q \neq p}}^{n} \{f_p(e_q G_{pq} + f_q B_{pq}) - e_p(f_q G_{pq} - e_q B_{pq})\} \tag{8.3.16}$$

$$p = 1, 2, \ldots, n-1$$

Differentiating, the off-diagonal elements of J_3 are

$$\frac{\partial Q_p}{\partial e_q} = e_p B_{pq} + f_p G_{pq} \qquad q \neq p$$

and the diagonal elements of J_3 are

$$\frac{\partial Q_p}{\partial e_p} = f_p G_{pp} - f_p G_{pp} + 2e_p B_{pp} - \sum_{\substack{q=1 \\ q \neq p}}^{n} (f_q G_{pq} - e_q B_{pq}) \tag{8.3.17}$$

The imaginary component of current from equation (8.3.14) is substituted in equation (8.3.17) to obtain

$$\frac{\partial Q_p}{\partial e_p} = e_p B_{pp} + f_p G_{pp} - d_p$$

From equation (8.3.16), the off-diagonal elements of J_4 are

$$\frac{\partial Q_p}{\partial f_q} = -e_p G_{pq} + f_p B_{pq} \qquad q \neq p$$

and the diagonal elements of J_4 are

$$\frac{\partial Q_p}{\partial f_p} = e_pG_{pp} + 2f_pB_{pp} - e_pG_{pp} + \sum_{\substack{q=1 \\ q \neq p}}^{n} (e_qG_{pq} + f_qB_{pq}) \tag{8.3.18}$$

The real component of current from equation (8.3.14) is substituted in equation (8.3.18) to obtain

$$\frac{\partial Q_p}{\partial f_p} = -e_pG_{pp} + f_pB_{pp} + c_p$$

Given an initial set of bus voltages, the real and reactive powers are calculated from equations (8.3.10). The changes in power are the differences between the scheduled and calculated values

$$\Delta P_p^k = P_{p\,(\text{scheduled})} - P_p^k$$
$$\Delta Q_p^k = Q_{p\,(\text{scheduled})} - Q_p^k \qquad p = 1, 2, \ldots, n-1$$

The estimated bus voltages and calculated powers are used to compute bus currents in order to evaluate the elements of the Jacobian. The linear set of equations (8.3.11) can be solved for Δe_p and Δf_p, $p = 1, 2, \ldots, n-1$, by a direct or an iterative method. Then, the new estimates for bus voltages are

$$e_p^{k+1} = e_p^k + \Delta e_p^k$$
$$f_p^{k+1} = f_p^k + \Delta f_p^k$$

The process is repeated until ΔP_p^k and ΔQ_p^k for all buses are within a specified tolerance. The sequence of steps for the load flow solution by the Newton-Raphson method is shown in Fig. 8.5.

The Newton-Raphson method can be applied also to solve the load flow problem when the equations are expressed in polar coordinates. In polar coordinates

$$E_p = |E_p|\epsilon^{j\delta_p} \qquad \text{and} \qquad Y_{pq} = |Y_{pq}|\epsilon^{-j\theta_{pq}}$$

Substituting in equation (8.3.9), the power at bus p is

$$P_p - jQ_p = \sum_{q=1}^{n} |E_pE_qY_{pq}|\epsilon^{-j(\theta_{pq}+\delta_p-\delta_q)}$$

Since $e^{-j(\theta_{pq}+\delta_p-\delta_q)} = \cos(\theta_{pq} + \delta_p - \delta_q) - j\sin(\theta_{pq} + \delta_p - \delta_q)$, the real and imaginary components of power are

$$P_p = \sum_{q=1}^{n} |E_pE_qY_{pq}| \cos(\theta_{pq} + \delta_p - \delta_q)$$
$$Q_p = \sum_{q=1}^{n} |E_pE_qY_{pq}| \sin(\theta_{pq} + \delta_p - \delta_q) \qquad p = 1, 2, \ldots, n-1 \tag{8.3.19}$$

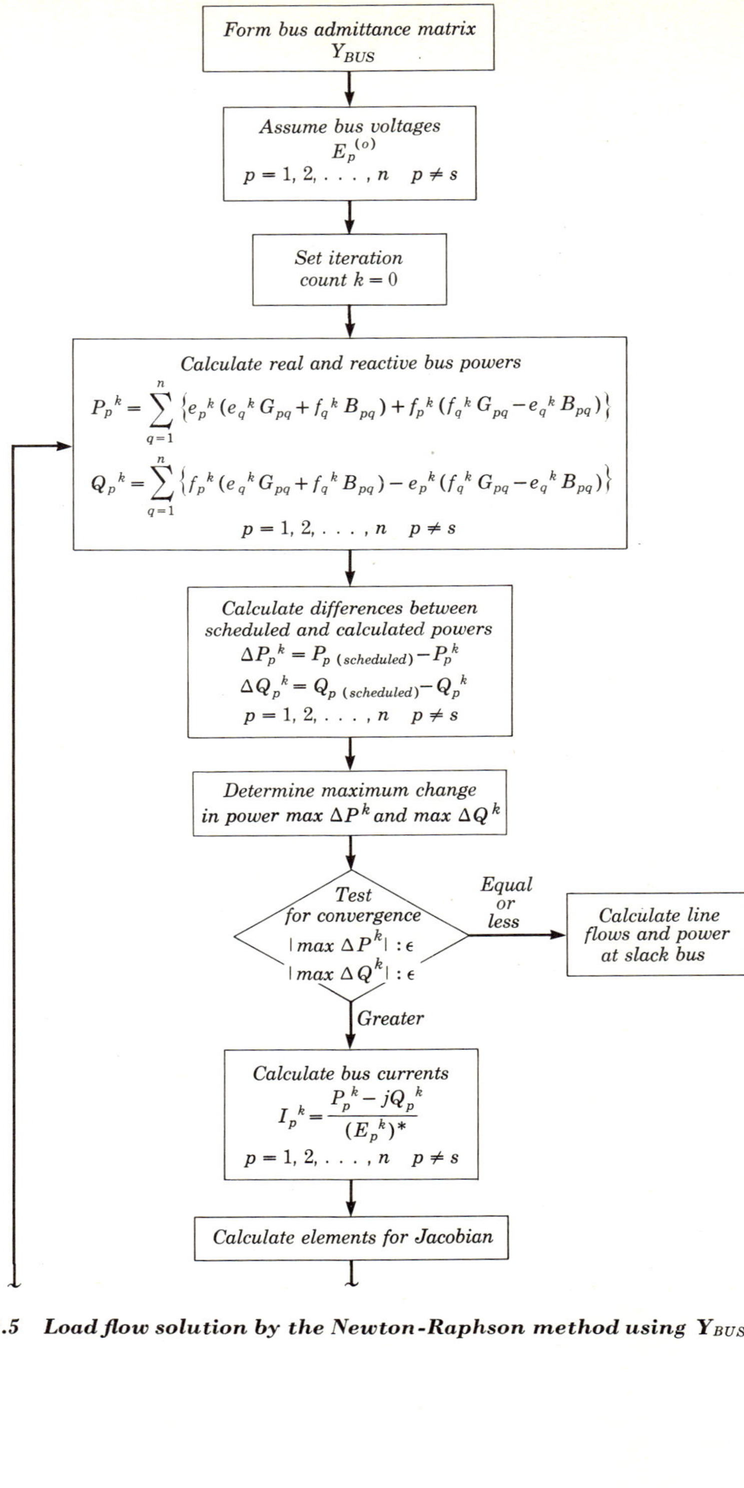

Fig. 8.5 Load flow solution by the Newton-Raphson method using Y_{BUS}.

The elements of the Jacobian are calculated from equations (8.3.19) and are

For J_1:

$$\frac{\partial P_p}{\partial \delta_q} = |E_p E_q Y_{pq}| \sin(\theta_{pq} + \delta_p - \delta_q) \qquad q \neq p$$

$$\frac{\partial P_p}{\partial \delta_p} = -\sum_{\substack{q=1\\q\neq p}}^{n} |E_p E_q Y_{pq}| \sin(\theta_{pq} + \delta_p - \delta_q)$$

For J_2:

$$\frac{\partial P_p}{\partial |E_q|} = |E_p Y_{pq}| \cos(\theta_{pq} + \delta_p - \delta_q) \qquad q \neq p$$

$$\frac{\partial P_p}{\partial |E_p|} = 2|E_p Y_{pp}| \cos\theta_{pp} + \sum_{\substack{q=1\\q\neq p}}^{n} |E_q Y_{pq}| \cos(\theta_{pq} + \delta_p - \delta_q)$$

For J_3:

$$\frac{\partial Q_p}{\partial \delta_q} = -|E_p E_q Y_{pq}| \cos(\theta_{pq} + \delta_p - \delta_q) \qquad q \neq p$$

$$\frac{\partial Q_p}{\partial \delta_p} = \sum_{\substack{q=1\\q\neq p}}^{n} |E_p E_q Y_{pq}| \cos(\theta_{pq} + \delta_p - \delta_q)$$

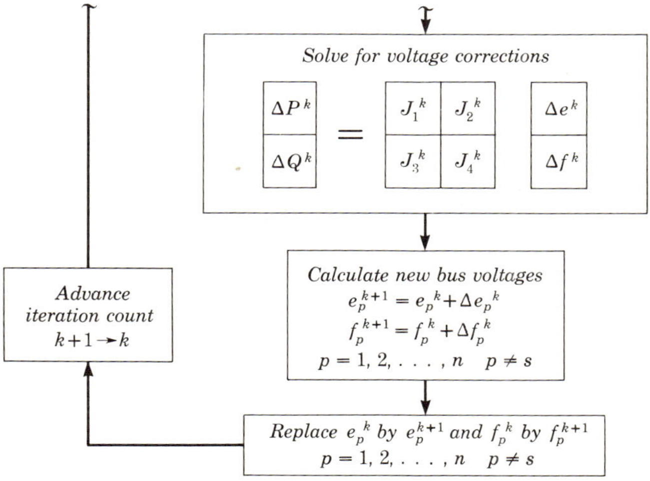

For J_4:

$$\frac{\partial Q_p}{\partial |E_q|} = |E_p Y_{pq}| \sin (\theta_{pq} + \delta_p - \delta_q) \qquad q \neq p$$

$$\frac{\partial Q_p}{\partial |E_p|} = 2|E_p Y_{pp}| \sin \theta_{pp} + \sum_{\substack{q=1 \\ q \neq p}}^{n} |E_q Y_{pq}| \sin (\theta_{pq} + \delta_p - \delta_q)$$

Then the equation relating the changes in power to the changes in the voltage magnitudes and phase angles for the Newton-Raphson method is

$$\begin{bmatrix} \Delta P \\ \hline \Delta Q \end{bmatrix} = \left[\begin{array}{c|c} J_1 & J_2 \\ \hline J_3 & J_4 \end{array}\right] \begin{bmatrix} \Delta \delta \\ \hline \Delta |E| \end{bmatrix}.$$

Approximations to Newton-Raphson method

In general, for a small change in the magnitude of bus voltage the real power at the bus does not change appreciably. Likewise, for a small change in the phase angle of the bus voltage the reactive power does not change appreciably. Therefore, using polar coordinates, a solution for the load flow problem can be obtained assuming the elements of the submatrices J_2 and J_3 are zero (Carpentier, 1963). The simplified matrix equation is

$$\begin{bmatrix} \Delta P \\ \hline \Delta Q \end{bmatrix} = \left[\begin{array}{c|c} J_1 & 0 \\ \hline 0 & J_4 \end{array}\right] \begin{bmatrix} \Delta \delta \\ \hline \Delta |E| \end{bmatrix}$$

Successful solutions can be obtained also by reevaluating the Jacobian in only the first few iterations.

When using rectangular coordinates, a solution to the load flow problem also can be obtained by neglecting the off-diagonal elements of the submatrices J_1, J_2, J_3, and J_4 of the Jacobian (Ward and Hale, 1956). This results in the following equations for the changes in real and reactive power at bus p:

$$\begin{aligned} \Delta P_p &= \frac{\partial P_p}{\partial e_p} \Delta e_p + \frac{\partial P_p}{\partial f_p} \Delta f_p \\ &= \Delta e_p(e_p G_{pp} - f_p B_{pp} + c_p) + \Delta f_p(e_p B_{pp} + f_p G_{pp} + d_p) \\ \Delta Q_p &= \frac{\partial Q_p}{\partial e_p} \Delta e_p + \frac{\partial Q_p}{\partial f_p} \Delta f_p \\ &= \Delta e_p(e_p B_{pp} + f_p G_{pp} - d_p) + \Delta f_p(-e_p G_{pp} + f_p B_{pp} + c_p) \end{aligned}$$

$$p = 1, 2, \ldots, n-1$$

These equations can be solved using the Gauss-Seidel iterative method.

Gauss iterative method using Z_{BUS}

Selecting an initial set of bus voltages, bus currents can be calculated from

$$I_p = \frac{P_p - jQ_p}{E_p^*} - y_p E_p \qquad \begin{matrix} p = 1, 2, \ldots, n \\ p \neq s \end{matrix} \tag{8.3.20}$$

where the shunt connections are treated as current sources. A new estimate of voltages can be obtained, then, from the bus impedance network equation

$$\bar{E}_{BUS} = Z_{BUS}\bar{I}_{BUS} + \bar{E}_R \tag{8.3.21}$$

where $\bar{E}_R$ is the vector whose elements are all equal to the voltage of the slack bus and the bus impedance matrix formed by using the slack bus as reference is of dimension $n - 1 \times n - 1$. The new voltage estimates are used in equation (8.3.20) to recalculate bus currents. The process is repeated until changes in all bus voltages are within a specified tolerance. This method of solving the load flow problem uses the Gauss iterative method since the new bus currents are recalculated only after the completion of an iteration.

Applying this method to the system given in Fig. 8.1 with bus 2 as slack (reference), the formula obtained from equation (8.3.21) is

$$E_p^{k+1} = E_2 + \sum_{\substack{q=1 \\ q \neq 2}}^{6} Z_{pq} I_q^k \qquad p = 1, 3, 4, 5, 6$$

where

$$I_q^k = \frac{P_q - jQ_q}{(E_q^k)^*} - y_q E_q^k$$

Gauss-Seidel iterative method using Z_{BUS}

The Gauss-Seidel iterative method can be applied also for the solution of the load flow problem using the bus impedance network equation (El-Abiad, Watson, and Stagg, 1961). The bus voltage equations (8.3.21) are solved one at a time in the order established by the bus coding. After each equation is solved to obtain a new estimate of bus voltage, the corresponding bus current is recalculated. Then the formula for the system given in Fig. 8.1 is

$$E_p^{k+1} = E_2 + \sum_{\substack{q=1 \\ q \neq 2}}^{p-1} Z_{pq} I_q^{k+1} + \sum_{\substack{q=p \\ q \neq 2}}^{6} Z_{pq} I_q^k \qquad p = 1, 3, 4, 5, 6$$

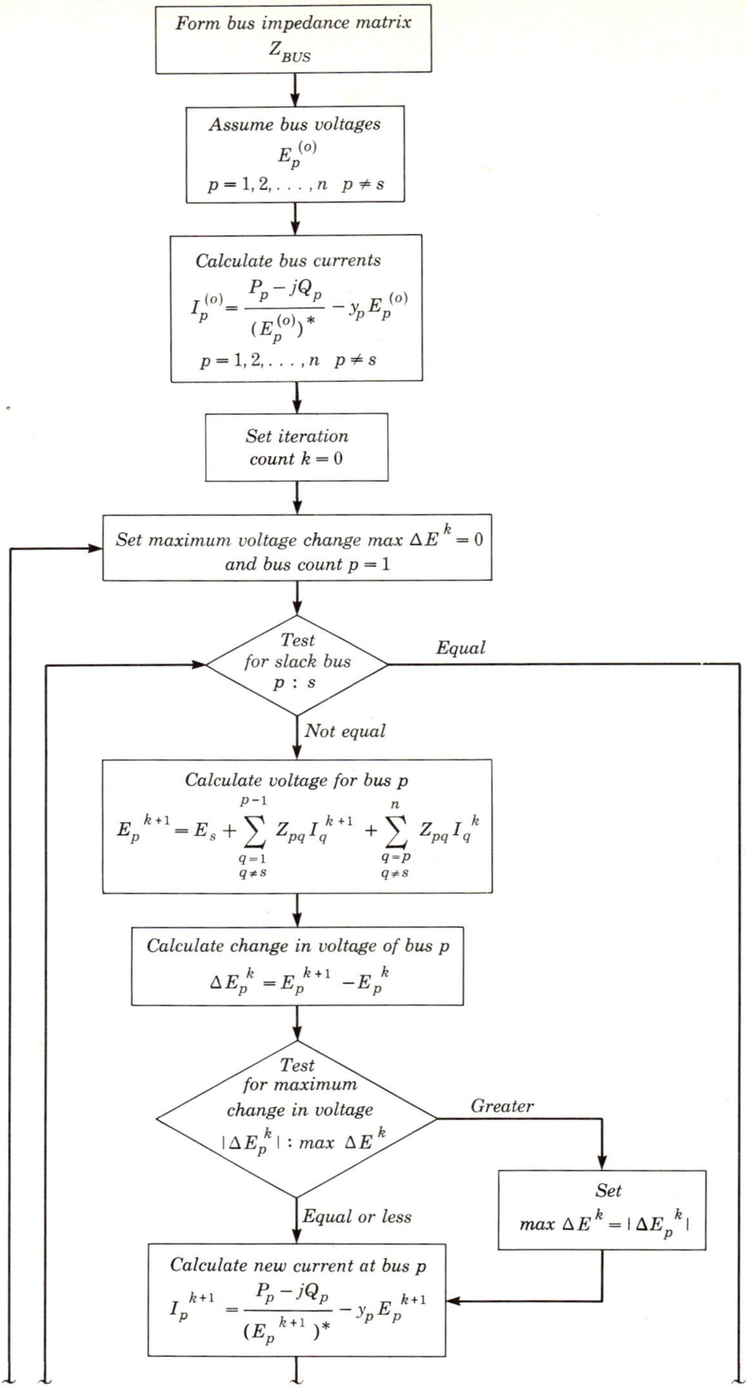

Fig. 8.6 Load flow solution by the Gauss-Seidel iterative method using Z_{BUS}.

where

$$I_q^{k+1} = \frac{P_q - jQ_q}{(E_q^{k+1})^*} - y_q E_q^{k+1}$$

The sequence of steps for the load flow solution by this method is shown in Fig. 8.6.

Gauss iterative method using Y_{LOOP}

The performance equations of the network in the loop frame of reference are not in terms of the bus quantities required for a load flow solution. In using loop matrices, therefore, it is necessary to convert from the initial assumed bus quantities to the loop quantities required for the solution technique and then to relate the changes in loop quantities back to bus quantities.

As in the previous methods, initial voltages are assumed for all buses except the slack bus. Then, bus currents can be calculated from

$$I_p = \frac{P_p - jQ_p}{E_p^*} - y_p E_p \qquad \begin{array}{l} p = 1, 2, \ldots, n \\ p \neq s \end{array}$$

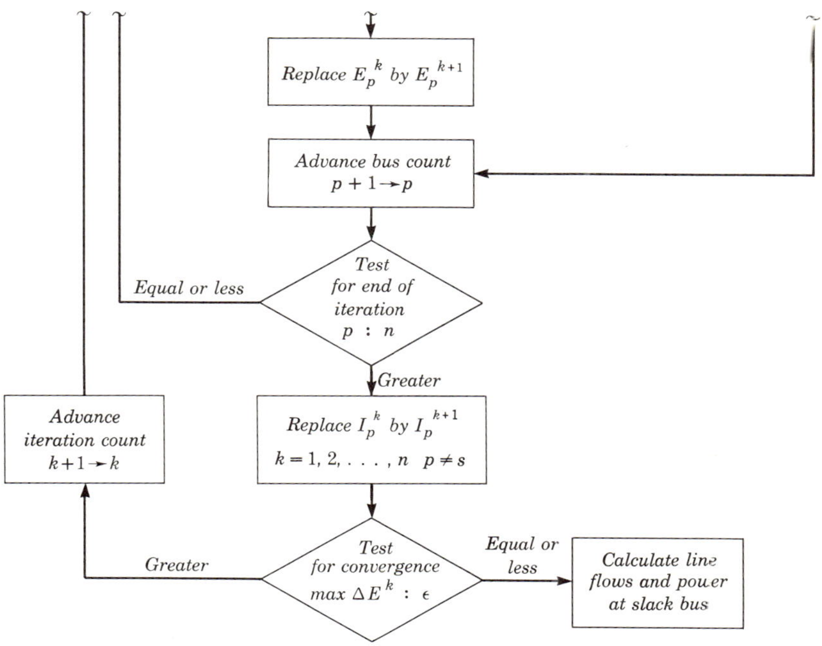

In order to use loop quantities it is necessary to estimate a flow of current through the network resulting from the calculated bus currents. An initial estimate of current flow can be obtained by assuming that the bus currents flow only through the tree of the network. Then the initial link currents are zero and the branch currents are

$$\bar{\imath}_b = K\bar{I}_{BUS}$$

where K is the branch-path incidence matrix. With the assumed current flow, the voltage across each element can be calculated from

$$\bar{v} = [z]\bar{\imath} \tag{8.3.22}$$

where $[z]$ is the primitive impedance matrix. The loop voltages $\bar{E}_{LOOP}$ and $\bar{v}$ are related by

$$\bar{E}_{LOOP} = -C^t\bar{v} \tag{8.3.23}$$

where C^t is the transpose of the basic loop incidence matrix. Substituting for $\bar{v}$ in equation (8.3.23) from equation (8.3.22),

$$\bar{E}_{LOOP} = -C^t[z]\bar{\imath} \tag{8.3.24}$$

where $\bar{E}_{LOOP}$ represents the errors in the loop voltages. The loop voltages must be zero for a load flow solution since the network excluding all shunt elements does not contain any sources.

To obtain an improved estimate of current flow in each element, loop balancing currents can be calculated from

$$\bar{I}_{LOOP} = Y_{LOOP}\bar{E}_{LOOP} \tag{8.3.25}$$

where Y_{LOOP} is the loop admittance matrix and does not include the effects of shunt elements. The loop balancing currents are superimposed on the previously estimated flows to obtain new current estimates. The changes in branch currents can be obtained from

$$\overline{\Delta i_b} = C_b\bar{I}_{LOOP} \tag{8.3.26}$$

where C_b is the submatrix relating branches and basic loops in the basic loop incidence matrix. The changes in the link currents are equal to the loop balancing currents. The new branch currents are used to calculate new bus voltages from

$$\bar{E}_{BUS} - \bar{E}_R = K^t[z_{bb}]\bar{\imath}_b \tag{8.3.27}$$

where $[z_{bb}]$ is the primitive impedance matrix for branches only and $\bar{E}_R$ is the vector whose elements are all equal to the fixed voltage of the slack bus.

The new estimates of bus voltages are used to reestimate bus currents. The changes in the bus currents are

$$\Delta I_p{}^k = I_p{}^k - I_p^{k-1} \qquad \begin{matrix} p = 1, 2, \ldots, n \\ p \neq s \end{matrix}$$

Then the changes in branch currents due to the changes in bus currents can be calculated from

$$\overline{\Delta i_b} = K\overline{\Delta I}{}^k_{BUS}$$

These changes are added to the previous branch currents to obtain

$$\bar{i}_b = \bar{i}_b^{k-1} + \overline{\Delta i_b}$$

With the branch currents $\bar{i}_b$ and the previous link currents $\bar{i}_l^{k-1}$, the new loop voltages are obtained from equation (8.3.24). The loop balancing currents are calculated from equation (8.3.25). The changes in the branch currents due to the loop balancing currents are calculated from equation (8.3.26). Then the new estimated branch currents are

$$\bar{i}_b{}^k = \bar{i}_b + C_b\bar{I}^k_{LOOP}$$

and the new link currents are

$$\bar{i}_l{}^k = \bar{i}_l^{k-1} + \bar{I}^k_{LOOP}$$

New estimates of bus voltages are obtained from equation (8.3.27). The process is repeated until all loop voltages become negligible.

The sequence of steps for the load flow solution by this method is shown in Fig. 8.7.

A solution of this type was employed first for digital load flow studies (Dunstan, 1947). However, the early methods used real and reactive power by assuming line flows and calculating voltage drops and phase angle differences in order to determine balancing loop power flows. Transmission losses had to be estimated during the iterative process.

8.4 Acceleration of convergence

In some cases the rate of convergence for an iterative process can be increased by applying an acceleration factor to the approximate solution obtained from each iteration. Let α and β be the acceleration factors respectively for the real and imaginary components of voltage. The accelerated values are

$$e^{k+1}_{p\,(\text{accelerated})} = e_p{}^k + \alpha(e_p^{k+1} - e_p{}^k)$$
$$f^{k+1}_{p\,(\text{accelerated})} = f_p{}^k + \beta(f_p^{k+1} - f_p{}^k)$$

and replace the calculated e_p^{k+1} and f_p^{k+1}.

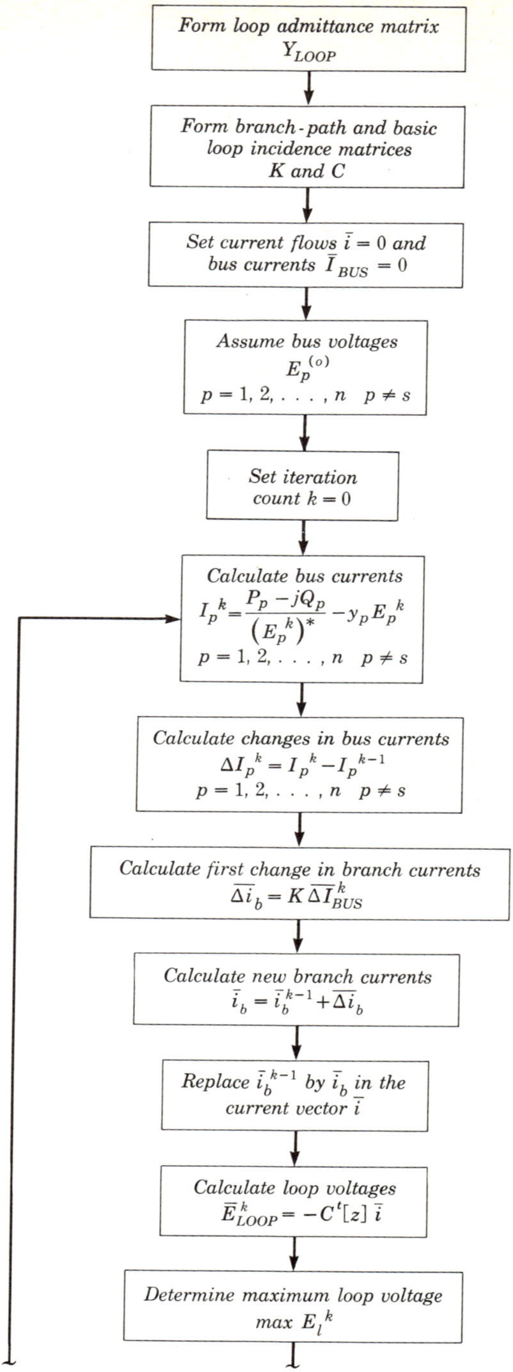

Fig. 8.7 Load flow solution by the Gauss iterative method using Y_{LOOP}.

8.5 Examples of load flow calculations

The methods for solving the load flow problem will be illustrated for the sample power system given in Fig. 8.8.

Problem

With bus 1 as the slack, use the following methods to obtain a load flow solution:

a. Gauss-Seidel using Y_{BUS}, with acceleration factors of 1.4 and 1.4 and tolerances of 0.0001 and 0.0001 per unit for the real and imaginary components of voltage

b. Newton-Raphson using Y_{BUS}, with tolerances of 0.01 per unit for the changes in the real and reactive bus powers

c. Gauss-Seidel using Z_{BUS}, with voltage tolerances of 0.001 and 0.001 per unit

d. Gauss using Y_{LOOP}, with loop voltage tolerances of 0.01 and 0.01 per unit

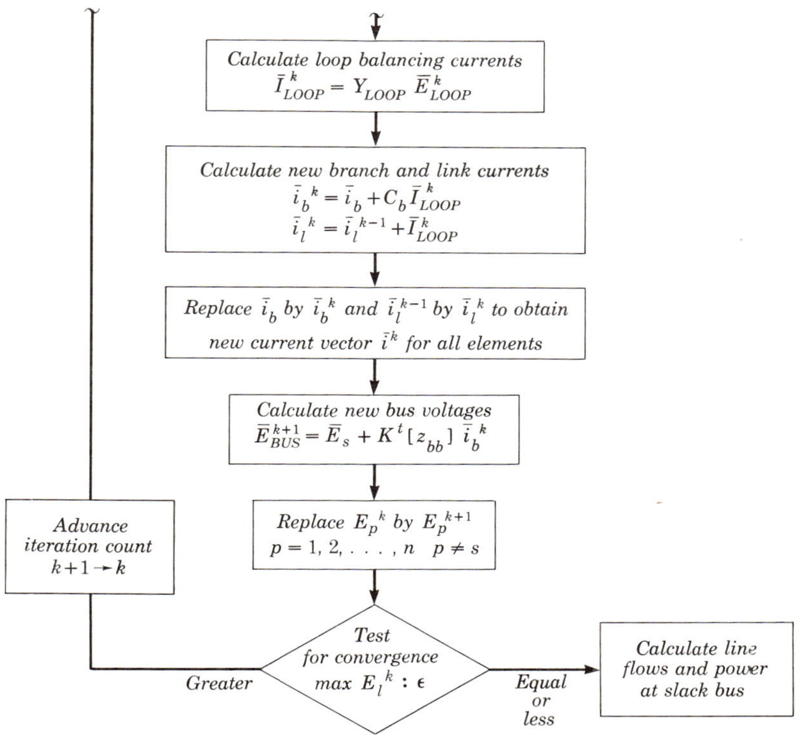

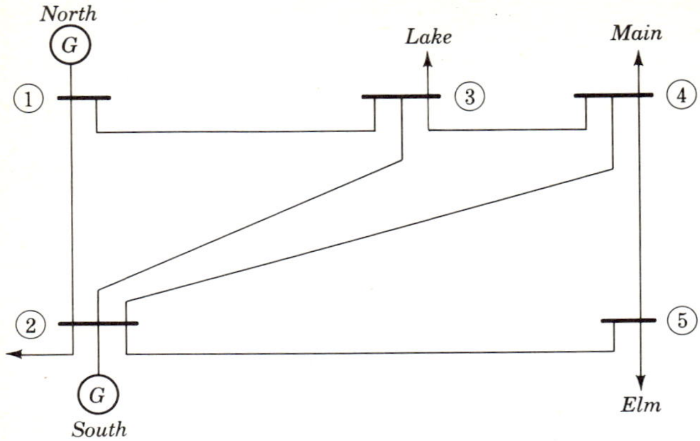

Fig. 8.8 Sample system for load flow solution.

Solution

The transmission line impedances and line charging admittances in per unit on a 100,000 kva base are given in Table 8.2. The scheduled generation and loads and the assumed per unit bus voltages are given in Table 8.3.

Table 8.2 Impedances and line charging for sample system

Bus code p–q	Impedance z_{pq}	Line charging $y'_{pq}/2$
1–2	$0.02 + j0.06$	$0.0 + j0.030$
1–3	$0.08 + j0.24$	$0.0 + j0.025$
2–3	$0.06 + j0.18$	$0.0 + j0.020$
2–4	$0.06 + j0.18$	$0.0 + j0.020$
2–5	$0.04 + j0.12$	$0.0 + j0.015$
3–4	$0.01 + j0.03$	$0.0 + j0.010$
4–5	$0.08 + j0.24$	$0.0 + j0.025$

Table 8.3 Scheduled generation and loads and assumed bus voltages for sample system

Bus code p	Assumed bus voltage	Generation		Load	
		Megawatts	*Megavars*	*Megawatts*	*Megavars*
1	$1.06 + j0.0$	0	0	0	0
2	$1.0 + j0.0$	40	30	20	10
3	$1.0 + j0.0$	0	0	45	15
4	$1.0 + j0.0$	0	0	40	5
5	$1.0 + j0.0$	0	0	60	10

a. The equations for the Gauss-Seidel iterative solution, using the bus code numbers given in Fig. 8.8, are

$$E_1 = 1.06 + j0.0$$

$$E_2^{k+1} = \frac{KL_2}{(E_2^k)^*} - YL_{21}E_1 - YL_{23}E_3^k - YL_{24}E_4^k - YL_{25}E_5^k$$

$$E_3^{k+1} = \frac{KL_3}{(E_3^k)^*} - YL_{31}E_1 - YL_{32}E_2^{k+1} - YL_{34}E_4^k$$

$$E_4^{k+1} = \frac{KL_4}{(E_4^k)^*} - YL_{42}E_2^{k+1} - YL_{43}E_3^{k+1} - YL_{45}E_5^k$$

$$E_5^{k+1} = \frac{KL_5}{(E_5^k)^*} - YL_{52}E_2^{k+1} - YL_{54}E_4^{k+1}$$

In order to calculate the parameters for these equations, it is necessary, first, to determine the elements of the bus admittance matrix from the transmission line and line charging admittances with ground as reference. The transmission line admittances, obtained by taking the reciprocal of the line impedances, are shown in Table 8.4 along with the total line charging admittance to ground at each bus. Since there is no mutual coupling in the representation of the system, the diagonal element of the bus admittance matrix for bus 1 is

$$Y_{11} = y_{12} + y_{13} + y_1$$

Table 8.4 Line admittances and admittances to ground for sample system

Bus code p–q	Line admittance y_{pq}
1–2	$5.00000 - j15.00000$
1–3	$1.25000 - j3.75000$
2–3	$1.66667 - j5.00000$
2–4	$1.66667 - j5.00000$
2–5	$2.50000 - j7.50000$
3–4	$10.00000 - j30.00000$
4–5	$1.25000 - j3.75000$

Bus code p	Admittance to ground y_p
1	$0.0 + j0.05500$
2	$0.0 + j0.08500$
3	$0.0 + j0.05500$
4	$0.0 + j0.05500$
5	$0.0 + j0.04000$

where y_1 is the sum of the line charging to ground at bus 1. Thus, Y_{11} is

$$
\begin{array}{l}
5.00000 - j15.00000 \\
1.25000 - j3.75000 \\
\underline{0.0 + j0.05500} \\
6.25000 - j18.69500
\end{array}
$$

The off-diagonal elements associated with bus 1 are

$$
\begin{aligned}
Y_{12} &= Y_{21} = -y_{12} = -5.00000 + j15.00000 \\
Y_{13} &= Y_{31} = -y_{13} = -1.25000 + j3.75000
\end{aligned}
$$

The bus admittance matrix with ground as reference for the sample system is

Y_{BUS} =

	①	②	③	④	⑤
①	$6.25000 - j18.69500$	$-5.00000 + j15.00000$	$- 1.25000 + j3.75000$		
②	$-5.00000 + j15.00000$	$10.83334 - j32.41500$	$-1.66667 + j5.00000$	$-1.66667 + j5.00000$	$-2.50000 + j7.50000$
③	$-1.25000 + j3.75000$	$-1.66667 + j5.00000$	$12.91667 - j38.69500$	$-10.00000 + j30.00000$	
④		$-1.66667 + j5.00000$	$-10.00000 + j30.00000$	$12.91667 - j38.69500$	$-1.25000 + j3.75000$
⑤		$-2.50000 + j7.50000$		$-1.25000 + j3.75000$	$3.75000 - j11.21000$

The KL_p's are obtained from the equation

$$KL_p = (P_p - jQ_p)L_p = (P_p - jQ_p)\frac{1}{Y_{pp}} \qquad p = 1, 2, \ldots, n$$

where $P_p - jQ_p$ is the net load in per unit at the pth bus. For bus 2

$$KL_2 = (0.20 - j0.20)\frac{1}{10.83334 - j32.41500}$$
$$= 0.00740 + j0.00370$$

The KL_p's for all buses are given in Table 8.5.

The YL_{pq}'s are obtained from the equation

$$YL_{pq} = Y_{pq}L_p = Y_{pq}\frac{1}{Y_{pp}} \qquad p, q = 1, 2, \ldots, n$$

For the element 1–2,

$$YL_{12} = (-5.00000 + j15.00000)\frac{1}{6.25000 - j18.69500}$$
$$= -0.80212 + j0.00071$$

The YL_{pq}'s for all elements are given in Table 8.6.

It is not necessary to calculate the parameters associated with the slack bus for the solution of a particular load flow. For actual planning and operating studies, however, the slack bus is frequently changed in subsequent load flow cases. This type of change can be made readily if the parameters of all buses are calculated and stored in the appropriate data lists.

The first step in the iterative solution is to calculate a new estimate of the voltage for bus 2. The new estimate from the equation

Table 8.5 Bus parameters for sample system

Bus code p	KL_p
1	0.0 + j0.0
2	0.00740 + j0.00370
3	−0.00698 − j0.00930
4	−0.00427 − j0.00891
5	−0.02413 − j0.04545

$$E_2^{(1)} = \frac{KL_2}{(E_2^{(0)})^*} - YL_{21}E_1 - YL_{23}E_3^{(0)} - YL_{24}E_4^{(0)} - YL_{25}E_5^{(0)}$$

is

$$\begin{aligned} E_2^{(1)} &= \frac{0.00740 + j0.00370}{1.0 - j0.0} - (-0.46263 + j0.00036)(1.06 + j0.0) \\ &\quad - (-0.15421 + j0.00012)(1.0 + j0.0) \\ &\quad - (-0.15421 + j0.00012)(1.0 + j0.0) \\ &\quad - (-0.23131 + j0.00018)(1.0 + j0.0) \\ &= 1.03752 + j0.00290 \end{aligned}$$

The change in voltage is

$$\Delta E_2^{(1)} = 0.03752 + j0.00290$$

The accelerated value of bus voltage from the equation

$$E_{2\,(\text{accelerated})}^{(1)} = E_2^{(0)} + \alpha \Delta E_2^{(1)}$$

is

$$\begin{aligned} E_{2\,(\text{accelerated})}^{(1)} &= 1.0 + j0.0 + 1.4(0.03752 + j0.00290) \\ &= 1.05253 + j0.00406 \end{aligned}$$

This value replaces the initial estimated value of voltage for bus 2 and is used in subsequent calculations of voltages for the remaining buses. The

Table 8.6 Line parameters for sample system

Bus code *p–q*	YL_{pq}
1–2	$-0.80212 + j0.00071$
1–3	$-0.20053 + j0.00018$
2–1	$-0.46263 + j0.00036$
2–3	$-0.15421 + j0.00012$
2–4	$-0.15421 + j0.00012$
2–5	$-0.23131 + j0.00018$
3–1	$-0.09690 + j0.00004$
3–2	$-0.12920 + j0.00006$
3–4	$-0.77518 + j0.00033$
4–2	$-0.12920 + j0.00006$
4–3	$-0.77518 + j0.00033$
4–5	$-0.09690 + j0.00004$
5–2	$-0.66881 + j0.00072$
5–4	$-0.33440 + j0.00036$

new estimate of the voltage for bus 3 from the equation

$$E_3^{(1)} = \frac{KL_3}{(E_3^{(0)})^*} - YL_{31}E_1 - YL_{32}E_2^{(1)} - YL_{34}E_4^{(0)}$$

is

$$\begin{aligned} E_3^{(1)} &= \frac{-0.00698 - j0.00930}{1.0 - j0.0} - (-0.09690 + j0.00004)(1.06 + j0.0) \\ &\quad - (-0.12920 + j0.00006)(1.05253 + j0.00406) \\ &\quad - (-0.77518 + j0.00033)(1.0 + j0.0) \\ &= 1.00690 - j0.00921 \end{aligned}$$

The change in voltage is

$$\Delta E_3^{(1)} = 0.00690 - j0.00921$$

The accelerated value of bus voltage from the equation

$$E_{3\,(\text{accelerated})}^{(1)} = E_3^{(0)} + \alpha \Delta E_3^{(1)}$$

is

$$\begin{aligned} E_{3\,(\text{accelerated})}^{(1)} &= 1.0 + j0.0 + 1.4(0.00690 - j0.00921) \\ &= 1.00966 - j0.01289 \end{aligned}$$

This value replaces the initial estimated value of voltage for bus 3. The process is continued for the remaining buses to complete one iteration. If the process has not converged, new estimates of voltage are calculated for all buses, starting again with bus 2. The bus voltages for all iterations are given in Table 8.7 and the changes in voltages in Table 8.8.

Table 8.7 Bus voltages from the Gauss-Seidel iterative solution using Y_{BUS}

Iteration count *k*	Bus voltages			
	Bus 2	*Bus 3*	*Bus 4*	*Bus 5*
0	1.0 + j0.0	1.0 + j0.0	1.0 + j0.0	1.0 + j0.0
1	1.05253 + j0.00406	1.00966 − j0.01289	1.01579 − j0.02635	1.02727 − j0.07374
2	1.04528 − j0.03015	1.02154 − j0.04227	1.02451 − j0.06353	1.01025 − j0.08932
3	1.04732 − j0.03618	1.02637 − j0.07153	1.02394 − j0.08326	1.01712 − j0.09826
4	1.04964 − j0.04730	1.02395 − j0.08289	1.02268 − j0.09079	1.01575 − j0.10787
5	1.04749 − j0.05016	1.02300 − j0.08693	1.02148 − j0.09393	1.01315 − j0.10782
6	1.04708 − j0.05057	1.02195 − j0.08877	1.02036 − j0.09473	1.01316 − j0.10873
7	1.04678 − j0.05127	1.02106 − j0.08901	1.01977 − j0.09493	1.01256 − j0.10908
8	1.04639 − j0.05120	1.02070 − j0.08913	1.01945 − j0.09501	1.01224 − j0.10893
9	1.04630 − j0.05123	1.02048 − j0.08918	1.01927 − j0.09502	1.01219 − j0.10904
10	1.04623 − j0.05126	1.02036 − j0.08917	1.01920 − j0.09504	1.01211 − j0.10904

Table 8.8 Changes in bus voltages from the Gauss-Seidel iterative solution using Y_{BUS}

Iteration count k	Changes in bus voltages			
	Bus 2	Bus 3	Bus 4	Bus 5
0	$0.0 + j0.0$	$0.0 + j0.0$	$0.0 + j0.0$	$0.0 + j0.0$
1	$0.05253 + j0.00406$	$0.00966 - j0.01289$	$0.01579 - j0.02635$	$0.02727 - j0.07374$
2	$-0.00724 - j0.03421$	$0.01188 - j0.02938$	$0.00872 - j0.03718$	$-0.07102 - j0.01558$
3	$0.00204 - j0.00603$	$0.00483 - j0.02926$	$-0.00057 - j0.01973$	$0.00687 - j0.00894$
4	$0.00232 - j0.01112$	$-0.00242 - j0.01136$	$-0.00126 - j0.00753$	$-0.00137 - j0.00961$
5	$-0.00215 - j0.00286$	$-0.00095 - j0.00404$	$-0.00120 - j0.00314$	$-0.00260 + j0.00005$
6	$-0.00041 - j0.00041$	$-0.00105 - j0.00184$	$-0.00112 - j0.00080$	$0.00001 - j0.00091$
7	$-0.00030 - j0.00070$	$-0.00089 - j0.00024$	$-0.00059 - j0.00020$	$-0.00060 - j0.00035$
8	$-0.00039 + j0.00007$	$-0.00036 - j0.00012$	$-0.00032 - j0.00008$	$-0.00032 + j0.00015$
9	$-0.00009 - j0.00003$	$-0.00022 - j0.00005$	$-0.00018 - j0.00001$	$-0.00005 - j0.00011$
10	$-0.00007 - j0.00003$	$-0.00012 + j0.00001$	$-0.00007 - j0.00002$	$-0.00008 + j0.00000$

Note: The changes in bus voltages given in this table are the differences between the accelerated values. The tolerance test was made on the unaccelerated voltage in determining convergence.

The line flows are calculated with the final bus voltages and the given line admittances and line charging. The flow in line 1–2 at bus 1 from the equation

$$P_{pq} - jQ_{pq} = E_p^*(E_p - E_q)y_{pq} + E_p^*E_p\frac{y'_{pq}}{2}$$

is

$$\begin{aligned} P_{12} - jQ_{12} &= (1.06 - j0.0)\{1.06 + j0.0 \\ &\quad - (1.04623 - j0.05126)\}(5.0 - j15.0) \\ &\quad + (1.06 - j0.0)(1.06 + j0.0)(0.0 + j0.03) \\ &= 0.888 + j0.086 \end{aligned}$$

The flow in megawatts and megavars is

$$P_{12} - jQ_{12} = 88.8 + j8.6$$

The flow in line 1–2 at bus 2 is

$$\begin{aligned} P_{21} - jQ_{21} &= (1.04623 + j0.05126)\{1.04623 - j0.05126 \\ &\quad - (1.06 + j0.0)\}(5.0 - j15.0) \\ &\quad + (1.04623 + j0.05126)(1.04623 - j0.05126)(0.0 + j0.3) \\ &= -0.874 - j0.062 \end{aligned}$$

or in megawatts and megavars,

$$P_{21} - jQ_{21} = -87.4 - j6.2$$

All line flows for the system are given in Table 8.9.

The slack bus power can be determined by summing the flows on the

Table 8.9 Calculated line flows for sample system

Bus code p–q	Line flows	
	Megawatts	Megavars
1–2	88.8	−8.6
1–3	40.7	1.1
2–1	−87.4	6.2
2–3	24.7	3.5
2–4	27.9	3.0
2–5	54.8	7.4
3–1	−39.5	−3.0
3–2	−24.3	−6.8
3–4	18.9	−5.1
4–2	−27.5	−5.9
4–3	−18.9	3.2
4–5	6.3	−2.3
5–2	−53.7	−7.2
5–4	−6.3	−2.8

lines terminating at the slack bus. The real slack bus power is 129.5 megawatts and the reactive power is −7.5 megavars.

b. The matrix equation for the solution of a load flow by the Newton-Raphson method is

$$\begin{bmatrix} \Delta P^k \\ \hline \Delta Q^k \end{bmatrix} = \begin{bmatrix} J_1^k & J_2^k \\ \hline J_3^k & J_4^k \end{bmatrix} \begin{bmatrix} \Delta e^k \\ \hline \Delta f^k \end{bmatrix}$$

This equation does not include the slack bus. The changes in bus powers are obtained from

$$\Delta P_p^k = P_{p\,(\text{scheduled})} - P_p^k$$
$$\Delta Q_p^k = Q_{p\,(\text{scheduled})} - Q_p^k$$

where $P_{p\,(\text{scheduled})}$ and $Q_{p\,(\text{scheduled})}$ are the net bus powers in per unit and are obtained from Table 8.3. The calculated bus powers are obtained from the equations

$$P_p^k = \sum_{q=1}^{n} \{e_p^k(e_q^k G_{pq} + f_q^k B_{pq}) + f_p^k(f_q^k G_{pq} - e_q^k B_{pq})\}$$

$$Q_p^k = \sum_{q=1}^{n} \{f_p^k(e_q^k G_{pq} + f_q^k B_{pq}) - e_p^k(f_q^k G_{pq} - e_q^k B_{pq})\}$$

using the initial bus voltages given in Table 8.3 and the elements of the bus admittance matrix.

The real and reactive power for bus 2 are

$$
\begin{aligned}
P_2^{(0)} = {} & 1.0\{1.06(-5.00000) + 0.0(-15.00000)\} \\
& + 0.0\{0.0(-5.00000) - 1.06(-15.00000)\} \\
& + 1.0\{1.0(10.83334) + 0.0(32.41500)\} \\
& + 0.0\{0.0(10.83334) - 1.0(32.41500)\} \\
& + 1.0\{1.0(-1.66667) + 0.0(-5.00000)\} \\
& + 0.0\{0.0(-1.66667) - 1.0(-5.00000)\} \\
& + 1.0\{1.0(-1.66667) + 0.0(-5.00000)\} \\
& + 0.0\{0.0(-1.66667) - 1.0(-5.00000)\} \\
& + 1.0\{1.0(-2.50000) + 0.0(-7.50000)\} \\
& + 0.0\{0.0(-2.50000) - 1.0(-7.50000)\} \\
= {} & -0.30000
\end{aligned}
$$

and

$$
\begin{aligned}
Q_2^{(0)} = {} & 0.0\{1.06(-5.00000) + 0.0(-15.00000)\} \\
& - 1.0\{0.0(-5.00000) - 1.06(-15.00000)\} \\
& + 0.0\{1.0(10.83334) + 0.0(32.41500)\} \\
& - 1.0\{0.0(10.83334) - 1.0(32.41500)\} \\
& + 0.0\{1.0(-1.66667) + 0.0(-5.00000)\} \\
& - 1.0\{0.0(-1.66667) - 1.0(-5.00000)\} \\
& + 0.0\{1.0(-1.66667) + 0.0(-5.00000)\} \\
& - 1.0\{0.0(-1.66667) - 1.0(-5.00000)\} \\
& + 0.0\{1.0(-2.50000) + 0.0(-7.50000)\} \\
& - 1.0\{0.0(-2.50000) - 1.0(-7.50000)\} \\
= {} & -0.98500.
\end{aligned}
$$

The powers for the remaining buses are

$$
\begin{aligned}
P_3^{(0)} &= -0.07500 \\
P_4^{(0)} &= 0.0 \\
P_5^{(0)} &= 0.0 \\
Q_3^{(0)} &= -0.28000 \\
Q_4^{(0)} &= -0.05500 \\
Q_5^{(0)} &= -0.04000
\end{aligned}
$$

The changes in the real and reactive power for bus 2 are

$$
\begin{aligned}
\Delta P_2^{(0)} &= 0.20000 - (-0.30000) = 0.50000 \\
\Delta Q_2^{(0)} &= 0.20000 - (-0.98500) = 1.18500
\end{aligned}
$$

The changes in powers for the remaining buses are

$$\begin{aligned}
\Delta P_3^{(0)} &= -0.37500\\
\Delta P_4^{(0)} &= -0.40000\\
\Delta P_5^{(0)} &= -0.60000\\
\Delta Q_3^{(0)} &= 0.13000\\
\Delta Q_4^{(0)} &= 0.00500\\
\Delta Q_5^{(0)} &= -0.06000
\end{aligned}$$

The bus currents used to determine the elements of the Jacobian can be computed from the equation

$$I_p{}^k = \frac{P_p{}^k - jQ_p{}^k}{(E_p{}^k)^*}$$

The current for bus 2 is

$$\begin{aligned}
I_2^{(0)} &= \frac{-0.30000 - j(-0.98500)}{1.0 - j0.0}\\
&= -0.30000 + j0.98500
\end{aligned}$$

The components of the current for bus 2 are, then,

$$\begin{aligned}
c_2^{(0)} &= -0.30000\\
d_2^{(0)} &= 0.98500
\end{aligned}$$

The components of currents for the remaining buses are

$$\begin{aligned}
c_3^{(0)} &= -0.07500\\
d_3^{(0)} &= 0.28000\\
c_4^{(0)} &= 0.0\\
d_4^{(0)} &= 0.05500\\
c_5^{(0)} &= 0.0\\
d_5^{(0)} &= 0.04000
\end{aligned}$$

The elements of the Jacobian are calculated using the bus voltages and currents and elements of the bus admittance matrix. The diagonal element in the first row of $J_1{}^k$ from the equation

$$\frac{\partial P_p}{\partial e_p} = e_p{}^k G_{pp} - f_p{}^k B_{pp} + c_p{}^k$$

is

$$\begin{aligned}
\frac{\partial P_2}{\partial e_2} &= 1.0(10.83334) - 0.0(32.41500) + (-0.30000)\\
&= 10.53334
\end{aligned}$$

and the off-diagonal elements from the equation

$$\frac{\partial P_p}{\partial e_q} = e_p^k G_{pq} - f_p^k B_{pq}$$

are

$$\frac{\partial P_2}{\partial e_3} = 1.0(-1.66667) - 0.0(-5.00000) = -1.66667$$

$$\frac{\partial P_2}{\partial e_4} = 1.0(-1.66667) - 0.0(-5.00000) = -1.66667$$

$$\frac{\partial P_2}{\partial e_5} = 1.0(-2.50000) - 0.0(-7.50000) = -2.50000$$

The diagonal element in the first row of J_2^k from the equation

$$\frac{\partial P_p}{\partial f_p} = e_p^k B_{pp} + f_p^k G_{pp} + d_p^k$$

is

$$\frac{\partial P_2}{\partial f_2} = 1.0(32.41500) + 0.0(10.83334) + 0.98500$$
$$= 33.40000$$

and the off-diagonal elements from the equation

$$\frac{\partial P_p}{\partial f_q} = e_p^k B_{pq} + f_p^k G_{pq}$$

are

$$\frac{\partial P_2}{\partial f_3} = 1.0(-5.00000) + 0.0(-1.66667) = -5.00000$$

$$\frac{\partial P_2}{\partial f_4} = 1.0(-5.00000) + 0.0(-1.66667) = -5.00000$$

$$\frac{\partial P_2}{\partial f_5} = 1.0(-7.50000) + 0.0(-2.50000) = -7.50000$$

The diagonal element in the first row of J_3^k from the equation

$$\frac{\partial Q_p}{\partial e_p} = e_p^k B_{pp} + f_p^k G_{pp} - d_p^k$$

is

$$\frac{\partial Q_2}{\partial e_2} = 1.0(32.41500) + 0.0(10.83334) - 0.98500$$
$$= 31.43000$$

and the off-diagonal elements from the equation

$$\frac{\partial Q_p}{\partial e_q} = f_p{}^k G_{pq} + e_p{}^k B_{pq}$$

are

$$\frac{\partial Q_2}{\partial e_3} = 0.0(-1.66667) + 1.0(-5.00000) = -5.00000$$
$$\frac{\partial Q_2}{\partial e_4} = 0.0(-1.66667) + 1.0(-5.00000) = -5.00000$$
$$\frac{\partial Q_2}{\partial e_5} = 0.0(-2.50000) + 1.0(-7.50000) = -7.50000$$

The diagonal element in the first row of $J_4{}^k$ from the equation

$$\frac{\partial Q_p}{\partial f_p} = -e_p{}^k G_{pp} + f_p{}^k B_{pp} + c_p{}^k$$

is

$$\frac{\partial Q_2}{\partial f_2} = -1.0(10.83334) + 0.0(32.41500) + (-0.30000)$$
$$= -11.13334$$

and the off-diagonal elements from the equation

$$\frac{\partial Q_p}{\partial f_q} = f_p{}^k B_{pq} - e_p{}^k G_{pq}$$

are

$$\frac{\partial Q_2}{\partial f_3} = 0.0(-5.00000) - 1.0(-1.66667) = 1.66667$$
$$\frac{\partial Q_2}{\partial f_4} = 0.0(-5.00000) - 1.0(-1.66667) = 1.66667$$
$$\frac{\partial Q_2}{\partial f_5} = 0.0(-7.50000) - 1.0(-2.50000) = 2.50000$$

Repeating the process to obtain the elements for the remaining rows, the Jacobian when $k = 0$ is

10.53334	− 1.66667	− 1.66667	− 2.50000	33.40000	− 5.00000	− 5.00000	− 7.50000
− 1.66667	12.84167	−10.00000	0.0	− 5.00000	38.97500	−30.00000	0.0
− 1.66667	−10.00000	12.91667	− 1.25000	− 5.00000	−30.00000	38.75000	− 3.75000
− 2.50000	0.0	− 1.25000	3.75000	− 7.50000	0.0	− 3.75000	11.25000
31.43000	− 5.00000	− 5.00000	− 7.50000	−11.13334	1.66667	1.66667	2.50000
− 5.00000	38.41500	−30.00000	0.0	1.66667	−12.99167	10.00000	0.0
− 5.00000	−30.00000	38.64000	− 3.75000	1.66667	10.00000	−12.91667	1.25000
− 7.50000	0.0	− 3.75000	11.17000	2.50000	0.0	1.25000	− 3.75000

The solution of the matrix equation for Δe_p and Δf_p, $p = 2, 3, 4, 5,$ can be obtained directly from

$$\begin{bmatrix} \Delta e^k \\ \hline \Delta f^k \end{bmatrix} = \begin{bmatrix} J^{-1} \end{bmatrix} \begin{bmatrix} \Delta P^k \\ \hline \Delta Q^k \end{bmatrix}$$

where the inverse of the Jacobian is

.01826	.01403	.01492	.01726	.05478	.04208	.04477	.05177
.01403	.03167	.02823	.01888	.04208	.09502	.08469	.05665
.01492	.02823	.03367	.02131	.04477	.08469	.10101	.06393
.01726	.01888	.02131	.04577	.05177	.05665	.06393	.13670
.04771	.03499	.03752	.04428	−.01627	−.01214	−.01302	−.01529
.03499	.08567	.07551	.04846	−.01214	−.02933	−.02597	−.01687
.03752	.07551	.09188	.05559	−.01302	−.02597	−.03148	−.01930
.04428	.04846	.05559	.12796	−.01529	−.01687	−.01930	−.04357

The voltage change vector is

$$\begin{bmatrix} \Delta e^{(0)} \\ \hline \Delta f^{(0)} \end{bmatrix} = \begin{bmatrix} 0.05505 \\ 0.03176 \\ 0.03136 \\ 0.02652 \\ \hline -0.05084 \\ -0.09123 \\ -0.09747 \\ -0.11284 \end{bmatrix}$$

The new bus voltages are obtained from the equation

$$E_p^{k+1} = E_p{}^k + \Delta E_p{}^k$$

and after the first iteration are

$$\begin{aligned} E_2^{(1)} &= 1.05505 - j0.05084 \\ E_3^{(1)} &= 1.03176 - j0.09123 \\ E_4^{(1)} &= 1.03136 - j0.09747 \\ E_5^{(1)} &= 1.02652 - j0.11284 \end{aligned}$$

These values are used to compute the bus powers and currents and elements of the Jacobian for the next iteration. The changes in bus voltages for all iterations are given in Table 8.10 and the bus voltages in Table 8.11. The process is terminated when the changes in both real and reactive power at each bus are less than 0.01. The changes in powers for all iterations are given in Table 8.12.

The line flows can be calculated as shown previously and are given in Table 8.9.

Table 8.10 Changes in bus voltages from the Newton-Raphson solution using Y_{BUS}

Iteration count k	Changes in bus voltages			
	Bus 2	Bus 3	Bus 4	Bus 5
0	0.0 + j0.0	0.0 + j0.0	0.0 + j0.0	0.0 + j0.0
1	0.05505 − j0.05084	0.03176 − j0.09123	0.03136 − j0.09747	0.02652 − j0.11284
2	−0.00876 − j0.00044	−0.01133 + j0.00201	−0.01206 + j0.00239	−0.01424 + j0.00375

Table 8.11 Bus voltages from the Newton-Raphson solution using Y_{BUS}

Iteration count k	Bus voltages			
	Bus 2	Bus 3	Bus 4	Bus 5
0	1.0 + j0.0	1.0 + j0.0	1.0 + j0.0	1.0 + j0.0
1	1.05505 − j0.05084	1.03176 − j0.09123	1.03136 − j0.09747	1.02652 − j0.11284
2	1.04629 − j0.05128	1.02043 − j0.08922	1.01930 − j0.09508	1.01228 − j0.10909

Table 8.12 Changes in bus powers from the Newton-Raphson solution using Y_{BUS}

Iteration count k	Changes in bus powers			
	Bus 2	Bus 3	Bus 4	Bus 5
0	0.50000 − j1.18500	−0.37500 − j0.13000	−0.40000 − j0.00500	−0.60000 + j0.06000
1	−0.09342 + j0.03857	−0.00103 + j0.03586	0.01171 + j0.03871	0.02244 + j0.06563
2	−0.00073 + j0.00037	−0.00010 + j0.00037	0.00003 + j0.00044	0.00006 + j0.00094

c. The bus impedance matrix for the sample system with bus 1 as reference is

Z_{BUS} =

	②	③	④	⑤
②	$0.0168571 + j0.0505714$	$0.0125714 + j0.0377143$	$0.0134286 + j0.0402857$	$0.0157143 + j0.0471429$
③	$0.0125714 + j0.0377143$	$0.0297143 + j0.0891429$	$0.0262857 + j0.0788571$	$0.0171429 + j0.0514286$
④	$0.0134286 + j0.0402857$	$0.0262857 + j0.0788571$	$0.0317143 + j0.0951429$	$0.0195238 + j0.0585714$
⑤	$0.0157143 + j0.0471429$	$0.0171429 + j0.0514286$	$0.0195238 + j0.0585714$	$0.0436508 + j0.1309524$

which was obtained by using the algorithm described in Sec. 4.2.

Prior to initiating the iterative process it is necessary to calculate the bus currents with the scheduled net bus powers and assumed initial bus voltages. From the values given in Table 8.3, the net bus powers in per unit are obtained and substituted in the equation

$$I_p{}^k = \frac{P_p - jQ_p}{(E_p{}^k)^*} - y_p E_p{}^k$$

The currents for all buses, except the reference, are

$$I_2^{(0)} = \frac{0.20 - j0.20}{1.0 - j0.0} - (0.0 + j0.085)(1.0 + j0.0) = 0.200 - j0.285$$

$$I_3^{(0)} = \frac{-0.45 + j0.15}{1.0 - j0.0} - (0.0 + j0.055)(1.0 + j0.0) = -0.450 + j0.095$$

$$I_4^{(0)} = \frac{-0.40 + j0.05}{1.0 - j0.0} - (0.0 + j0.055)(1.0 + j0.0) = -0.400 - j0.005$$

$$I_5^{(0)} = \frac{-0.60 + j0.10}{1.0 - j0.0} - (0.0 + j0.040)(1.0 + j0.0) = -0.600 + j0.060$$

The first step in the iterative process is to calculate a new estimate of the voltage for bus 2 by multiplying the first row of the bus impedance matrix by the vector of bus currents as follows:

$$\begin{aligned} E_2^{(1)} - E_1 &= Z_{22}I_2^{(0)} + Z_{23}I_3^{(0)} + Z_{24}I_4^{(0)} + Z_{25}I_5^{(0)} \\ &= (0.0168751 + j0.0505714)(0.200 - j0.285) \\ &\quad + (0.0125714 + j0.0377143)(-0.450 + j0.095) \\ &\quad + (0.0134286 + j0.0402857)(-0.400 - j0.005) \\ &\quad + (0.0157143 + j0.0471429)(-0.600 + j0.060) \\ &= -0.00888 - j0.05399 \end{aligned}$$

Since the voltage at the reference bus has been specified, as given in Table 8.3,

$$\begin{aligned} E_2^{(1)} &= (-0.00888 - j0.05399) + (1.060 + j0.0) \\ &= 1.05112 - j0.05399 \end{aligned}$$

The new current for bus 2 is

$$\begin{aligned} I_2^{(1)} &= \frac{P_2 - jQ_2}{(E_2^{(1)})^*} - y_2 E_2^{(1)} \\ &= \frac{0.20 - j0.20}{1.05112 + j0.05399} - (0.0 + j0.085)(1.05112 - j0.05399) \\ &= 0.17544 - j0.028887 \end{aligned}$$

This new value of bus current replaces the previously calculated value for bus 2 in subsequent calculations.

A new estimate of the voltage for bus 3 is obtained next by multiplying the second row of the bus impedance matrix by the vector of bus currents as follows:

$$E_3^{(1)} - E_1 = Z_{32}I_2^{(1)} + Z_{33}I_3^{(0)} + Z_{34}I_4^{(0)} + Z_{35}I_5^{(0)}$$
$$E_3^{(1)} = 1.02777 - j0.09581$$

The new current for bus 3 is

$$I_3^{(1)} = -0.42585 + j0.12863$$

The process is continued to obtain

$$E_4^{(1)} = 1.02521 - j0.09920$$
$$I_4^{(1)} = -0.38732 + j0.02933$$

and

$$E_5^{(1)} = 1.01913 - j0.11403$$
$$I_5^{(1)} = -0.57518 + j0.12120$$

A second iteration is performed by repeating the process, starting again with bus 2. The bus voltages for all iterations are given in Table

Table 8.13 Bus voltages from the Gauss-Seidel iterative solution using Z_{BUS}

Iteration count k	Bus voltages			
	Bus 2	*Bus 3*	*Bus 4*	*Bus 5*
0	1.0 + j0.0	1.0 + j0.0	1.0 + j0.0	1.0 + j0.0
1	1.05112 − j0.05399	1.02777 − j0.09581	1.02521 − j0.09920	1.01913 − j0.11403
2	1.04622 − j0.05086	1.02041 − j0.08837	1.01924 − j0.09454	1.01220 − j0.10841
3	1.04622 − j0.05129	1.02035 − j0.08924	1.01918 − j0.09508	1.01212 − j0.10908

Table 8.14 Changes in bus voltages from the Gauss-Seidel iterative solution using Z_{BUS}

Iteration count k	Changes in bus voltages			
	Bus 2	*Bus 3*	*Bus 4*	*Bus 5*
0	0.0 + j0.0	0.0 + j0.0	0.0 + j0.0	0.0 + j0.0
1	0.05112 − j0.05399	0.02777 − j0.09581	0.02521 − j0.09920	0.01913 − j0.11403
2	−0.00490 + j0.00313	−0.00736 + j0.00744	−0.00597 + j0.00466	−0.00693 + j0.00562
3	0.00000 − j0.00043	−0.00006 − j0.00087	−0.00006 − j0.00054	−0.00008 − j0.00067

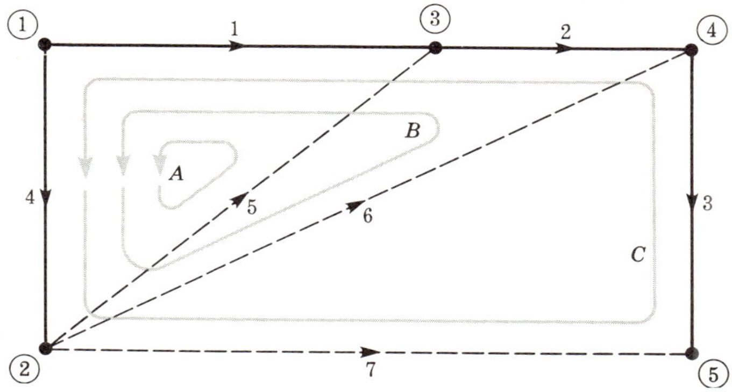

Fig. 8.9 Tree, cotree, and basic loops of the oriented connected graph for sample power system.

8.13 and the changes in bus voltages in Table 8.14. The process is terminated when the changes in both components of the voltage at each bus are less than 0.001. The line flows can be calculated as shown previously and are given in Table 8.9.

d. The tree, cotree, and basic loops of the oriented connected graph for the power system are shown in Fig. 8.9. The branch-path incidence matrix is

$K =$

b \ path	②	③	④	⑤
1		−1	−1	−1
2			−1	−1
3				−1
4	−1			

The basic loop incidence matrix is

$C =$

e \ l	A	B	C
1	−1	−1	−1
2		−1	−1
3			−1
4	1	1	1
5	1		
6		1	
7			1

The loop admittance matrix is

$Y_{LOOP} =$

	A	B	C
A	$1.07143 - j3.21429$	$-0.47619 + j1.42857$	$-0.23809 + j0.71429$
B	$-0.47619 + j1.42857$	$1.06349 - j3.19048$	$-0.30159 + j0.90476$
C	$-0.23809 + j0.71429$	$-0.30159 + j0.90476$	$0.68254 - j2.04762$

which was obtained by first forming the loop impedance matrix by singular transformation and then taking its inverse. The loop admittance matrix can be derived also from the bus impedance matrix by using the algorithm described in Sec. 4.5.

The first step in the iterative process is to calculate the bus currents with the scheduled bus powers and assumed initial bus voltages. The currents in per unit for all buses except the slack are determined from the equation

$$I_p^k = \frac{P_p - jQ_p}{(E_p^k)^*} - y_p E_p^k$$

These currents, identical to the initial bus currents calculated in the bus impedance method, are

$$
\begin{aligned}
I_2^{(0)} &= 0.200 - j0.285 \\
I_3^{(0)} &= -0.450 + j0.095 \\
I_4^{(0)} &= -0.400 - j0.005 \\
I_5^{(0)} &= -0.600 + j0.060
\end{aligned}
$$

The changes in bus currents are calculated from

$$\Delta I_p{}^k = I_p{}^k - I_p^{k-1}$$

Since the bus currents prior to initiating the iterative procedure were assumed to be zero, the changes in bus currents are

$$
\begin{aligned}
\Delta I_2^{(0)} &= 0.200 - j0.285 \\
\Delta I_3^{(0)} &= -0.450 + j0.095 \\
\Delta I_4^{(0)} &= -0.400 - j0.005 \\
\Delta I_5^{(0)} &= -0.600 + j0.060
\end{aligned}
$$

The first changes in branch currents from the equation

$$\overline{\Delta i_b} = K\overline{\Delta I}_{BUS}^{(0)}$$

are

$$
\overline{\Delta i_b} =
\begin{array}{c}
\\ 1 \\ 2 \\ 3 \\ 4
\end{array}
\begin{array}{c}
\begin{array}{cccc} ② & ③ & ④ & ⑤ \end{array} \\
\left[\begin{array}{cccc}
 & -1 & -1 & -1 \\
 & & -1 & -1 \\
 & & & -1 \\
-1 & & &
\end{array}\right]
\end{array}
\begin{array}{c}
② \\ ③ \\ ④ \\ ⑤
\end{array}
\left[\begin{array}{c}
0.200 - j0.285 \\
-0.450 + j0.095 \\
-0.400 - j0.005 \\
-0.600 + j0.060
\end{array}\right]
$$

$$
=
\begin{array}{c}
1 \\ 2 \\ 3 \\ 4
\end{array}
\left[\begin{array}{c}
1.450 - j0.150 \\
1.000 - j0.055 \\
0.600 - j0.060 \\
-0.200 + j0.285
\end{array}\right]
$$

Since the initial currents in the elements were assumed zero, the new currents are

$$\bar{\imath} = \begin{array}{c|c|} & \\ 1 & 1.450 - j0.150 \\ 2 & 1.000 - j0.055 \\ 3 & 0.600 - j0.060 \\ 4 & -0.200 + j0.285 \\ 5 & 0.0 \quad + j0.0 \\ 6 & 0.0 \quad + j0.0 \\ 7 & 0.0 \quad + j0.0 \end{array}$$

Then, from the equation

$$\bar{E}_{LOOP}^{k} = -C^t[z]\bar{\imath}$$

the loop voltages are

$$\bar{E}_{LOOP}^{(0)} = - \begin{array}{c} \\ A \\ B \\ C \end{array} \begin{bmatrix} 1 & 2 & 3 & 4 & 5 & 6 & 7 \\ -1 & & & 1 & 1 & & \\ -1 & -1 & & 1 & & 1 & \\ -1 & -1 & -1 & 1 & & & 1 \end{bmatrix} \begin{bmatrix} 0.08 + j0.24 & & & & & & \\ & 0.01 + j0.03 & & & & & \\ & & 0.08 + j0.24 & & & & \\ & & & 0.02 + j0.06 & & & \\ & & & & 0.06 + j0.18 & & \\ & & & & & 0.06 + j0.18 & \\ & & & & & & 0.04 + j0.12 \end{bmatrix} \begin{bmatrix} 1.450 - j0.150 \\ 1.000 - j0.055 \\ 0.600 - j0.060 \\ -0.200 + j0.285 \\ 0.0 + j0.0 \\ 0.0 + j0.0 \\ 0.0 + j0.0 \end{bmatrix}$$

$$= \begin{array}{c|c|} A & 0.17310 + j0.34230 \\ B & 0.18475 + j0.37175 \\ C & 0.24715 + j0.51095 \end{array}$$

From the loop performance equation the balancing loop currents are

$$\bar{I}^{(0)}_{LOOP} = \begin{array}{c|c|c|c|} & A & B & C \\ \hline A & 1.07143 - j3.21429 & -0.47619 + j1.42857 & -0.23809 + j0.71429 \\ \hline B & -0.47619 + j1.42857 & 1.06349 + j3.19048 & -0.30159 + j0.90476 \\ \hline C & -0.23809 + j0.71429 & -0.30159 + j0.90476 & 0.68254 - j2.04762 \\ \hline \end{array} \begin{array}{c|c|} \hline A & 0.17310 + j0.34230 \\ \hline B & 0.18475 + j0.37175 \\ \hline C & 0.24715 + j0.51095 \\ \hline \end{array}$$

$$= \begin{array}{c|c|} \hline A & 0.24286 - j0.04786 \\ \hline B & 0.27429 - j0.04029 \\ \hline C & 0.53714 - j0.06014 \\ \hline \end{array}$$

The new branch currents from the equation

$$\bar{i}_b^k = \bar{i}_b + C_b \bar{I}^k_{LOOP}$$

are

$$\bar{i}_b^{(0)} = \begin{array}{c} 1 \\ 2 \\ 3 \\ 4 \end{array} \begin{bmatrix} 1.450 - j0.150 \\ 1.000 - j0.055 \\ 0.600 - j0.060 \\ -0.200 + j0.285 \end{bmatrix} + \begin{array}{c} \\ 1 \\ 2 \\ 3 \\ 4 \end{array} \begin{array}{c} \begin{array}{ccc} A & B & C \end{array} \\ \begin{bmatrix} -1 & -1 & -1 \\ & -1 & -1 \\ & & -1 \\ 1 & 1 & 1 \end{bmatrix} \end{array} \begin{array}{c} A \\ B \\ C \end{array} \begin{bmatrix} 0.24286 - j0.04786 \\ 0.27429 - j0.04029 \\ 0.53714 - j0.06014 \end{bmatrix}$$

$$= \begin{array}{c} 1 \\ 2 \\ 3 \\ 4 \end{array} \begin{bmatrix} 0.39572 - j0.00171 \\ 0.18857 + j0.04543 \\ 0.06286 + j0.00014 \\ 0.85428 + j0.13671 \end{bmatrix}$$

The new link currents from the equation

$$\bar{i}_l^k = \bar{i}_l^{k-1} + \bar{I}_{LOOP}^k$$

are

$$\bar{i}_l^{(0)} = \begin{array}{r|c|} 5 & 0.24286 - j0.04786 \\ 6 & 0.27429 - j0.04029 \\ 7 & 0.53714 - j0.06014 \end{array}$$

The new currents in all elements are, then,

$$\bar{i}^{(0)} = \begin{array}{r|c|} 1 & 0.39572 - j0.00171 \\ 2 & 0.18857 + j0.04543 \\ 3 & 0.06286 + j0.00014 \\ 4 & 0.85428 + j0.13671 \\ 5 & 0.24286 - j0.04786 \\ 6 & 0.27429 - j0.04029 \\ 7 & 0.53714 - j0.06014 \end{array}$$

These values replace the previous estimated flows.

The new bus voltages from the equation

$$\bar{E}_{BUS}^{k+1} = \bar{E}_s + K^t[z_{bb}]\bar{i}_b^{\,k}$$

are

$$\bar{E}_{BUS}^{(1)} = \begin{array}{c} ② \\ ③ \\ ④ \\ ⑤ \end{array}\begin{bmatrix} 1.06 + j0.0 \\ 1.06 + j0.0 \\ 1.06 + j0.0 \\ 1.06 + j0.0 \end{bmatrix} + \begin{array}{c} \\ ② \\ ③ \\ ④ \\ ⑤ \end{array}\begin{array}{cccc} 1 & 2 & 3 & 4 \\ \hline & & & -1 \\ -1 & & & \\ -1 & -1 & & \\ -1 & -1 & -1 & \end{array} \quad \begin{array}{c} \\ 1 \\ 2 \\ 3 \\ 4 \end{array}\begin{array}{cccc} 1 & 2 & 3 & 4 \\ \hline 0.08 + j0.24 & & & \\ & 0.01 + j0.03 & & \\ & & 0.08 + j0.24 & \\ & & & 0.02 + j0.06 \end{array} \quad \begin{array}{c} 1 \\ 2 \\ 3 \\ 4 \end{array}\begin{bmatrix} 0.39572 - j0.00171 \\ 0.18857 + j0.04543 \\ 0.06286 + j0.00014 \\ 0.85428 + j0.13671 \end{bmatrix}$$

$$= \begin{array}{c} ② \\ ③ \\ ④ \\ ⑤ \end{array}\begin{bmatrix} 1.05112 - j0.05399 \\ 1.02793 - j0.09483 \\ 1.02741 - j0.10095 \\ 1.02241 - j0.11604 \end{bmatrix}$$

The new bus voltages replace the initially assumed voltages. If all loop voltages are within a specified tolerance, the line flows are calculated. If any loop voltage is greater than the tolerance, the new bus voltages are used to calculate new estimates of bus currents. The remaining steps of the iterative calculation are repeated to obtain a second estimate for the bus voltages. The bus voltages for each iteration are shown in Table 8.15 and changes in bus voltages are shown in Table 8.16. The loop voltages for each iteration are shown in Table 8.17.

Table 8.15 Bus voltages from the Gauss iterative solution using Y_{LOOP}

Iteration count k	Bus voltages			
	Bus 2	Bus 3	Bus 4	Bus 5
0	1.0 + j0.0	1.0 + j0.0	1.0 + j0.0	1.0 + j0.0
1	1.05112 − j0.05399	1.02793 − j0.09483	1.02741 − j0.10095	1.02241 − j0.11604
2	1.04626 − j0.05073	1.02046 − j0.08826	1.01932 − j0.09405	1.01229 − j0.10778
3	1.04624 − j0.05131	1.02038 − j0.08925	1.01925 − j0.09511	1.01221 − j0.10912

Table 8.16 Changes in bus voltages from the Gauss iterative solution using Y_{LOOP}

Iteration count k	Changes in bus voltages			
	Bus 2	Bus 3	Bus 4	Bus 5
0	0.0 + j0.0	0.0 + j0.0	0.0 + j0.0	0.0 + j0.0
1	0.05112 − j0.05399	0.02793 − j0.09483	0.02741 − j0.10095	0.02241 − j0.11604
2	−0.00486 + j0.00326	−0.00747 + j0.00657	−0.00809 + j0.00690	−0.01012 + j0.00826
3	−0.00002 − j0.00058	−0.00008 − j0.00099	−0.00007 − j0.00106	−0.00008 − j0.00134

Table 8.17 Loop voltages from the Gauss iterative solution using Y_{LOOP}

Iteration count k	Loop voltages		
	Loop A	Loop B	Loop C
0	0.17310 + j0.34230	0.18475 + j0.37175	0.24715 + j0.51095
1	0.02559 − j0.02738	0.02806 − j0.02956	0.04065 − j0.04094
2	0.00034 + j0.00381	0.00035 + j0.00415	0.00036 + j0.00596

8.6 Voltage controlled buses

Voltage control at the terminal of a reactive power source

A modification of, or deviation from, the normal computational procedures for the solution of the load flow problem is required to take into account voltage controlled buses. At these buses the voltage magnitude and the real power are specified.

In the Gauss and Gauss-Seidel methods using Y_{BUS}, the reactive power at a voltage controlled bus p must be calculated before proceeding with the calculation of voltage at that bus. Separating the real and imaginary parts of the bus power equation

$$P_p - jQ_p = E_p^* \sum_{q=1}^{n} Y_{pq}E_q$$

the reactive bus power is

$$Q_p = e_p{}^2 B_{pp} + f_p{}^2 B_{pp} + \sum_{\substack{q=1 \\ q \neq p}}^{n} \{f_p(e_q G_{pq} + f_q B_{pq}) - e_p(f_q G_{pq} - e_q B_{pq})\} \quad (8.6.1)$$

where e_p and f_p are the components of voltage at bus p. The values of e_p and f_p must satisfy the relation

$$e_p{}^2 + f_p{}^2 = \{|E_p|_{(\text{scheduled})}\}^2 \quad (8.6.2)$$

in order to calculate the reactive bus power required to provide the scheduled bus voltage. The present estimates of $e_p{}^k$ and $f_p{}^k$ must be adjusted, therefore, to satisfy equation (8.6.2).

The phase angle of the estimated bus voltage is

$$\delta_p{}^k = \arctan \frac{f_p{}^k}{e_p{}^k}$$

Assuming that the angles of the estimated and scheduled voltages are equal, then adjusted estimates for $e_p{}^k$ and $f_p{}^k$ are

$$e^k_{p\ (\text{new})} = |E_p|_{(\text{scheduled})} \cos \delta_p{}^k$$
$$f^k_{p\ (\text{new})} = |E_p|_{(\text{scheduled})} \sin \delta_p{}^k$$

Substituting $e^k_{p\ (\text{new})}$ and $f^k_{p\ (\text{new})}$ in equation (8.6.1), the reactive power $Q_p{}^k$ is obtained and is used with $E^k_{p\ (\text{new})}$ for calculating the new voltage estimate E_p^{k+1}.

In actual practice the limits of reactive power source at the voltage controlled bus must be taken into account. If the calculated $Q_p{}^k$ exceeds the maximum capability $Q_{p(\text{max})}$ of the source the maximum value is taken as the reactive power at that bus. If the calculated value is less than minimum capability $Q_{p(\text{min})}$ the minimum value is used. In either case it is impossible to obtain a solution with the specified scheduled voltage and therefore $E^k_{p\,(\text{new})}$ cannot be used in the calculation of E_p^{k+1}.

The sequence of steps required to include the effects of voltage controlled buses in the Gauss-Seidel iterative method using Y_{BUS} is shown in Fig. 8.10.

In the relaxation method using Y_{BUS} the reactive power at a voltage controlled bus must be calculated prior to obtaining the new current I_p^{k+1}.

In the Newton-Raphson method the equations for a voltage controlled bus p are

$$P_p = \sum_{q=1}^{n} \{e_p(e_q G_{pq} + f_q B_{pq}) + f_p(f_q G_{pq} - e_q B_{pq})\}$$

and

$$|E_p|^2 = e_p^2 + f_p^2 \tag{8.6.3}$$

where equation (8.6.3) replaces the equation for the reactive power. The matrix equation relating the changes in bus powers and the square of voltage magnitudes to changes in real and imaginary components of voltage is

$$\begin{bmatrix} \Delta P \\ \Delta Q \\ \Delta |E|^2 \end{bmatrix} = \begin{bmatrix} J_1 & J_2 \\ J_3 & J_4 \\ J_5 & J_6 \end{bmatrix} \begin{bmatrix} \Delta e \\ \Delta f \end{bmatrix}$$

The elements of the submatrices J_1, J_2, J_3, and J_4 are calculated as shown in Sec. 8.3. The off-diagonal elements of J_5, from equation (8.6.3), are

$$\frac{\partial |E_p|^2}{\partial e_q} = 0 \qquad q \neq p$$

and the diagonal elements are

$$\frac{\partial |E_p|^2}{\partial e_p} = 2e_p$$

Similarly, the off-diagonal elements of J_6 are

$$\frac{\partial |E_p|^2}{\partial f_q} = 0 \qquad q \neq p$$

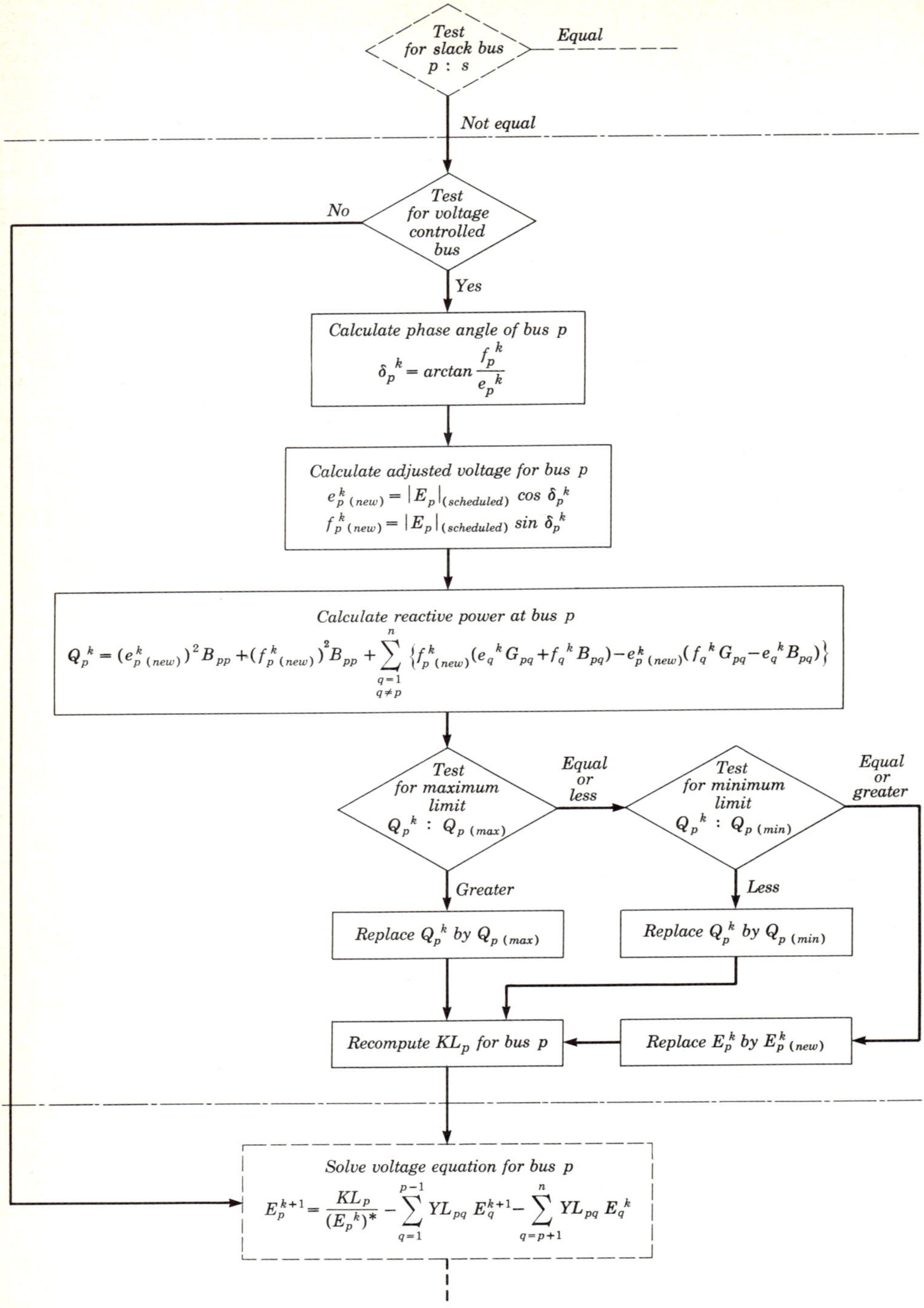

Fig. 8.10 Calculation of reactive power at voltage controlled buses in the Gauss-Seidel iterative method using Y_{BUS}.

and the diagonal elements are

$$\frac{\partial|E_p|^2}{\partial f_p} = 2f_p$$

The change in the square of the voltage magnitude at bus p is

$$\Delta|E_p^k|^2 = \{|E_p|_{(\text{scheduled})}\}^2 - |E_p^k|^2$$

If sufficient reactive capability is not available to hold the desired magnitude of bus voltage the reactive power must be fixed at a limit. In this case the bus is treated as a load bus with fixed reactive power.

In the Gauss-Seidel method using Z_{BUS} a correction to the reactive bus power can be calculated in order to provide the scheduled voltage (Brown, Carter, Happp, and Person, 1963). From the performance equation, the voltage at bus p is

$$e_p^{k+1} + jf_p^{k+1} = Z_{p1}I_1^{k+1} \cdots + Z_{pp}I_p^k \cdots + Z_{pn}I_n^k$$

The current at bus p can be corrected by ΔI_p^k to obtain

$$e_p + jf_p = Z_{p1}I_1^{k+1} \cdots + Z_{pp}(I_p^k + \Delta I_p^k) \cdots + Z_{pn}I_n^k$$

where e_p and f_p satisfy the equation (8.6.2). Subtracting the two voltage equations,

$$\Delta I_p^k = \frac{(e_p + jf_p) - (e_p^{k+1} + jf_p^{k+1})}{Z_{pp}}$$

or

$$\Delta I_p^k = \frac{(e_p^{k+1} + jf_p^{k+1})}{Z_{pp}} \left\{ \frac{e_p + jf_p}{e_p^{k+1} + jf_p^{k+1}} - 1 \right\}$$

Assuming that the phase angles of the scheduled voltage and E_p^{k+1} are equal, then,

$$\Delta I_p^k = \frac{(e_p^{k+1} + jf_p^{k+1})}{Z_{pp}} \left\{ \frac{|E_p|_{(\text{scheduled})}}{|E_p^{k+1}|} - 1 \right\}$$

The corresponding correction in the reactive power is

$$\Delta Q_p^k = -\text{Im}\{(E_p^{k+1})^* \Delta I_p^k\}$$

If the new reactive power

$$Q_p^{k+1} = Q_p^k + \Delta Q_p^k$$

is within the capability limits of the reactive source, then the new bus current is

$$I_p^{k+1} = \frac{P_p - jQ_p^{k+1}}{(E_p^{k+1})^*}$$

If Q_p^{k+1} is not within the reactive capability, the appropriate limit replaces the calculated value in the determination of the bus current. The new bus current is used in subsequent bus voltage calculations.

Voltage control at a remote bus

It is the practice in the operation of many power systems to control the voltage at a bus other than the terminal of a reactive source. This makes it necessary in the load flow solution to determine a reactive power at bus p that will hold the voltage magnitude specified for bus q, as shown in Fig. 8.11.

A procedure developed to accomplish this assumes a scheduled voltage magnitude for bus p. A reasonable first approximation is

$$\{|E_p|_{(\text{scheduled})}\}^{(0)} = |E_q|_{(\text{scheduled})}$$

During the iterative solution the reactive power at bus p is calculated in the usual manner using this assumed scheduled voltage. After the calculation of the voltage at bus q, however, the deviation from the scheduled voltage magnitude is determined from

$$\Delta|E_q|^k = |E_q|_{(\text{scheduled})} - |E_q^k|$$

where E_q^k is the calculated bus voltage. If the value of $\Delta|E_q|^k$ is greater than a specified tolerance, the scheduled voltage for bus p is reestimated from

$$\{|E_p|_{(\text{scheduled})}\}^{k+1} = \{|E_p|_{(\text{scheduled})}\}^k + \Delta|E_q|^k$$

This procedure has been employed in the Gauss-Seidel method using Y_{BUS}. During the iterative solution a change in the assumed scheduled voltage at bus p does not affect immediately the calculated voltage at bus q. It is necessary, therefore, to complete a number of iterations before reestimating the scheduled voltage for bus p. Tests have shown that five iterations are required to obtain sufficiently accurate changes in the calculated voltage at bus q for reestimating a new scheduled voltage for bus p. A voltage tolerance of 0.005 per unit provides acceptable results.

An alternate procedure is to change $\{|E_p|_{(\text{scheduled})}\}^{k+1}$ by a small specified amount each iteration until the magnitude of $\Delta|E_q|^k$ is less than the tolerance.

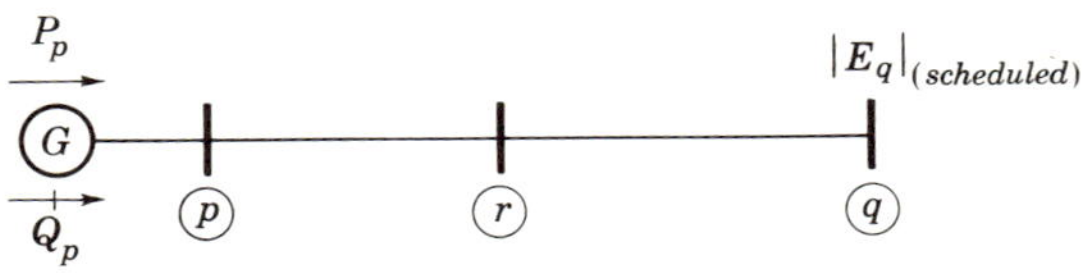

Fig. 8.11 Single line diagram of reactive power source and remote voltage controlled bus.

8.7 Representation of transformers

Fixed tap setting transformers

A transformer with off-nominal turns ratios can be represented by its impedance, or admittance, connected in series with an ideal autotransformer as shown in Fig. 8.12*a*. An equivalent π circuit can be obtained from this representation to be used in load flow studies. The elements of the equivalent π circuit, then, can be treated in the same manner as line elements.

The parameters of the equivalent π circuit, shown in Fig. 8.12*b*, can be derived by equating the terminal currents of the transformer with corresponding currents of the equivalent π circuit. At bus p the terminal current I_p of the transformer, shown in Fig. 8.12*a*, is

$$I_p = \frac{i_{tq}}{a}$$

where a is the turns ratio of the ideal autotransformer and i_{tq}, the current flowing from t to q, is

$$i_{tq} = (E_t - E_q)y_{pq}$$

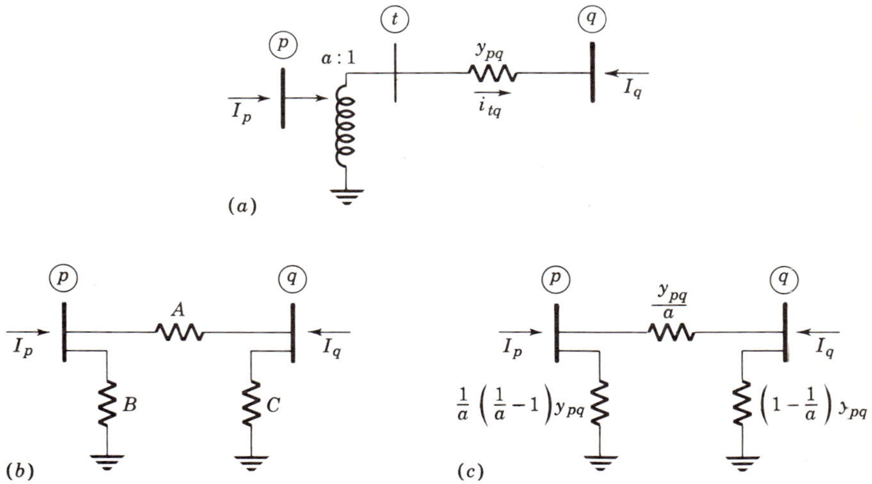

Fig. 8.12 Transformer representations. (a) Equivalent circuit; (b) equivalent π circuit; (c) equivalent π circuit with parameters expressed in terms of admittance and off-nominal turns ratios.

Therefore,

$$I_p = (E_t - E_q)\frac{y_{pq}}{a} \tag{8.7.1}$$

Since

$$E_t = \frac{E_p}{a}$$

equation (8.7.1) becomes

$$I_p = (E_p - aE_q)\frac{y_{pq}}{a^2} \tag{8.7.2}$$

Similarly, the terminal current I_q at bus q is

$$I_q = (E_q - E_t)y_{pq} \tag{8.7.3}$$

Substituting for E_t, equation (8.7.3) becomes

$$I_q = (aE_q - E_p)\frac{y_{pq}}{a} \tag{8.7.4}$$

The corresponding terminal currents for the equivalent π circuit shown in Fig. 8.12*b* are

$$I_p = (E_p - E_q)A + E_pB \tag{8.7.5}$$
$$I_q = (E_q - E_p)A + E_qC \tag{8.7.6}$$

Letting $E_p = 0$ and $E_q = 1$ in equation (8.7.2),

$$I_p = -\frac{y_{pq}}{a}$$

Letting $E_p = 0$ and $E_q = 1$ in equation (8.7.5),

$$I_p = -A$$

Since the terminal currents for the transformer and its equivalent π circuit must be equal,

$$A = \frac{y_{pq}}{a} \tag{8.7.7}$$

Similarly, substituting $E_p = 0$ and $E_q = 1$ in both equations (8.7.4) and (8.7.6),

$$I_q = y_{pq} \qquad \text{and} \qquad I_q = A + C$$

Again, since the terminal currents for the transformer and its equivalent must be equal,

$$y_{pq} = A + C$$

Substituting for A from equation (8.7.7) and solving for C,

$$C = y_{pq} - \frac{y_{pq}}{a}$$

$$= \left(1 - \frac{1}{a}\right) y_{pq}$$

Equating the current from equations (8.7.2) and (8.7.5) and substituting for A from (8.7.7),

$$(E_p - aE_q)\frac{y_{pq}}{a^2} = (E_p - E_q)\frac{y_{pq}}{a} + E_p B$$

Solving for B,

$$B = \frac{(E_p - aE_q)\dfrac{y_{pq}}{a^2} - (E_p - E_q)\dfrac{y_{pq}}{a}}{E_p}$$

$$= \frac{y_{pq}}{a^2} - \frac{y_{pq}}{a}$$

$$= \frac{1}{a}\left(\frac{1}{a} - 1\right) y_{pq}$$

The equivalent π circuit with its parameters expressed in terms of the turns ratio a and the transformer admittance are shown in Fig. 8.12c.

When the off-nominal turns ratio is represented at bus p for a transformer connecting p and q, the self-admittance at bus p is

$$Y_{pp} = y_{p1} \cdots + \frac{y_{pq}}{a} \cdots + y_{pn} + \frac{1}{a}\left(\frac{1}{a} - 1\right) y_{pq}$$

$$= y_{p1} + y_{p2} \cdots + \frac{y_{pq}}{a^2} \cdots + y_{pn}$$

The mutual admittance from p to q is

$$Y_{pq} = -\frac{y_{pq}}{a}$$

The self-admittance at bus q is

$$Y_{qq} = y_{q1} \cdots + \frac{y_{qp}}{a} \cdots + y_{qn} + \left(1 - \frac{1}{a}\right) y_{qp}$$

$$= y_{q1} \cdots + y_{qp} \cdots + y_{qn}$$

and is unchanged. The mutual admittance from q to p is

$$Y_{qp} = -\frac{y_{qp}}{a}$$

The equivalent π circuit shown in Fig. 8.12c can be used also in the methods employing the bus impedance matrix. The elements of Z_{BUS} are calculated with the equivalent impedance of a/y_{pq} from p to q. If in load flow calculations the elements of Z_{BUS} do not include the effect of shunt connections to ground, the total currents at buses p and q, respectively, are

$$I_p = \frac{P_p - jQ_p}{E_p^*} - y_pE_p - \frac{1}{a}\left(\frac{1}{a} - 1\right) y_{pq}E_p$$

$$I_q = \frac{P_q - jQ_q}{E_q^*} - y_qE_q - \left(1 - \frac{1}{a}\right) y_{pq}E_q$$

Tap changing under load transformers

In the representation of tap changing under load (TCUL) transformers, it is necessary to change the turns ratio to obtain the desired magnitude of voltage at a specified bus. This can be accomplished by changing the turns ratio by a small increment Δa once in any iteration when the voltage magnitude of bus q is such that

$$||E_q^k| - |E_q|_{(\text{scheduled})}| > \epsilon$$

The standard change in tap setting of TCUL transformers is $\pm 5/8$ percent per step. This value has proved satisfactory for Δa since it obviates, in general, additional iterations to obtain a voltage solution. It is not necessary to check the voltage magnitude of those buses controlled by TCUL transformers every iteration. Performing this check in alternate iterations has proved sufficient. A voltage magnitude tolerance ϵ of 0.01 per unit has given acceptable results.

The self-admittance Y_{pp} and the mutual admittances $Y_{pq} = Y_{qp}$ must be recalculated for every change in the tap setting of the transformer connecting buses p and q. In the Gauss and Gauss-Seidel iterative methods using Y_{BUS} the parameters L_p, YL_{pq}, YL_{qp}, and KL_p must be recomputed also. These calculations must be made before continuing the iterative solution.

Also, the elements of Z_{BUS} must be modified for every change in tap setting. These changes can be effected by adding a new element from bus p to q such that the series impedance of the π equivalent is

$$(a + \Delta a)z_{pq}$$

where a is the original turns ratio and Δa is the change. Let bz_{pq} be the impedance of the element to be added, as shown in Fig. 8.13. Then,

$$(a + \Delta a)z_{pq} = \frac{abz_{pq}^2}{az_{pq} + bz_{pq}}$$

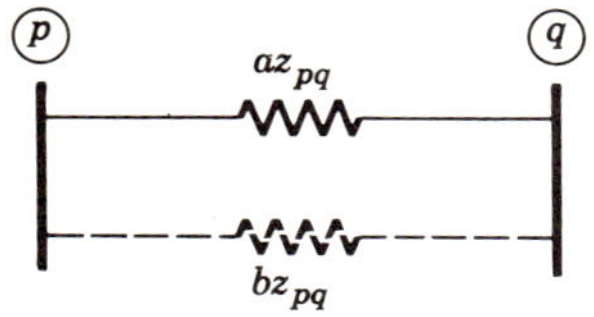

Fig. 8.13 Element added to network to reflect change in transformer tap setting

Solving for bz_{pq},

$$bz_{pq} = -\frac{a(a + \Delta a)}{\Delta a} z_{pq}$$

The change in tap setting of any transformer requires that every element of the Z_{BUS} matrix be recomputed. To avoid these extensive calculations an alternative equivalent can be used, in which the series impedance is made equal to the original transformer impedance and the shunt elements are varied to correspond to tap changes (Gupta and Humphrey-Davies, 1961).

Letting $A = y_{pq}$ and equating the corresponding terminal currents from equations (8.7.2) and (8.7.5) for the transformer and its equivalent, respectively, then

$$(E_p - E_q)y_{pq} + E_pB = (E_p - aE_q)\frac{y_{pq}}{a^2}$$

Solving for B,

$$\begin{aligned} B &= \left\{(E_p - aE_q)\frac{y_{pq}}{a^2} - (E_p - E_q)y_{pq}\right\}\frac{1}{E_p} \\ &= \left\{\left(\frac{1}{a^2} - 1\right) - \left(\frac{1}{a} - 1\right)\frac{E_q}{E_p}\right\} y_{pq} \\ &= \left(\frac{1}{a} - 1\right)\left\{\left(\frac{1}{a} + 1\right) - \frac{E_q}{E_p}\right\} y_{pq} \end{aligned} \tag{8.7.8}$$

Similarly, equating the terminal currents I_q from equations (8.7.4) and (8.7.6) with $A = y_{pq}$,

$$(E_q - E_p)y_{pq} + E_qC = (aE_q - E_p)\frac{y_{pq}}{a}$$

Solving for C,

$$\begin{aligned} C &= \left\{(aE_q - E_p)\frac{y_{pq}}{a} - (E_q - E_p)y_{pq}\right\}\frac{1}{E_q} \\ &= \left(1 - \frac{1}{a}\right)\frac{y_{pq}E_p}{E_q} \end{aligned} \tag{8.7.9}$$

The shunt admittances, (8.7.8) and (8.7.9), at buses p and q, respectively, are a function of the voltages E_p and E_q. The bus loading equations are, then,

$$I_p = \frac{P_p - jQ_p}{E_p^*} - y_p E_p - \left(\frac{1}{a} - 1\right)\left\{\left(\frac{1}{a} + 1\right) - \frac{E_q}{E_p}\right\} y_{pq} E_p$$

$$I_q = \frac{P_q - jQ_q}{E_q^*} - y_q E_q - \left(1 - \frac{1}{a}\right) y_{pq} E_p$$

Phase shifting transformers

A phase shifting transformer can be represented in load flow studies by its impedance, or admittance, connected in series with an ideal autotransformer having a complex turns ratio, as shown in Fig. 8.14. Then the terminal voltages E_p and E_s are related by

$$\frac{E_p}{E_s} = a_s + jb_s \tag{8.7.10}$$

Since there is no power loss in an ideal autotransformer,

$$E_p^* i_{pr} = E_s^* i_{sq} \tag{8.7.11}$$

It follows from equations (8.7.10) and (8.7.11) that

$$\frac{i_{pr}}{i_{sq}} = \frac{E_s^*}{E_p^*} = \frac{1}{a_s - jb_s}$$

Since

$$i_{sq} = (E_s - E_q) y_{pq}$$

then

$$i_{pr} = (E_s - E_q) \frac{y_{pq}}{a_s - jb_s}$$

Substituting for E_s from equation (8.7.10),

$$i_{pr} = \{E_p - (a_s + jb_s)E_q\} \frac{y_{pq}}{a_s^2 + b_s^2} \tag{8.7.12}$$

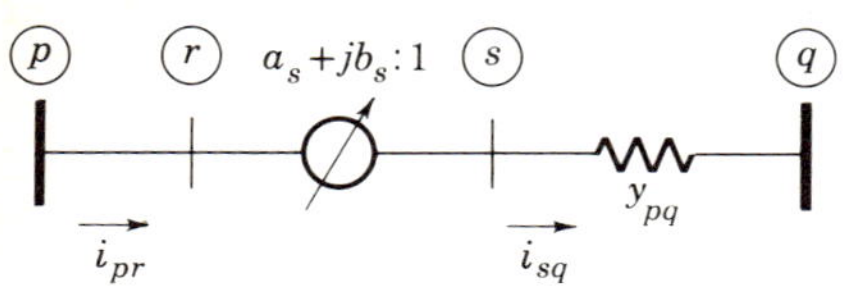

Fig. 8.14 Phase shifting transformer representation.

Similarly, the transformer current at bus q, i_{qs}, is

$$i_{qs} = (E_q - E_s)y_{pq}$$

Substituting again for E_s,

$$i_{qs} = \{(a_s + jb_s)E_q - E_p\} \frac{y_{pq}}{a_s + jb_s}$$

When a phase shifting transformer is connected between buses p and q, the self-admittance at bus p can be determined by letting E_p equal to one per unit and short circuiting all other network buses. Then

$$Y_{pp} = i_{p1} + i_{p2} \cdots + i_{pr} \cdots + i_{pn}$$

Substituting for i_{pr} from equation (8.7.12), and since

$$i_{p1} = y_{p1}$$
$$i_{p2} = y_{p2}$$
$$\cdots\cdots$$
$$i_{pn} = y_{pn}$$

then

$$Y_{pp} = y_{p1} + y_{p2} \cdots + \frac{y_{pq}}{a_s^2 + b_s^2} \cdots + y_{pn}$$

The current flowing out of bus q to bus p is $-i_{sq}$. Therefore the mutual admittance is

$$Y_{qp} = -i_{sq}$$

Then

$$Y_{qp} = -y_{pq}E_s$$

and from equation (8.7.10),

$$Y_{qp} = -\frac{y_{pq}}{a_s + jb_s}$$

Similarly, letting E_q equal one per unit and short circuiting all other network buses, the self-admittance at bus q is

$$Y_{qq} = i_{q1} + i_{q2} \cdots + i_{qs} \cdots + i_{qn}$$

or

$$Y_{qq} = y_{q1} + y_{q2} \cdots + y_{qp} \cdots + y_{qn}$$

The current flowing out of bus p to bus q is i_{pr}. Therefore the mutual admittance is

$$Y_{pq} = i_{pr}$$

Then

$$Y_{pq} = \frac{i_{sq}}{a_s - jb_s}$$

and therefore

$$Y_{pq} = \frac{y_{pq}}{a_s - jb_s}(E_s - E_q)$$

Since $E_s = 0$, then

$$Y_{pq} = -\frac{y_{pq}}{a_s - jb_s}$$

The complex turns ratio for a specified angular displacement and tap setting can be calculated from

$$a_s + jb_s = a(\cos\theta + j\sin\theta)$$

where

$$|E_p| = a|E_s|$$

If the phase shift from bus p to bus s is positive, that is, if the sign of θ is plus, then the voltage at bus p leads the voltage at bus s.

8.8 Tie line control

In studies involving several interconnected power systems, the load flow solution must satisfy a specified net power interchange for each system. The first step in the procedure of solving the problem is to calculate a voltage solution for the entire system, with an assumed generation schedule for each system. Next, using this voltage solution, the individual tie line flows are calculated and algebraically summed by system to determine the actual net power interchanges. Then, the actual and scheduled power interchanges for each system are compared to determine the adjustments that must be made in the assumed generation schedules.

A practical means of effecting the necessary changes in system generation is to select one generator in each system as a regulating generator. Each regulating generator is adjusted to satisfy the specified net power interchange. Thus, for system A, shown in Fig. 8.15, the actual net power interchange is

$$P_T{}^k = P_{F1}^k + P_{F2}^k - P_{F3}^k + P_{F4}^k$$

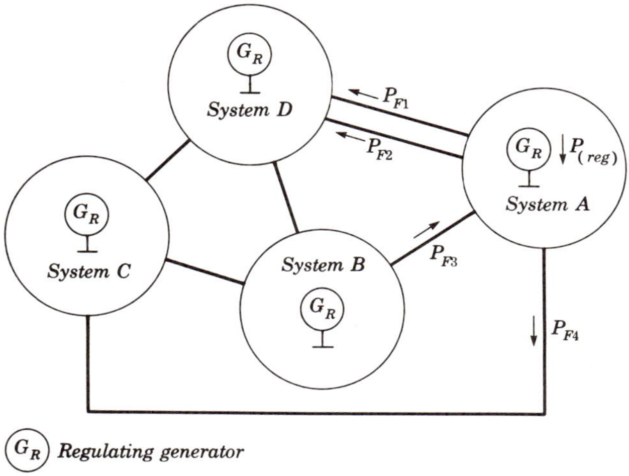

Fig. 8.15 Simplified representation of interconnected power systems.

The difference between the actual and scheduled interchanges is

$$\Delta P_T{}^k = P_{T\,(\text{scheduled})} - P_T{}^k$$

The new estimate of the power output for the regulating generator of system A is

$$P^{k+1}_{(\text{reg})} = P^{k}_{(\text{reg})} + \Delta P_T{}^k$$

Similar calculations are made for the other systems and a new iterative voltage solution is obtained. The process is repeated until all $\Delta P_T{}^k$ are less than or equal to a specified tolerance. A tolerance of 5 megawatts is usually acceptable.

8.9 Comparison of methods

An evaluation of the methods for obtaining a load flow solution must include the following:

1. The computing time required to process system input data in order to obtain the parameters for the iterative calculation
2. Computer programming and storage requirements
3. Iterative solution time
4. The computing time required to modify network data and to effect system operating changes

The first step in all load flow methods is the coding of the network and the formation of the appropriate network matrix. In the bus frame of reference the assignment of numbers to buses and to the corresponding terminals of the network elements provides adequate information to describe the network connections. In the loop frame of reference it is necessary also to identify the basic loops of the network.

The bus admittance matrix can be formed by a simple and straightforward procedure because mutual coupling is not involved. A diagonal element Y_{pp} of this matrix is equal to the sum of the admittances of the network elements connected to bus p. An off-diagonal element Y_{pq} is equal to the negative of the admittance of the network element connecting bus p to bus q. Moreover, since the bus admittance matrix is sparse, that is, a large number of elements are zero, relatively few elements have to be calculated.

With the bus admittance matrix it is possible to conserve computer storage because it is not necessary to store the zero elements. One way in which this can be accomplished is to store the nonzero elements along with a list of bus numbers corresponding to the rows and columns of the matrix.

The formation of the bus impedance matrix requires either a matrix inversion, nonsingular transformations, or the use of the algorithm. Unlike the bus admittance matrix, the bus impedance matrix is a full matrix that has no zero elements except in the row and column associated with the reference bus. Consequently, for a 101 bus system, of which one bus is the reference, 20,000 words of computer storage would be required to store the entire complex matrix. Since the bus impedance matrix is symmetrical, only the diagonal elements and half the off-diagonal elements need to be stored. This reduces the storage requirement to 10,100 words. In contrast, a 101 bus system with an average of four lines per bus would require only 1,000 words of computer memory to store all nonzero complex elements of the bus admittance matrix. Taking advantage of symmetry would reduce the storage to 600 words. In addition, space for the bus numbering list would be required.

The formation of the loop admittance matrix involves a matrix inversion, using either the loop impedance matrix obtained by a singular transformation or the augmented loop admittance matrix obtained by a nonsingular transformation. As an alternative the bus impedance matrix can be formed and then the algorithm can be used to obtain the loop admittance matrix. The loop admittance matrix is a full matrix.

Test computer programs were developed in order to evaluate the effectiveness of the methods presented for the load flow solution. These programs were used to obtain load flow solutions on actual power systems and to obtain relative solution times.

The computer time required to perform the iterative solution depends on the following:

1. The number of logical and arithmetic operations required to complete an iteration
2. The rate of convergence of the solution technique
3. The size and characteristics of the power system

As compared with the Gauss-Seidel method, the Gauss method using either the bus admittance matrix or the bus impedance matrix requires additional iterations to obtain a solution. Since the time per iteration for these two methods is about the same, the Gauss method was not evaluated in detail. The relaxation method using the bus admittance matrix also required more iterations plus additional time per iteration and therefore was not studied in detail.

In the development of the computer programs for the Gauss-Seidel and Newton-Raphson methods using the bus admittance matrix, advantage was taken of the sparsity of the network matrix in order to reduce the number of arithmetic operations per iteration. The Gauss-Seidel method was programmed using rectangular coordinates, and the Newton-Raphson method was programmed using polar coordinates.

The times per iteration obtained for the principal methods presented are shown in Fig. 8.16. The Gauss-Seidel method using the bus admittance matrix requires the fewest number of arithmetic operations to complete an iteration. This is because of the sparsity of the network matrix and the simplicity of the solution technique. Consequently, this method requires the least time per iteration. The Newton-Raphson method using the bus admittance matrix also takes advantage of the sparsity of the network matrix in order to reduce the number of arithmetic operations. However, the computation of the elements of the Jacobian for each iteration requires additional computer time. The time per iteration in both these methods increases directly as the number of buses of the network, because the number of nonzero elements added to the network matrix for each new bus is approximately the same.

The Gauss-Seidel method using the bus impedance matrix requires a relatively simple solution procedure. However, the time per iteration for this method is greater and varies approximately with the square of the number of buses, because the bus impedance matrix is a full matrix.

The Gauss method using the loop admittance matrix requires additional arithmetic and logical operations to relate bus and loop quantities during the iterative solution. The time per iteration also varies approximately with the square of the number of buses, because the loop admittance matrix is a full matrix.

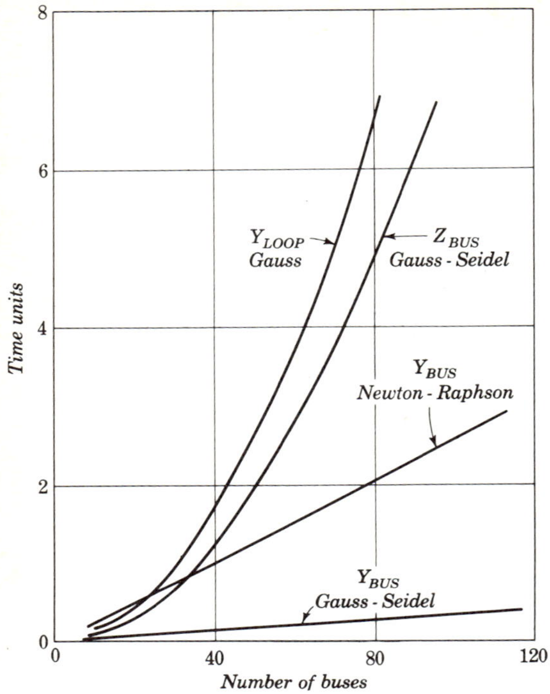

Fig. 8.16 ***Time per iteration for load flow methods.***

The rate of convergence of the Gauss-Seidel method using the bus admittance matrix is slow, requiring a relatively greater number of iterations to obtain a solution than the Newton-Raphson method and the methods using the bus impedance or loop admittance matrices. In addition, the number of iterations for the Gauss-Seidel method increased directly as the number of buses of the network, whereas the number of iterations for the other methods remained relatively constant, independent of system size. A significant increase in the rate of convergence can be obtained for the Gauss-Seidel method using the bus admittance matrix by applying acceleration factors.

The optimum values of acceleration factors for a load flow solution are difficult to calculate; however, they can be determined empirically. The selection of values for α and β, the acceleration factors for the real and imaginary components of voltage, depends on the characteristics of the network and the method of solution. The effectiveness of different acceleration factors on the rate of convergence for the principal methods presented is shown in Fig. 8.17. A system of 30 buses and 41 lines was used for this analysis.

The tolerance required to obtain a solution varies with the different methods. The slower converging Gauss-Seidel method using the bus admittance matrix required relatively smaller voltage tolerances to obtain comparable accuracy with that obtained by the methods using the bus impedance or loop admittance matrices. A voltage tolerance of 0.0001 per unit for both the real and imaginary components of voltage was used in the tests. For the Gauss-Seidel method using the bus impedance matrix a voltage tolerance of 0.001 per unit provided comparable results. A voltage tolerance of 0.01 per unit produced the required accuracy for the Gauss method using the loop admittance matrix.

The Newton-Raphson method using the bus admittance matrix has the advantage that the tolerances are specified for the net real and reactive powers at a bus. The tolerances, therefore, are given directly in quantities that are meaningful to the engineer who specifies the desired accuracy. Tolerances of 0.001 per unit for the real and reactive bus powers were used in the test calculations and produced comparable results. The number of iterations for different size systems along with the acceleration factors and tolerances used for each method are summarized in Table 8.18. The initial bus voltages were assumed equal to $1.0 + j0$ for all tests performed.

The time required for the iterative solution was least for the Newton-

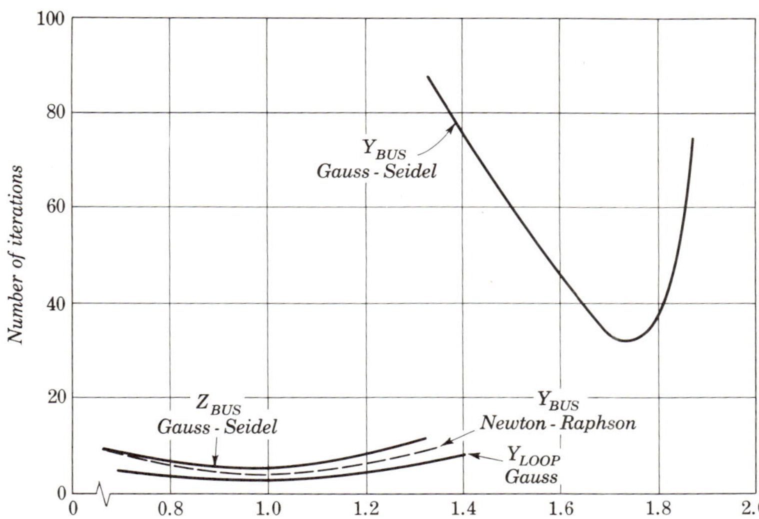

Fig. 8.17 Effect of acceleration factors on the rate of convergence for load flow solutions.

Table 8.18 Number of iterations for load flow solutions

Number of buses	Y_{BUS} Gauss-Seidel†	Y_{BUS} Newton-Raphson‡	Z_{BUS} Gauss-Seidel§	Y_{LOOP} Gauss¶
14	24	4	5	4
30	33	4	5	4
57	59	4	6	6
92	80	4	5	7
113	92	4	5	–

† Acceleration factors of 1.7 and 1.7 and tolerances of 0.0001 and 0.0001 per unit used for real and imaginary components of voltage.

‡ Tolerances of 0.001 and 0.001 per unit used for real and reactive bus powers. No acceleration.

§ Tolerances of 0.001 and 0.001 per unit used for real and imaginary components of voltage. No acceleration.

¶ Tolerances of 0.01 and 0.01 per unit used for real and imaginary components of voltage. No acceleration.

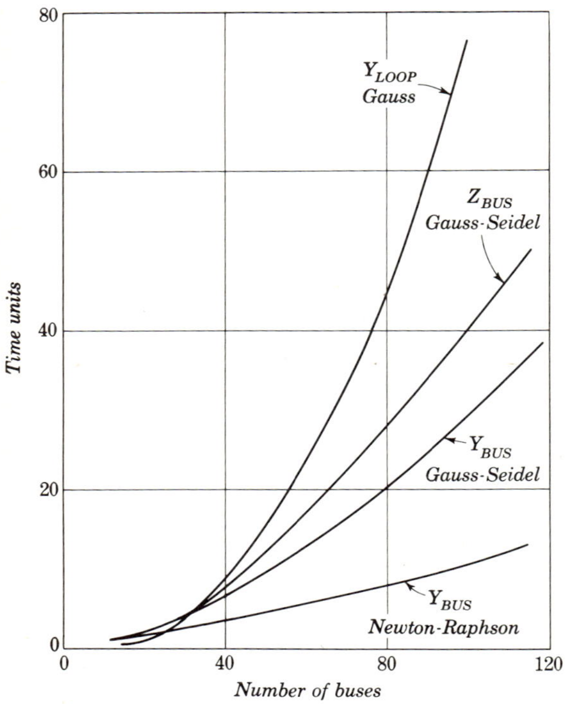

Fig. 8.18 Time for iterative solution not including the effects of voltage controlled buses.

Raphson method using the bus admittance matrix. The total iterative solution times for the principal methods are shown in Fig. 8.18. Voltage controlled buses were not represented in these initial tests.

When voltage controlled buses are represented, the Gauss-Seidel method using the bus admittance matrix usually requires fewer iterations to obtain a solution. However, a few more iterations usually are required for the Newton-Raphson method using the bus admittance matrix and the Gauss-Seidel method using the bus impedance matrix. The time per iteration for the Gauss-Seidel method using the bus admittance or bus impedance matrix increases as a result of the added computations. The time per iteration for the Newton-Raphson method decreases slightly because the number of arithmetic operations is reduced for the voltage controlled buses. The total iterative solution times for the principal methods when automatic voltage controlled buses are represented are shown in Fig. 8.19.

When a series of load flow solutions representing various system conditions are to be obtained, it is necessary to revise system data before proceeding from case to case. Network changes such as the addition

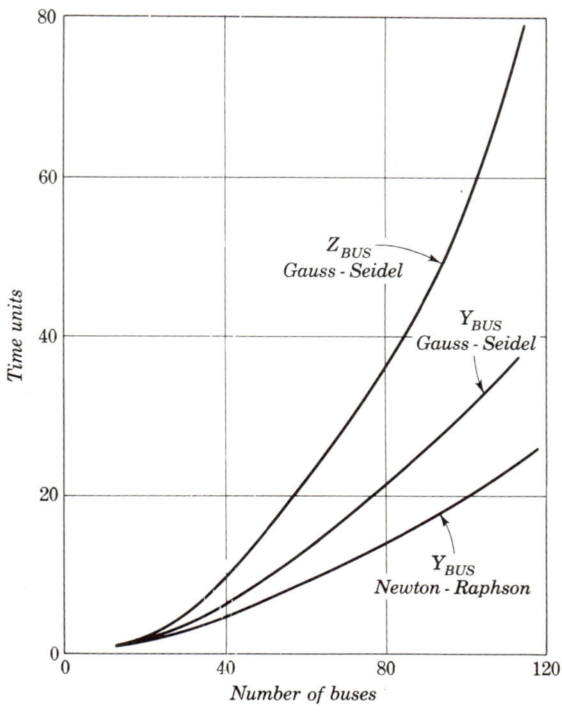

***Fig. 8.19** Time for iterative solution including the effects of voltage controlled buses.*

and removal of lines and transformers require that the network matrix be modified.

When the bus admittance matrix is used, it is necessary to recompute only those elements of the matrix that are associated with the terminals of the lines or transformers that are being changed. Because relatively few matrix elements are associated with any one bus, network changes can be effected simply and quickly. However, in order to modify the bus impedance matrix it is necessary to use the algorithm. Network changes that result in adding a new link make it necessary that all elements of the network matrix be modified. An algorithm would have to be developed in order to provide a means of modifying the loop admittance matrix.

The selection of initial values for bus voltages can have a marked effect on solution time. When a series of load flow calculations are performed, the usual procedure is to use the final calculated bus voltages of each case as the initial voltages for the next case. This tends to reduce the number of iterations, particularly when there are only minor changes in system conditions.

The actual computer time required for a load flow solution is dependent also on the speed of the digital computer and the efficiency of the program. The time units used in the comparison, therefore, would differ considerably from one digital computer to another. In general, however, each time unit is equal to about 1 sec for a medium size computer and to 0.1 sec or less for a large-scale computer.

8.10 Description of load flow program

Large-scale load flow computer programs incorporate many automatic features to facilitate their use in power system planning, operating, and interconnection studies. The principal objectives of these features are to make maximum use of the computer's capability and to minimize the number of manual operations required by the engineer in specifying and maintaining system data for the initial and subsequent load flow cases.

The American Electric Power Load Flow Program consists of an integrated set of computer programs to perform load flow calculations and associated data processing. The principal components of this program are:

Input

The input program provides the ability to read into the computer the power system data for a load flow calculation. This data is converted to the proper computer representation and stored in memory in the specified locations.

The information required for a load flow solution is divided into three parts. First is the base data which describes completely the network and operating conditions of the power system. This data includes line and transformer impedances, generation, loads, transformer taps, static capacitor and shunt reactor admittances, as well as information pertaining to the swing machine and the voltage regulating capability of the system. To facilitate data preparation all power system facilities are identified by actual power plant and substation names. For interconnection studies involving two or more systems, tie line information and interchange schedules are required. Second are the study title, case number, and control statements which govern the sequence of operations for the calculation of a series of load flows. Finally, there is the data required to effect changes in the system representation and operating conditions for the calculation of subsequent cases.

Data assembly

The data assembly program prepares and checks data and performs all computations preliminary to the iterative calculation. In the preparation of data this program assigns bus numbers and identifies the various facilities with each system in an interconnection study.

Voltage and power flow calculation

The program for the voltage and power flow calculation performs the iterative calculation to obtain bus voltages and then uses these voltages to compute line and transformer loadings.

Output

The output program uses system and station names together with assigned bus numbers to identify the load flow results. The first information printed includes the study title, load level, case number, remarks, and study totals. Next, the tie line flows and totals are printed for each system represented in the study. An example of the output listing of this information is shown in Fig. 8.20.

Detailed results are printed then for each system. First, transformer tap settings and voltage data are listed for the system, followed by the static capacitor and shunt reactor information. Finally, station conditions and line flows are printed. For line flows a plus value indicates a flow out of a bus and a minus value indicates a flow into a bus. An example of the output of this information is shown in Fig. 8.21.

System changes

The system change program provides a means of automatically effecting data changes in the calculation of a series of load flow cases. The data changes that can be made are divided into two types, network changes

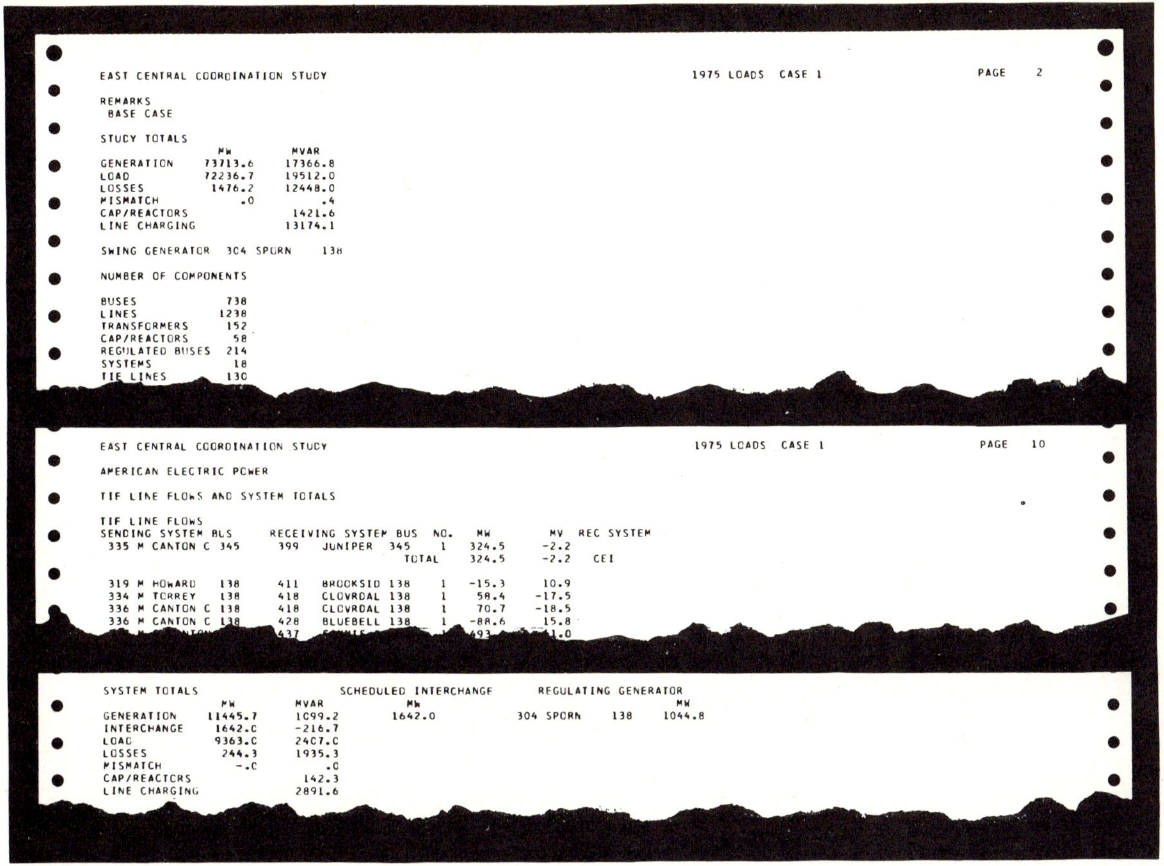

```
EAST CENTRAL COORDINATION STUDY                          1975 LOADS  CASE 1          PAGE   2

REMARKS
 BASE CASE

STUDY TOTALS
                    MW         MVAR
GENERATION      73713.6     17366.8
LOAD            72236.7     19512.0
LOSSES           1476.2     12448.0
MISMATCH             .0          .4
CAP/REACTORS                 1421.6
LINE CHARGING               13174.1

SWING GENERATOR   304 SPORN     138

NUMBER OF COMPONENTS

BUSES               738
LINES              1238
TRANSFORMERS        152
CAP/REACTORS         58
REGULATED BUSES     214
SYSTEMS              18
TIE LINES           130
```

```
EAST CENTRAL COORDINATION STUDY                          1975 LOADS  CASE 1          PAGE   10

AMERICAN ELECTRIC POWER

TIE LINE FLOWS AND SYSTEM TOTALS

TIE LINE FLOWS
SENDING SYSTEM BUS        RECEIVING SYSTEM BUS  NO.   MW         MV  REC SYSTEM
 335 M CANTON C 345        399   JUNIPER  345    1   324.5      -2.2
                                          TOTAL      324.5      -2.2   CEI

 319 M HOWARD   138        411   BROOKSID 138    1   -15.3      10.9
 334 M TORREY   138        418   CLOVRDAL 138    1    58.4     -17.5
 336 M CANTON C 138        418   CLOVRDAL 138    1    70.7     -18.5
 336 M CANTON C 138        428   BLUEBELL 138    1   -88.6      15.8
```

```
SYSTEM TOTALS                          SCHEDULED INTERCHANGE      REGULATING GENERATOR
                    MW         MVAR           MW                                   MW
GENERATION      11445.7      1099.2        1642.0            304 SPORN    138   1044.8
INTERCHANGE      1642.0      -216.7
LOAD             9363.0      2407.0
LOSSES            244.3      1935.3
MISMATCH            -.0          .0
CAP/REACTORS                  142.3
LINE CHARGING                2891.6
```

Fig. 8.20 Sample output of the American Electric Power Load Flow Program showing study totals and system totals and tie flows.

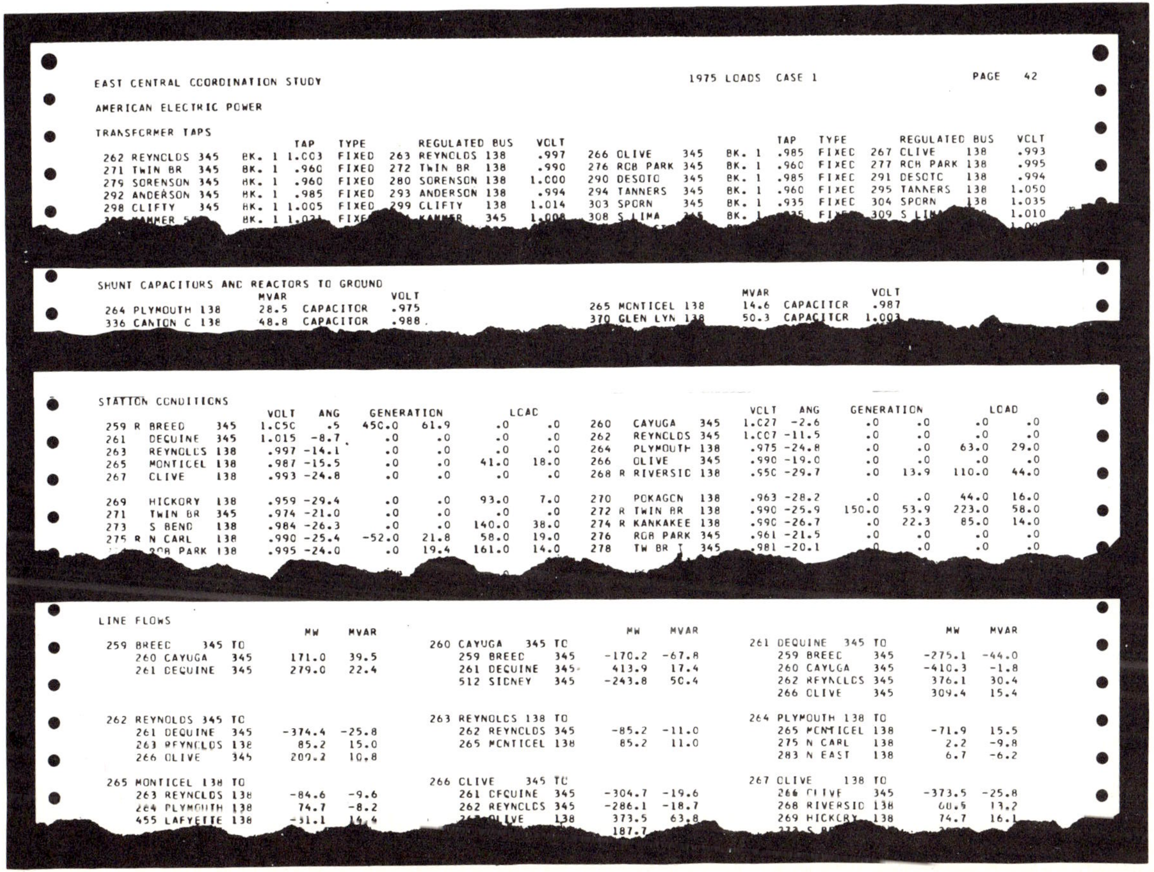

```
EAST CENTRAL COORDINATION STUDY                                   1975 LOADS  CASE 1                    PAGE   42

AMERICAN ELECTRIC POWER

TRANSFORMER TAPS
                           TAP    TYPE       REGULATED BUS    VOLT                                   TAP    TYPE       REGULATED BUS    VOLT
 262 REYNOLDS 345     BK. 1 1.003  FIXED  263 REYNOLDS 138     .997   266 OLIVE     345   BK. 1  .985  FIXED  267 CLIVE     138     .993
 271 TWIN BR  345     BK. 1  .960  FIXED  272 TWIN BR  138     .990   276 RCB PARK 345   BK. 1  .960  FIXED  277 RCH PARK 138     .995
 279 SORENSON 345     BK. 1  .960  FIXED  280 SORENSON 138    1.000   290 DESOTO    345   BK. 1  .985  FIXED  291 DESOTC   138     .994
 292 ANDERSON 345     BK. 1  .985  FIXED  293 ANDERSON 138     .994   294 TANNERS   345   BK. 1  .960  FIXED  295 TANNERS  138    1.050
 298 CLIFTY   345     BK. 1 1.005  FIXED  299 CLIFTY   138    1.014   303 SPORN     345   BK. 1  .935  FIXED  304 SPORN    138    1.035
                                                                                                                  309 S LIMA         1.010

SHUNT CAPACITORS AND REACTORS TO GROUND
                  MVAR               VOLT                                          MVAR               VOLT
 264 PLYMOUTH 138  28.5  CAPACITOR   .975             265 MONTICEL 138     14.6  CAPACITOR   .987
 336 CANTON C 138  48.8  CAPACITOR   .988             370 GLEN LYN 138     50.3  CAPACITOR  1.003

STATION CONDITIONS
                         VOLT   ANG    GENERATION       LOAD                                 VOLT   ANG    GENERATION        LOAD
 259 R BREED    345     1.050    .5   450.0   61.9     .0     .0    260   CAYUGA   345   1.027  -2.6      .0     .0      .0     .0
 261   DEQUINE  345     1.015  -8.7      .0     .0     .0     .0    262   REYNOLDS 345   1.007 -11.5      .0     .0      .0     .0
 263   REYNOLDS 138      .997 -14.1      .0     .0     .0     .0    264   PLYMOUTH 138    .975 -24.8      .0     .0    63.0   29.0
 265   MONTICEL 138      .987 -15.5      .0     .0   41.0   18.0    266   OLIVE    345    .990 -19.0      .0     .0      .0     .0
 267   CLIVE    138      .993 -24.8      .0     .0     .0     .0    268 R RIVERSIC 138    .550 -29.7      .0   13.9   110.0   44.0

 269   HICKORY  138      .959 -29.4      .0     .0   93.0    7.0    270   POKAGON  138    .963 -28.2      .0     .0    44.0   16.0
 271   TWIN BR  345      .974 -21.0      .0     .0     .0     .0    272 R TWIN BR  138    .990 -25.9   150.0   53.9   223.0   58.0
 273   S BEND   138      .984 -26.3      .0     .0  140.0   38.0    274 R KANKAKEE 138    .990 -26.7      .0   22.3    85.0   14.0
 275 R N CARL   138      .990 -25.4   -52.0   21.8   58.0   19.0    276   RCB PARK 345    .961 -21.5      .0     .0      .0     .0
       RCB PARK 138      .995 -24.0      .0   19.4  161.0   14.0    278   TW BR T  345    .981 -20.1      .0     .0      .0     .0

LINE FLOWS
                              MW     MVAR                                MW     MVAR                                MW     MVAR
 259 BREED      345 TO                        260 CAYUGA   345 TC                         261 DEQUINE  345 TO
     260 CAYUGA     345     171.0    39.5         259 BREED     345    -170.2   -67.8         259 BREED     345    -275.1   -44.0
     261 DEQUINE    345     279.0    22.4         261 DEQUINE   345     413.9    17.4         260 CAYUGA    345    -410.3    -1.8
                                                  512 SIDNEY    345    -243.8    50.4         262 REYNOLDS  345     376.1    30.4
                                                                                              266 OLIVE     345     309.4    15.4

 262 REYNOLDS 345 TO                          263 REYNOLDS 138 TO                         264 PLYMOUTH 138 TO
     261 DEQUINE    345    -374.4   -25.8         262 REYNOLDS  345     -85.2   -11.0         265 MONTICEL  138     -71.9    15.5
     263 REYNOLDS   138      85.2    15.0         265 MONTICEL  138      85.2    11.0         275 N CARL    138       2.2    -9.8
     266 OLIVE      345     209.2    10.8                                                     283 N EAST    138       6.7    -6.2

 265 MONTICEL 138 TO                          266 OLIVE    345 TO                         267 CLIVE    138 TO
     263 REYNOLDS   138     -84.6    -9.6         261 DEQUINE   345    -304.7   -19.6         266 CLIVE     345    -373.5   -25.8
     264 PLYMOUTH   138      74.7    -8.2         262 REYNOLDS  345    -286.1   -18.7         268 RIVERSIC  138      60.5    13.2
     455 LAFAYETTE  138     -31.1    14.4             OLIVE     138     373.5    63.8         269 HICKORY   138      74.7    16.1
                                                                        187.7
```

Fig. 8.21 ***Sample output of the American Electric Power Load Flow Program showing system power flows, voltages, and transformer tap settings.***

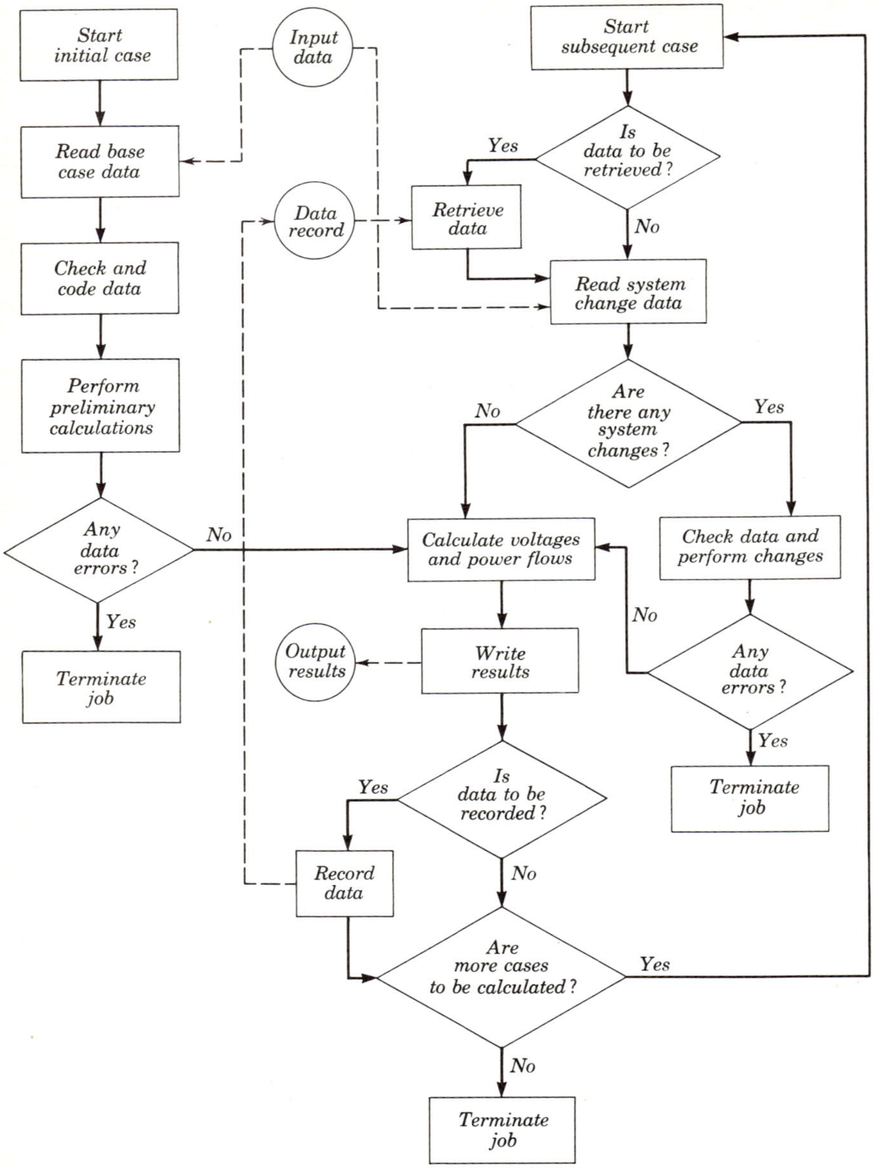

Fig. 8.22 Sequence of computer operations for the American Electric Power Load Flow Program.

and operating changes. Network changes include line and transformer additions, removals, or impedance modifications. Operating changes include changes in generation, loads, tap settings, capacitor, and reactors as well as revisions in the voltage schedule, regulating capability, and interchange schedule.

Data recording and retrieval

The data recording and retrieval program provides a means of preserving selected case data on an auxiliary storage device for later retrieval and use in studying alternate system conditions.

The sequence in which these programs are used to calculate a series of load flow cases is shown in Fig. 8.22.

Digital computers have provided the ability to obtain load flow solutions for very large interconnected systems. Interconnection studies involving the generation and transmission facilities of a dozen or more electric companies supplying major portions of twelve or more states are a continuing and important part of power system planning. Studies of this magnitude require representation of 1,000 or more buses and 2,000 or more lines and transformers.

Problems

8.1 The load flow data for the sample power system shown in Fig. 8.23 is given in Tables 8.19 and 8.20. The voltage magnitude at bus 2 is to be maintained at 1.03 per unit. The maximum and minimum reactive power limits of the generator at bus 2 are 35 and 0 megavars, respectively. With bus 1 as the slack use the following methods to obtain a load flow solution:

a. Gauss-Seidel using Y_{BUS} with acceleration factors of 1.4 and 1.4 and voltage tolerances of 0.001 and 0.001 per unit

b. Newton-Raphson using Y_{BUS} in rectangular coordinates with

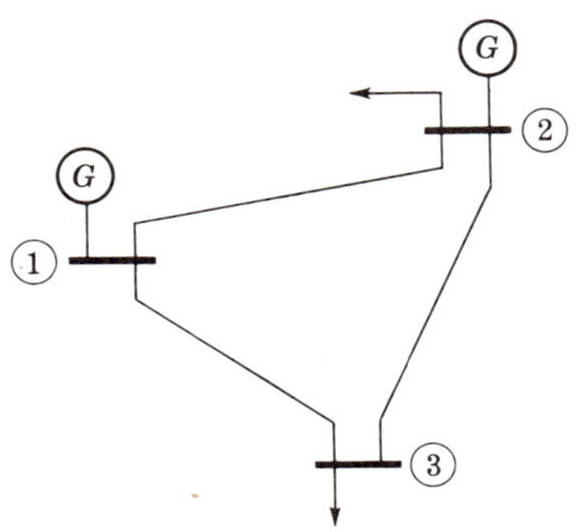

***Fig. 8.23** Sample power system for Prob. 8.1.*

tolerances of 0.01 per unit for the changes in real and reactive bus powers

c. Gauss-Seidel using Z_{BUS} with voltage tolerances of 0.01 and 0.01 per unit

8.2 With tolerances of 0.01 per unit for the changes in real and reactive bus powers, obtain a load flow solution for the sample system shown in Fig. 8.23 using the following methods:

a. Newton-Raphson using Y_{BUS} in rectangular coordinates. Assume the off-diagonal elements of the submatrices J_1, J_2, J_3, and J_4 of the Jacobian to be zero.

b. Newton-Raphson using Y_{BUS} in polar coordinates.

c. Newton-Raphson using Y_{BUS} in polar coordinates. Assume the submatrices J_2 and J_3 of the Jacobian to be zero.

Compare the convergence characteristics of these techniques and that of the method used in Prob. 8.1, part *b*.

8.3 Add to the sample system shown in Fig. 8.23 a second circuit from bus 1 to bus 3 with an impedance of $0.02 + j0.06$. Assume a fixed reactive generation of 25 megavars at bus 2 instead of maintaining voltage at that bus. Using the data given in Tables 8.19 and 8.20, obtain a load flow solution by the Gauss method using Y_{LOOP}. Let the loop voltage tolerances be 0.01 and 0.01.

Table 8.19 Impedances for sample system for Prob. 8.1

Bus code p–q	*Impedance* z_{pq}	*Line charging* $y'_{pq}/2$
1–2	$0.08 + j0.24$	0
1–3	$0.02 + j0.06$	0
2–3	$0.06 + j0.18$	0

Table 8.20 Scheduled generation and loads and assumed bus voltages for sample system for Prob. 8.1

Bus code	*Assumed*	*Generation*		*Load*	
p	*bus voltage*	*Megawatts*	*Megavars*	*Megawatts*	*Megavars*
1	$1.05 + j0$	0	0	0	0
2	$1.0 + j0$	20	0	50	20
3	$1.0 + j0$	0	0	60	25

8.4 The sample power system shown in Fig. 8.24 is composed of a tap changing under load transformer with an impedance of $0 + j0.03$. The load at bus 2 is 200 megawatts and 100 megavars. Let bus 1 be the slack and its voltage be $1.0 + j0$ per unit. Assume an initial tap setting of 1.0 and determine the required tap setting to hold a voltage magnitude of 1.0 per unit, within ± 0.005 per unit, at bus 2. Use the Gauss-Seidel method with Y_{BUS} and a ⅝ percent step per iteration for the tap change.

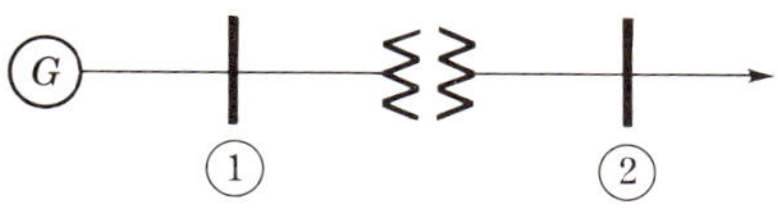

***Fig. 8.24** Sample power system for Prob. 8.4.*

8.5 The load flow data for the sample power system shown in Fig. 8.25 is given in Tables 8.21 and 8.22. The voltage magnitude at bus 2 is to be held at 1.0 per unit by means of the synchronous condenser at bus 3. The maximum and minimum reactive power limits of the condenser are 50 and -10 megavars, respectively. With bus 1 as the slack, use the Gauss-Seidel method and the bus admittance matrix to obtain a load flow solution. Use voltage tolerances of 0.001 and 0.001 per unit and acceleration factors of 1.4 and 1.4.

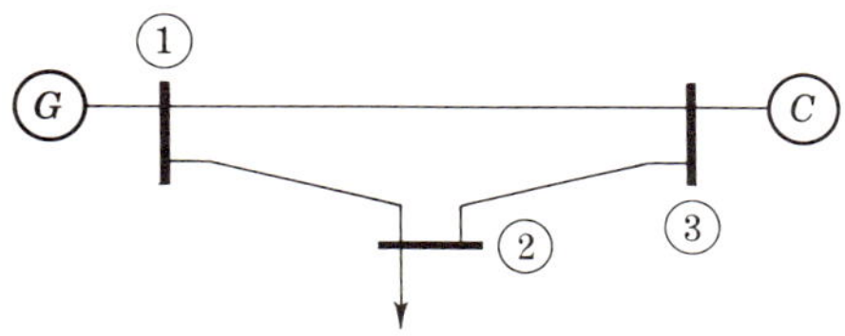

***Fig. 8.25** Sample power system for Prob. 8.5.*

***Table 8.21** Impedances for sample system for Prob. 8.5*

Bus code p–q	*Impedance* z_{pq}	*Line charging* $y'_{pq}/2$
1–2	$0 + j0.05$	0
1–3	$0 + j0.10$	0
2–3	$0 + j0.05$	0

Table 8.22 Scheduled generation and loads and assumed bus voltages for sample system for Prob. 8.5

Bus code p	Assumed bus voltage	Generation		Load	
		Megawatts	Megavars	Megawatts	Megavars
1	$1.03 + j0$	0	0	0	0
2	$1.00 + j0$	0	0	200	100
3	$1.00 + j0$	0	0	0	0

Bibliography

Bennett, J. M.: Digital Computers and the Load Flow Program, *Proc. Brit. Inst. Elec. Engrs.*, vol. 103, pt. B, suppl. 1, pp. 16–25, 1956.

Brown, H. E., G. K. Carter, H. H. Happ, and C. E. Person: Power Flow Solution by Impedance Matrix Iterative Method, *Trans. IEEE on Power Apparatus and Systems,* vol. 82, pp. 1–10, 1963.

Brown, Rodney J., and William F. Tinney: Digital Solutions for Large Power Networks, *Trans. AIEE,* vol. 76, pt. III, pp. 347–355, 1957.

Carpentier, M.: Application de la Methode de Newton au Calcul des Reseaux Mailles, Power Systems Computation Conference, Queen Mary College, London, 1963.

Cronin, J. H., and M. B. Newman: Digital Load Flow Program for 1,000 Bus Systems, *Trans. IEEE on Power Apparatus and Systems,* vol. 83, pp. 718–720, 1964.

Dunstan, L. A.: Machine Computation of Power Network Performance, *Trans. AIEE,* vol. 66, pp. 610–624, 1947.

Dunstan, L. A.: Digital Load Flow Studies, *Trans. AIEE,* vol. 73, pt. III-A, pp. 825–832, 1954.

Dyrkacz, M. S., and F. J. Maginniss: A New Automatic Program for Load Flow Studies on the IBM 704, *Trans. AIEE*, vol. 78, pt. III-A, pp. 52–62, 1959.

El-Abiad, A. H., M. Watson, and G. W. Stagg: The Load Flow Problem—Its Formulation and Solution—Part I, *Conf. Paper, AIEE Fall General Meeting,* 1961.

Glimn, A. F., and G. W. Stagg: Automatic Calculation of Load Flows, *Trans. AIEE,* vol. 76, pt. III, pp. 817–828, 1957.

Gupta, P. P., and M. W. Humphrey-Davies: Digital Computers in Power System Analysis, *Proc. Brit. Inst. Elec. Engrs.*, vol. 108, pt. A, pp. 383–404, 1961.

Hale, H. W., and R. W. Goodrich: Digital Computation of Power Flow—

Some New Aspects, *Trans. AIEE*, vol. 78, pt. IIIA, pp. 919–924, 1959.

Henderson, J. M.: Automatic Digital Computer Solution of Load Flow Studies, *Trans. AIEE*, vol. 73, pt. III-B, pp. 1696–1702, 1954.

Jordan, R. H.: Rapidly Converging Digital Load Flow, *Trans. AIEE*, vol. 76, pt. III, pp. 1433–1438, 1957.

Rindt, L. J., R. W. Long, and C. W. King: Load Flow Using Mesh Analysis, *Conf. Paper, AIEE Power Ind. Computer Appl. Conf.*, 1960.

St. Clair, H. P., and G. W. Stagg: Experience in Computation of Load Flow Studies Using High-speed Computers, *Trans. AIEE*, vol. 77, pt. III, pp. 1275–1282, 1958.

Van Ness, James E., and John H. Griffin: Elimination Methods for Load Flow Studies, *Trans. AIEE*, vol. 80, pt. III, pp. 299–304, 1961.

Ward, J. B., and H. W. Hale: Digital Computer Solution of Power-Flow Problems, *Trans. AIEE*, vol. 75, pt. III, pp. 398–404, 1956.

chapter 9
Numerical solution of differential equations

9.1 Introduction

Many complex physical systems are described by differential equations for which a solution cannot be determined in analytical form. However, techniques are available to obtain approximate solutions of such differential equations, or sets of equations, by numerical methods. Thus, the solution of differential equations is another important phase in numerical analysis.

In general, methods of numerical integration employ a step-by-step process to determine a series of values for each dependent variable corresponding to a selected set of values of the independent variable. The usual procedure is to select values of the independent variable at fixed intervals. The accuracy of a solution by numerical integration depends both on the method chosen and the size of the interval. Some of the methods frequently used are described in the following sections.

9.2 Numerical methods for solution of differential equations

Euler's method

Given the first-order differential equation

$$\frac{dy}{dx} = f(x,y) \tag{9.2.1}$$

where x is the independent variable and y is the dependent variable, the

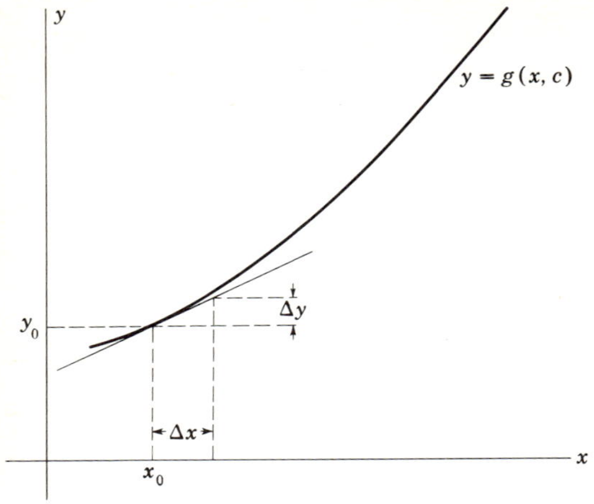

Fig. 9.1 Graph of the function for solution of a differential equation.

solution of equation (9.2.1) will be in the form

$$y = g(x,c) \tag{9.2.2}$$

where c is a constant determined from specified initial conditions. The curve representing equation (9.2.2) is shown in Fig. 9.1. Since this is a smooth curve, short segments can be assumed to be straight lines. Thus, at a particular point (x_0,y_0) on this curve,

$$\Delta y \approx \left.\frac{dy}{dx}\right|_0 \Delta x$$

where $\left.\frac{dy}{dx}\right|_0$ is the slope of the curve at the point (x_0,y_0) and is obtained by substituting x_0 and y_0 in equation (9.2.1). Hence, given the initial values x_0 and y_0 a new value of y can be determined for a specified Δx. Letting $h = \Delta x$, then

$$y_1 = y_0 + \Delta y \qquad \text{or} \qquad y_1 = y_0 + \left.\frac{dy}{dx}\right|_0 h$$

where Δy is the increment of y corresponding to the increment of x. In turn, a second value of y can be determined by

$$y_2 = y_1 + \left.\frac{dy}{dx}\right|_1 h$$

where

$$\left.\frac{dy}{dx}\right|_1 = f(x_1,y_1)$$

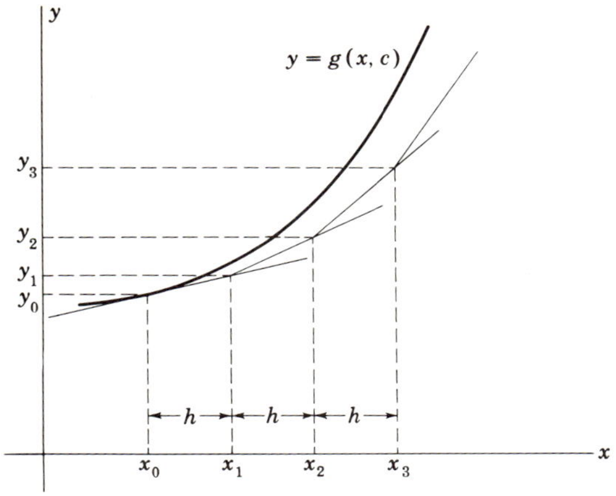

Fig. 9.2 Graph of approximate solution of a differential equation by Euler's method.

This process can be continued using

$$y_3 = y_2 + \left.\frac{dy}{dx}\right|_2 h$$

$$y_4 = y_3 + \left.\frac{dy}{dx}\right|_3 h$$

. . . .

to provide a table of x and y values for the integral solution of equation (9.2.1). This method is illustrated in Fig. 9.2.

The modified Euler method

In the application of Euler's method, a value of dy/dx calculated at the beginning of the interval is assumed to apply over the entire interval. An improvement can be obtained by calculating the new value of y for x_1 as before:

$$x_1 = x_0 + h$$

$$y_1^{(0)} = y_0 + \left.\frac{dy}{dx}\right|_0 h$$

and using these new values x_1 and $y_1^{(0)}$ in equation (9.2.1) to calculate the approximate value of $\left.\frac{dy}{dx}\right|_1$ at the end of the interval, i.e.,

$$\left.\frac{dy}{dx}\right|_1^{(0)} = f(x_1,y_1^{(0)})$$

Then, an improved value $y_1^{(1)}$ can be found by using the average of $\left.\frac{dy}{dx}\right|_0$ and $\left.\frac{dy}{dx}\right|_1^{(0)}$ as follows:

$$y_1^{(1)} = y_0 + \left(\frac{\left.\frac{dy}{dx}\right|_0 + \left.\frac{dy}{dx}\right|_1^{(0)}}{2} \right) h$$

Using x_1 and $y_1^{(1)}$, a third approximation $y_1^{(2)}$ can be obtained by the same process:

$$y_1^{(2)} = y_0 + \left(\frac{\left.\frac{dy}{dx}\right|_0 + \left.\frac{dy}{dx}\right|_1^{(1)}}{2} \right) h$$

And a fourth:

$$y_1^{(3)} = y_0 + \left(\frac{\left.\frac{dy}{dx}\right|_0 + \left.\frac{dy}{dx}\right|_1^{(2)}}{2} \right) h$$

This process can be continued until two consecutive estimates for y are equal within the desired tolerance. The entire process is then repeated to obtain y_2. The improved accuracy obtained with this modification of Euler's method is illustrated in Fig. 9.3.

Euler's method can be applied for the solution of simultaneous differential equations. Given the two equations

$$\frac{dy}{dx} = f_1(x,y,z)$$

$$\frac{dz}{dx} = f_2(x,y,z)$$

with the initial values x_0, y_0, and z_0, the new value y_1 will be

$$y_1 = y_0 + \left.\frac{dy}{dx}\right|_0 h$$

where

$$\left.\frac{dy}{dx}\right|_0 = f_1(x_0,y_0,z_0)$$

Similarly

$$z_1 = z_0 + \left.\frac{dz}{dx}\right|_0 h$$

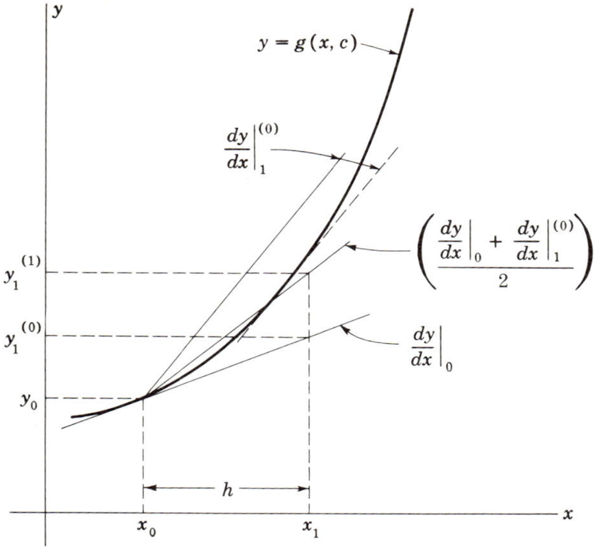

Fig. 9.3 Graph of approximate solution of a differential equation by the modified Euler method.

where

$$\left.\frac{dz}{dx}\right|_0 = f_2(x_0,y_0,z_0)$$

For the next increment, the values $x_1 = x_0 + h$, y_1, and z_1 are used to determine y_2 and z_2. In the modified Euler method y_1 and z_1 are used to evaluate the derivatives at x_1 for estimating the second approximations $y_1^{(1)}$ and $z_1^{(1)}$.

Picard's method of successive approximations

The basis of Picard's method is to determine a solution by approximating y as a function of x over a given range of x values, that is,

$$y \approx g(x)$$

This expression is evaluated by directly substituting values of x to obtain corresponding values for y. Given the differential equation (9.2.1),

$$dy = f(x,y)\,dx$$

and integrating between corresponding limits for x and y,

$$\int_{y_0}^{y_1} dy = \int_{x_0}^{x_1} f(x,y)\,dx$$

then

$$y_1 - y_0 = \int_{x_0}^{x_1} f(x,y)\,dx$$

or

$$y_1 = y_0 + \int_{x_0}^{x_1} f(x,y)\,dx \tag{9.2.3}$$

The integral term represents the change in y resulting from the change in x from x_0 to x_1. A solution can be obtained by evaluating the indicated integral by a method of successive approximations.

The first approximation for y as a function of x can be obtained by replacing y under the integral with y_0, the initial given value, that is,

$$y_1^{(1)} = y_0 + \int_{x_0}^{x_1} f(x,y_0)\,dx$$

and performing the indicated integration. This new value of y may now be substituted in equation (9.2.3) to obtain a second approximation for y, that is,

$$y_1^{(2)} = y_0 + \int_{x_0}^{x_1} f(x,y_1^{(1)})\,dx$$

The process may be repeated as many times as necessary to obtain the desired accuracy.

However, evaluating the integral may be complicated even though one of the variables is assumed fixed. This difficulty and the need to perform many integrations restrict the application of this method.

Picard's method can be applied to the solution of simultaneous equations such as

$$\frac{dy}{dx} = f_1(x,y,z)$$

$$\frac{dz}{dx} = f_2(x,y,z)$$

by using the formulas

$$y_1 = y_0 + \int_{x_0}^{x_1} f_1(x,y_0,z_0)\,dx$$

$$z_1 = z_0 + \int_{x_0}^{x_1} f_2(x,y_0,z_0)\,dx$$

The Runge-Kutta method

In the Runge-Kutta method the changes in the values of the dependent variable are calculated from a given set of formulas that are expressed in terms of the derivative evaluated at predetermined points. Since each value of y is uniquely determined by the formulas, this method does not require repeated approximations as in the modified Euler method or successive integrations as in Picard's method.

The formulas are derived by using an approximation to replace a truncated Taylor's series expansion. The Runge-Kutta second-order approximation can be written in the form

$$y_1 = y_0 + a_1 k_1 + a_2 k_2 \tag{9.2.4}$$

where

$$k_1 = f(x_0,y_0)h$$
$$k_2 = f(x_0 + b_1 h,\ y_0 + b_2 k_1)h$$

and the coefficients a_1, a_2, b_1, and b_2 are to be determined. First, expand $f(x_0 + b_1 h,\ y_0 + b_2 k_1)$ in a Taylor's series about (x_0,y_0); then,

$$k_2 = \left\{ f(x_0,y_0) + b_1 \frac{\partial f}{\partial x}\bigg|_0 h + b_2 k_1 \frac{\partial f}{\partial y}\bigg|_0 + \cdots \right\} h$$

Substituting for k_1 and two terms of the series for k_2 in equation (9.2.4), the approximation becomes

$$y_1 = y_0 + (a_1 + a_2) f(x_0,y_0)h + a_2 b_1 \frac{\partial f}{\partial x}\bigg|_0 h^2 + a_2 b_2 f(x_0,y_0) \frac{\partial f}{\partial y}\bigg|_0 h^2 \tag{9.2.5}$$

The Taylor's series expansion of y about (x_0,y_0) is

$$y_1 = y_0 + \frac{dy}{dx}\bigg|_0 h + \frac{d^2y}{dx^2}\bigg|_0 \frac{h^2}{2} + \cdots \tag{9.2.6}$$

Since

$$\frac{dy}{dx}\bigg|_0 = f(x_0,y_0) \qquad \text{and} \qquad \frac{d^2y}{dx^2}\bigg|_0 = \frac{\partial f}{\partial x}\bigg|_0 + \frac{\partial f}{\partial y}\bigg|_0 f(x_0,y_0)$$

equation (9.2.6) becomes

$$y_1 = y_0 + f(x_0,y_0)h + \frac{\partial f}{\partial x}\bigg|_0 \frac{h^2}{2} + \frac{\partial f}{\partial y}\bigg|_0 f(x_0,y_0) \frac{h^2}{2} + \cdots \tag{9.2.7}$$

Equating coefficients of equations (9.2.5) and (9.2.7), then

$$a_1 + a_2 = 1$$
$$a_2 b_1 = 1/2$$
$$a_2 b_2 = 1/2$$

Selecting an arbitrary value for a_1,

$$a_1 = 1/2$$

then

$$a_2 = \tfrac{1}{2}$$
$$b_1 = 1$$
$$b_2 = 1$$

Substituting these values in equation (9.2.4), the Runge-Kutta second-order approximation formula is

$$y_1 = y_0 + \tfrac{1}{2}k_1 + \tfrac{1}{2}k_2$$

where $k_1 = f(x_0,y_0)h$
$\quad\quad k_2 = f(x_0 + h,\ y_0 + k_1)h$

Hence,

$$\Delta y = \tfrac{1}{2}(k_1 + k_2)$$

The application of the Runge-Kutta method for a second-order approximation requires the computation of k_1 and k_2. The error in this approximation is of the order h^3 because the series was truncated after the second-order terms.

The general Runge-Kutta fourth-order approximation formula is

$$y_1 = y_0 + a_1k_1 + a_2k_2 + a_3k_3 + a_4k_4 \tag{9.2.8}$$

where $k_1 = f(x_0,y_0)h$
$\quad\quad k_2 = f(x_0 + b_1h,\ y_0 + b_2k_1)h$
$\quad\quad k_3 = f(x_0 + b_3h,\ y_0 + b_4k_2)h$
$\quad\quad k_4 = f(x_0 + b_5h,\ y_0 + b_6k_3)h$

Following the same procedure used for the second-order approximation, the coefficients in equation (9.2.8) are determined:

$$a_1 = \tfrac{1}{6}$$
$$a_2 = \tfrac{2}{6}$$
$$a_3 = \tfrac{2}{6}$$
$$a_4 = \tfrac{1}{6}$$

and

$$b_1 = \tfrac{1}{2}$$
$$b_2 = \tfrac{1}{2}$$
$$b_3 = \tfrac{1}{2}$$
$$b_4 = \tfrac{1}{2}$$
$$b_5 = 1$$
$$b_6 = 1$$

Substituting these values in equation (9.2.8), the Runge-Kutta fourth-order approximation becomes

$$y_1 = y_0 + \tfrac{1}{6}(k_1 + 2k_2 + 2k_3 + k_4)$$

where
$$\begin{aligned}
k_1 &= f(x_0,y_0)h \\
k_2 &= f\left(x_0 + \frac{h}{2}, y_0 + \frac{k_1}{2}\right)h \\
k_3 &= f\left(x_0 + \frac{h}{2}, y_0 + \frac{k_2}{2}\right)h \\
k_4 &= f(x_0 + h, y_0 + k_3)h
\end{aligned}$$

Thus, the calculation of Δy with this formula requires the computation of k_1, k_2, k_3, and k_4 and

$$\Delta y = \tfrac{1}{6}(k_1 + 2k_2 + 2k_3 + k_4)$$

The error in this approximation is of the order h^5.

Runge-Kutta fourth-order approximation formulas for simultaneous differential equations of the form

$$\frac{dy}{dx} = f(x,y,z)$$

$$\frac{dz}{dx} = g(x,y,z)$$

are

$$\begin{aligned}
y_1 &= y_0 + \tfrac{1}{6}(k_1 + 2k_2 + 2k_3 + k_4) \\
z_1 &= z_0 + \tfrac{1}{6}(l_1 + 2l_2 + 2l_3 + l_4)
\end{aligned}$$

where
$$\begin{aligned}
k_1 &= f(x_0,y_0,z_0)h \\
k_2 &= f\left(x_0 + \frac{h}{2}, y_0 + \frac{k_1}{2}, z_0 + \frac{l_1}{2}\right)h \\
k_3 &= f\left(x_0 + \frac{h}{2}, y_0 + \frac{k_2}{2}, z_0 + \frac{l_2}{2}\right)h \\
k_4 &= f(x_0 + h, y_0 + k_3, z_0 + l_3)h \\
l_1 &= g(x_0,y_0,z_0)h \\
l_2 &= g\left(x_0 + \frac{h}{2}, y_0 + \frac{k_1}{2}, z_0 + \frac{l_1}{2}\right)h \\
l_3 &= g\left(x_0 + \frac{h}{2}, y_0 + \frac{k_2}{2}, z_0 + \frac{l_2}{2}\right)h \\
l_4 &= g(x_0 + h, y_0 + k_3, z_0 + l_3)h
\end{aligned}$$

Predictor-corrector methods

Methods that are based on extrapolation, or integration ahead, and iteration for the solution of the differential equation

$$\frac{dy}{dx} = f(x,y) \tag{9.2.9}$$

are called predictor-corrector methods. The general procedure in a predictor-corrector method is to advance from the point (x_n,y_n) to the point (x_{n+1},y_{n+1}) by means of a formula that does not include the unknown derivative at the latter point; then, to determine $\left.\frac{dy}{dx}\right|_{n+1}$ from the differential equation and to correct y_{n+1} by the application of a more accurate formula.

A simple type of predictor formula is that in Euler's method, i.e.,

$$y_{n+1} = y_n + y'_n h \tag{9.2.10}$$

where

$$y'_n = \left.\frac{dy}{dx}\right|_n$$

A corrector formula is not used in Euler's method. However, in the modified Euler method an approximate value of y_{n+1} is obtained from the predictor formula (9.2.10) and this value is substituted in the differential equation (9.2.9) to determine y'_{n+1}. Then, a more accurate value for y_{n+1} is obtained by means of the corrector formula

$$y_{n+1} = y_n + (y'_{n+1} + y'_n)\frac{h}{2} \tag{9.2.11}$$

This value is substituted in the differential equation (9.2.9) to obtain a better estimate for y'_{n+1}, which in turn is substituted in equation (9.2.11) for a more accurate y_{n+1}. This iterative process is continued until two successive calculations of y_{n+1} from equation (9.2.11) yield the same value within the desired tolerance.

The classic predictor-corrector method is that of Milne. Milne's predictor and corrector formulas, respectively, are

$$y_{n+1}^{(0)} = y_{n-3} + \frac{4h}{3}(2y'_{n-2} - y'_{n-1} + 2y'_n)$$

and

$$y_{n+1} = y_{n-1} + \frac{h}{3}(y'_{n-1} + 4y'_n + y'_{n+1})$$

where

$$y'_{n+1} = f(x_{n+1}, y_{n+1}^{(0)})$$

The start of the computation requires four consecutive known values of y. These may be calculated by the Runge-Kutta or some other numerical method before applying Milne's predictor-corrector formulas. The error in this method is of the order h^5.

In general, it is desirable to choose h sufficiently small so that only a few iterations are required to obtain any y_{n+1} to the desired accuracy.

These methods can be extended for the numerical solution of simultaneous differential equations. The predictor and corrector formulas are applied independently to each differential equation as though it were a single equation. However, substitution of values for all the dependent variables into each differential equation is required to estimate the derivatives at (x_{n+1}, y_{n+1}).

9.3 Solution of higher-order differential equations

The techniques previously described for the solution of first-order differential equations can be applied also to the solution of higher-order differential equations by the introduction of auxiliary variables. For example, given the second-order differential equation

$$a \frac{d^2y}{dx^2} + b \frac{dy}{dx} + cy = 0$$

and the initial conditions, x_0, y_0, and $\left.\frac{dy}{dx}\right|_0$, this equation can be written as two simultaneous first-order differential equations:

$$\frac{dy}{dx} = y'$$

$$\frac{d^2y}{dx^2} = \frac{dy'}{dx} = -\frac{by' + cy}{a}$$

One of the methods previously described can be employed to solve these two first-order differential equations simultaneously.

In a similar manner, any equation or system of equations of higher order can be reduced to a system of equations of the first order.

9.4 Examples of numerical solution of differential equations

The solution of a differential equation will be illustrated by calculating the current for a series RL circuit.

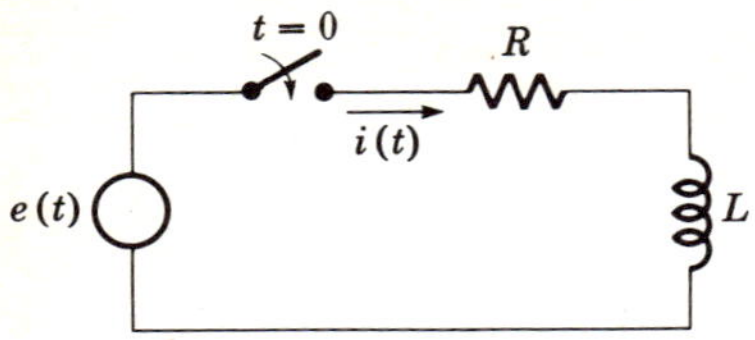

Fig. 9.4 Representation of an RL circuit.

Problem

Given the RL circuit in Fig. 9.4 where the voltage applied by closing the switch is

$$e(t) = 5t \qquad 0 \leq t \leq 0.2$$
$$e(t) = 1 \qquad t > 0.2$$

the resistance, in ohms, is

$$R = 1 + 3i^2$$

and the inductance, in henrys, is

$$L = 1$$

Find the current in the circuit by each of the following methods:

a. Euler's
b. The modified Euler
c. Runge-Kutta fourth-order approximation
d. Milne's
e. Picard's

Solution

The differential equation of the circuit is

$$L\frac{di}{dt} + Ri = e(t)$$

Substituting for R and L, then,

$$\frac{di}{dt} + (1 + 3i^2)i = e(t)$$

The initial conditions at $t = 0$ are $e_0 = 0$ and $i_0 = 0$. The interval selected for the independent variable is $\Delta t = 0.025$.
a. The equations for Euler's method are

$$\Delta i_n = \left.\frac{di}{dt}\right|_n \Delta t$$

$$i_{n+1} = i_n + \Delta i_n$$

where

$$\left.\frac{di}{dt}\right|_n = e_n - (1 + 3i_n^2)i_n$$

Substituting the initial values in the differential equation, $\left.\frac{di}{dt}\right|_0 = 0$ and $\Delta i_0 = 0$. Hence, the current $i_1 = 0$. At $t_1 = 0.025$, $e_1 = 0.125$ and

$$\left.\frac{di}{dt}\right|_1 = 0.125 - \{1 + 3(0)^2\}0 = 0.125$$

$$\Delta i_1 = (0.125)0.025 = 0.00313$$

Then,

$$i_2 = 0 + 0.00313 = 0.00313$$

The tabulated results for this solution are shown in Table 9.1.

b. The equations for the modified Euler method are

$$\Delta i_n^{(0)} = \left.\frac{di}{dt}\right|_n \Delta t$$

$$i_{n+1}^{(0)} = i_n + \Delta i_n^{(0)}$$

$$\Delta i_n^{(1)} = \left(\frac{\left.\frac{di}{dt}\right|_n + \left.\frac{di}{dt}\right|_{n+1}^{(0)}}{2}\right)\Delta t$$

$$i_{n+1}^{(1)} = i_n + \Delta i_n^{(1)}$$

Table 9.1 Solution by Euler's method

n	*Time* t_n	*Voltage* e_n	*Current* $i_n = i_{n-1} + \left.\frac{di}{dt}\right\|_{n-1} \Delta t$	$\left.\frac{di}{dt}\right\|_n = e_n - (1 + 3i_n^2)i_n$
0	0.000	0.000	0.00000	0.00000
1	0.025	0.125	0.00000	0.12500
2	0.050	0.250	0.00313	0.24687
3	0.075	0.375	0.00930	0.36570
4	0.100	0.500	0.01844	0.48154
5	0.125	0.625	0.03048	0.59444
6	0.150	0.750	0.04534	0.70438
7	0.175	0.875	0.06295	0.81130
8	0.200	1.000	0.08323	0.91504
9	0.225	1.000	0.10611	0.89031
10	0.250	1.000	0.12837	0.86528
11	0.275	1.000	0.15000	0.83988
12	0.300	1.000	0.17100	

where

$$\left.\frac{di}{dt}\right|_{n+1}^{(0)} = e_{n+1} - \{1 + 3(i_{n+1}^{(0)})^2\}i_{n+1}^{(0)}$$

Substituting the initial values $e_0 = 0$ and $i_0 = 0$ in the differential equation,

$$\left.\frac{di}{dt}\right|_0 = 0$$

and, therefore,

$$\Delta i_0^{(0)} = 0$$
$$i_1^{(0)} = 0$$

Substituting in the differential equation $i_1^{(0)} = 0$ and $e_1 = 0.125$,

$$\left.\frac{di}{dt}\right|_1^{(0)} = 0.125 - \{1 + 3(0)^2\}0 = 0.125$$

and

$$\Delta i_0^{(1)} = \left(\frac{0.125 + 0}{2}\right)0.025 = 0.00156$$

Then,

$$i_1^{(1)} = 0 + 0.00156 = 0.00156$$

In solving the example by this method, no iteration was performed; hence $i_{n+1}^{(1)} = i_{n+1}$. The solution obtained by this method is shown in Table 9.2.

Table 9.2 Solution by the modified Euler method

n	*Time* t_n	*Voltage* e_n	*Current* i_n	$\left.\frac{di}{dt}\right\|_n$	$\Delta i_n^{(0)}$	e_{n+1}	$i_{n+1}^{(0)}$	$\left.\frac{di}{dt}\right\|_{n+1}^{(0)}$	$\Delta i_n^{(1)}$
0	0.000	0.000	0.00000	0.00000	0.00000	0.125	0.00000	0.12500	0.00156
1	0.025	0.125	0.00156	0.12344	0.00309	0.250	0.00465	0.24535	0.00461
2	0.050	0.250	0.00617	0.24383	0.00610	0.375	0.01227	0.36272	0.00758
3	0.075	0.375	0.01375	0.36124	0.00903	0.500	0.02278	0.47718	0.01048
4	0.100	0.500	0.02423	0.47573	0.01189	0.625	0.03612	0.58874	0.01331
5	0.125	0.625	0.03754	0.58730	0.01468	0.750	0.05222	0.69735	0.01606
6	0.150	0.750	0.05360	0.69594	0.01740	0.875	0.07100	0.80293	0.01874
7	0.175	0.875	0.07234	0.80152	0.02004	1.000	0.09238	0.90525	0.02133
8	0.200	1.000	0.09367	0.90386	0.02260	1.000	0.11627	0.87901	0.02229
9	0.225	1.000	0.11596	0.87936	0.02198	1.000	0.13794	0.85419	0.02167
10	0.250	1.000	0.13763	0.85455	0.02136	1.000	0.15899	0.82895	0.02104
11	0.275	1.000	0.15867	0.82935	0.02073	1.000	0.17940	0.80328	0.02041
12	0.300	1.000	0.17908						

c. The equations used for the Runge-Kutta method to solve

$$\frac{di}{dt} = e(t) - (1 + 3i^2)i$$

are

$$k_1 = \{e(t_n) - (1 + 3i_n^2)i_n\}\,\Delta t$$

$$k_2 = \left\{e\,(t_n + \Delta t/2) - \left[1 + 3\left(i_n + \frac{k_1}{2}\right)^2\right]\left(i_n + \frac{k_1}{2}\right)\right\}\Delta t$$

$$k_3 = \left\{e\,(t_n + \Delta t/2) - \left[1 + 3\left(i_n + \frac{k_2}{2}\right)^2\right]\left(i_n + \frac{k_2}{2}\right)\right\}\Delta t$$

$$k_4 = \{e(t_n + \Delta t) - [1 + 3(i_n + k_3)^2](i_n + k_3)\}\,\Delta t$$

$$\Delta i_n = \tfrac{1}{6}(k_1 + 2k_2 + 2k_3 + k_4)$$

$$i_{n+1} = i_n + \Delta i_n$$

where $e(t_n) = e_n$

$$e\,(t_n + \Delta t/2) = \frac{e_n + e_{n+1}}{2}$$

$$e(t_n + \Delta t) = e_{n+1}$$

Substituting the initial values to solve for k_1,

$$k_1 = 0$$

Solving for k_2,

$$k_2 = \left\{\frac{0 + 0.125}{2} - [1 + 3(0)^2]0\right\} 0.025 = 0.00156$$

Solving for k_3,

$$k_3 = \left\{\frac{0 + 0.125}{2} - \left[1 + 3\left(\frac{0.00156}{2}\right)^2\right]\frac{0.00156}{2}\right\} 0.025 = 0.00154$$

Solving for k_4,

$$k_4 = \{0 + 0.125 - [1 + 3(0.00154)^2]0.00154\}0.025 = 0.00309$$

Then,

$$\Delta i_0 = \tfrac{1}{6}(0 + 0.00312 + 0.00308 + 0.00309) = 0.00155$$

and

$$i_1 = i_0 + \Delta i_0 = 0 + 0.00155 = 0.00155$$

The solution obtained by this method is shown in Table 9.3.

Table 9.3 **Solution by the Runge-Kutta method**

n	Time t_n	Voltage e_n	Current i_n	k_1	$\frac{e_n + e_{n+1}}{2}$	$i_n + \frac{k_1}{2}$	k_2	$i_n + \frac{k_2}{2}$	k_3	e_{n+1}	$i_n + k_3$	k_4	Δi_n
0	0.000	0.000	0.00000	0.00000	0.0625	0.00000	0.00156	0.00078	0.00154	0.125	0.00154	0.00309	0.00155
1	0.025	0.125	0.00155	0.00309	0.1875	0.00310	0.00461	0.00386	0.00459	0.250	0.00614	0.00610	0.00460
2	0.050	0.250	0.00615	0.00610	0.3125	0.00920	0.00758	0.00994	0.00756	0.375	0.01371	0.00903	0.00757
3	0.075	0.375	0.01372	0.00903	0.4375	0.01824	0.01048	0.01896	0.01046	0.500	0.02418	0.01189	0.01047
4	0.100	0.500	0.02419	0.01189	0.5625	0.03014	0.01331	0.03084	0.01329	0.625	0.03748	0.01468	0.01330
5	0.125	0.625	0.03749	0.01468	0.6875	0.04483	0.01606	0.04552	0.01604	0.750	0.05353	0.01740	0.01605
6	0.150	0.750	0.05354	0.01740	0.8125	0.06224	0.01874	0.06291	0.01872	0.875	0.07226	0.02004	0.01873
7	0.175	0.875	0.07227	0.02004	0.9375	0.08229	0.02134	0.08294	0.02132	1.000	0.09359	0.02260	0.02133
8	0.200	1.000	0.09360	0.02260	1.0000	0.10490	0.02229	0.10475	0.02230	1.000	0.11590	0.02199	0.02230
9	0.225	1.000	0.11590	0.02199	1.0000	0.12690	0.02167	0.12674	0.02168	1.000	0.13758	0.02137	0.02168
10	0.250	1.000	0.13758	0.02137	1.0000	0.14827	0.02105	0.14811	0.02105	1.000	0.15863	0.02073	0.02105
11	0.275	1.000	0.15863	0.02073	1.0000	0.16900	0.02041	0.16884	0.02042	1.000	0.17905	0.02009	0.02041
12	0.300	1.000	0.17904										

d. The predictor and corrector formulas for Milne's method are

$$i_{n+1}^{(0)} = i_{n-3} + \frac{4\Delta t}{3} (2i'_{n-2} - i'_{n-1} + 2i'_n)$$

$$i_{n+1} = i_{n-1} + \frac{\Delta t}{3} (i'_{n-1} + 4i'_n + i'_{n+1})$$

where

$$i'_n = \left. \frac{di}{dt} \right|_n$$

and

$$\left. \frac{di}{dt} \right|_n = e_n - (1 + 3i_n{}^2)i_n$$

The required starting values were obtained from the Runge-Kutta solution, where

$$\begin{aligned} i_0 &= 0 \\ i_1 &= 0.00155 \\ i_2 &= 0.00615 \\ i_3 &= 0.01372 \end{aligned}$$

Substituting these in the differential equation, then

$$\begin{aligned} i'_0 &= 0 \\ i'_1 &= 0.12345 \\ i'_2 &= 0.24385 \\ i'_3 &= 0.36127 \end{aligned}$$

Starting at $t_4 = 0.100$ and substituting in the predictor formula, the first estimate for i_4 is

$$i_4^{(0)} = 0 + \tfrac{4}{3}(0.025)[2(0.12345) - 0.24385 + 2(0.36127)] = 0.02418$$

Substituting $e_4 = 0.500$ and $i_4 = 0.02418$ in the differential equation,

$$i'_4 = 0.500 - [1 + 3(0.02418)^2]0.02418 = 0.47578$$

Substituting in the corrector formula,

$$i_4 = 0.00615 + \frac{0.025}{3} [0.24385 + 4(0.36127) + 0.47578] = 0.02419$$

The predicted and corrected values differ only by one in the fifth decimal place and thus no iteration is required. The results for subsequent steps are given in Table 9.4. At t_9, the predicted value of current was 0.11742, whereas the corrected value was 0.11639. An iteration was performed

Table 9.4 Solution by Milne's method

n	Time t_n	Voltage e_n	Current (predicted) i_n	i'_n	Current (corrected) i_n
4	0.100	0.500	0.02418	0.47578	0.02419
5	0.125	0.625	0.03748	0.58736	0.03748
6	0.150	0.750	0.05353	0.69601	0.05353
7	0.175	0.875	0.07226	0.80161	0.07226
8	0.200	1.000	0.09359	0.90395	0.09358
9	0.225	1.000	0.11742	0.87772	0.11639
				0.87888	0.11640†
10	0.250	1.000	0.13543	0.85712	0.13755
				0.85464	0.13753†
11	0.275	1.000	0.16021	0.82745	0.15911
				0.82881	0.15912†
12	0.300	1.000	0.17894	0.80387	0.17898
				0.80382	0.17898†

† Second corrected value obtained by iteration.

by substituting this corrected value in the differential equation to obtain $i'_9 = 0.87888$. This in turn was used in the corrector formula to obtain the second estimate for $i_9 = 0.11640$, which checks the previous corrected value. An iteration was performed in all subsequent steps to assure the desired accuracy.

e. The equation used for Picard's method to generate an approximating function for i, near $i_0 = 0$, is

$$i = i_0 + \int_0^t (e(t) - i - 3i^3)\,dt$$

Substituting $e(t) = 5t$ and the initial value $i_0 = 0$,

$$i^{(1)} = \int_0^t 5t\,dt = \frac{5t^2}{2}$$

Then, substituting $i^{(1)}$ for i in the integral equation,

$$i^{(2)} = \int_0^t \left(5t - \frac{5t^2}{2} - \frac{375t^6}{8}\right) dt$$

$$= \frac{5t^2}{2} - \frac{5t^3}{6} - \frac{375t^7}{56}$$

Continuing,

$$i^{(3)} = \int_0^t \left(5t - \frac{5t^2}{2} + \frac{5t^3}{6} - \frac{375t^6}{8} + \frac{375t^7}{7} - \frac{125t^8}{8} + \cdots\right) dt$$

$$= \frac{5t^2}{2} - \frac{5t^3}{6} + \frac{5t^4}{24} - \frac{375t^7}{56} + \cdots$$

$$i^{(4)} = \int_0^t \left(5t - \frac{5t^2}{2} + \frac{5t^3}{6} - \frac{5t^4}{24} - \frac{375t^6}{8} + \frac{375t^7}{7} + \cdots\right) dt$$

$$= \frac{5t^2}{2} - \frac{5t^3}{6} + \frac{5t^4}{24} - \frac{t^5}{24} - \frac{375t^7}{56} + \cdots$$

Terminating this series after the fourth-degree term, then

$$i = \frac{5t^2}{2} - \frac{5t^3}{6} + \frac{5t^4}{24}$$

If this function is used to approximate i correct to four decimal places with the first neglected term as an approximation of the truncation error, then

$$\frac{t^5}{24} \leq 0.00005$$

$$5 \log t \leq \log 0.00120$$

$$\log t \leq 9.415836 - 10$$

$$t \leq 0.2605$$

This is the limiting value of t for which the approximating function is valid. However, in this example the function can be used to obtain y only for the range $0 \leq t \leq 0.2$, because for $t > 0.2$, $e(t) = 1$. Consequently, another approximating function must be determined for the range $0.2 \leq t \leq 0.3$ as follows:

$$i = 0.09367 + \int_{0.2}^t (1 - i - 3i^3)\, dt$$

$$i^{(1)} = 0.09367 + \int_{0.2}^t \{1 - 0.09367 - 3(0.09367)^3\}\, dt$$

$$= 0.09367 + 0.90386(t - 0.2)$$

$$i^{(2)} = 0.09367 + \int_{0.2}^t \{1 - 0.09367 - 0.90386(t - 0.2) - 3[0.09367 + 0.90386(t - 0.2)]^3\}\, dt$$

$$= 0.09367 + 0.90386 \int_{0.2}^t \{1 - 1.07897(t - 0.2) - 0.76198(t - 0.2)^2 - 2.45089(t - 0.2)^3\}\, dt$$

$$= 0.09367 + 0.90386 \left\{(t - 0.2) - 1.07897 \frac{(t - 0.2)^2}{2} - 0.76198 \frac{(t - 0.2)^3}{3} - 2.45089 \frac{(t - 0.2)^4}{4}\right\}$$

Table 9.5 Solution by Picard's method

n	Time t_n	Voltage e_n	Current i_n
0	0	0	0
1	0.025	0.125	0.00155
2	0.050	0.250	0.00615
3	0.075	0.375	0.01372
4	0.100	0.500	0.02419
5	0.125	0.625	0.03749
6	0.150	0.750	0.05354
7	0.175	0.875	0.07229
8	0.200	1.000	0.09367
9	0.225	1.000	0.11596
10	0.250	1.000	0.13764
11	0.275	1.000	0.15868
12	0.300	1.000	0.17910

Finally,

$$\begin{aligned} i^{(3)} = {} & 0.09367 + 0.90386(t - 0.2) - 0.48762(t - 0.2)^2 \\ & - 0.05420(t - 0.2)^3 - 0.30611(t - 0.2)^4 + 0.86646(t - 0.2)^5 \cdots \end{aligned}$$

Terminating the series, the approximating function is

$$\begin{aligned} i = {} & 0.09367 + 0.90386(t - 0.2) - 0.48762(t - 0.2)^2 \\ & - 0.05420(t - 0.2)^3 - 0.30611(t - 0.2)^4 \end{aligned}$$

For i correct to four decimal places, since

$$\begin{aligned} 0.86646(t - 0.2)^5 &\leq 0.00005 \\ (t - 0.2) &\leq 0.14198 \end{aligned}$$

this function is valid for the range $0.2 \leq t < 0.342$.

The values obtained by Picard's method are shown in Table 9.5.

9.5 Comparison of methods

In the solution of a differential equation a functional relation between the dependent variable y and the independent variable x is sought to satisfy the differential equation. An analytical solution is difficult and for some problems impossible to obtain. Numerical methods are used to obtain a solution by (1) expressing y as some function of the independent variable x from which approximate values of y can be obtained by direct substitution, or (2) expressing an approximate relation between successive values of y determined for selected values of x. Picard's method is a numerical method of the first type. The methods of Euler, Runge-Kutta, and Milne are examples of the second type.

The principal difficulties that arise from methods approximating y by a function, such as Picard's method, occur in the repeated explicit integrations that must be performed to obtain a satisfactory function. Hence these methods are impractical in most cases and are seldom used.

The methods of the second type require simple arithmetic operations and thus are applicable for a computer solution of differential equations. In general, the simpler relations require the use of smaller intervals for the independent variable whereas the more complex methods can employ relatively larger intervals without sacrificing the accuracy of the solution. Euler's method is the simplest, but unless a very small interval is used it is too inaccurate to be practical. The modified method of Euler is also simple to apply and has the additional advantage that systematic checking is inherent in the process of obtaining improved estimates for y. This method is of limited accuracy, however, and requires the use of small intervals for the independent variable. The Runge-Kutta method requires a larger number of arithmetic operations, but the results are more accurate.

Milne's predictor-corrector method is less laborious than is the Runge-Kutta method and has comparable accuracy of order h^5. However, Milne's method requires four starting values for the dependent variable that must be obtained by some other method, such as the modified Euler or Runge-Kutta method, that is self-starting. For a computer application this requires programming a numerical method for starting the solution as well as Milne's method for continuing the solution. The use of different formulas for predicting and then correcting a value of y provides a systematic process for checking as well as correcting the initial estimate. If the difference between the predicted and corrected values is significant, the interval can be reduced. This capability in the Milne method is not available in the Runge-Kutta method.

Problems

9.1 Solve the differential equation

$$\frac{dy}{dx} = x^2 - y$$

for $0 \leq x \leq 0.3$, with the interval equal to 0.05 and initial values $x_0 = 0$ and $y_0 = 1$, by the following numerical methods:

a. Euler's

b. The modified Euler

c. Picard's

d. Runge-Kutta fourth-order approximation

e. Milne's, using starting values obtained from the Runge-Kutta method

9.2 Solve by the modified Euler method the simultaneous differential equations

$$\frac{dx}{dt} = 2y$$

$$\frac{dy}{dt} = -\frac{x}{2}$$

for $0 \leq t \leq 1.0$, with the interval equal to 0.2 and initial values $t_0 = 0$, $x_0 = 0$, and $y_0 = 1$.

9.3 Solve by the Runge-Kutta fourth-order approximation the second-order differential equation

$$y'' = y + xy'$$

for $0 \leq x \leq 0.4$, with the interval equal to 0.1 and initial values $x_0 = 0$, $y_0 = 1$, and $y'_0 = 0$.

Bibliography

Hildebrand, F. B.: "Introduction to Numerical Analysis," McGraw-Hill Book Company, New York, 1956.

Kunz, Kaiser S.: "Numerical Analysis," McGraw-Hill Book Company, New York, 1957.

Levy, H., and E. A. Baggott: "Numerical Solutions of Differential Equations," Dover Publications, Inc., New York, 1950.

McCormick, John M., and Mario G. Salvadori: "Numerical Methods in FORTRAN," Prentice-Hall, Inc., Englewood Cliffs, N.J., 1964.

McCracken, Daniel D., and William S. Dorn: "Numerical Methods and FORTRAN Programming," John Wiley & Sons, Inc., New York, 1964.

Milne, William Edmund: "Numerical Calculus," Princeton University Press, Princeton, N.J., 1949.

Milne, William Edmund: "Numerical Solution of Differential Equations," John Wiley & Sons, Inc., New York, 1953.

Salvadori, Mario G., and Melvin L. Baron: "Numerical Methods in Engineering," Prentice-Hall, Inc., Englewood Cliffs, N.J., 1952.

Scarborough, James B.: "Numerical Mathematical Analysis," The Johns Hopkins Press, Baltimore, 1950.

chapter 10
Transient stability studies

10.1 Introduction

Transient stability studies provide information related to the capability of a power system to remain in synchronism during major disturbances resulting from either the loss of generating or transmission facilities, sudden or sustained load changes, or momentary faults. Specifically, these studies provide the changes in the voltages, currents, powers, speeds, and torques of the machines of the power system, as well as the changes in system voltages and power flows, during and immediately following a disturbance. The degree of stability of a power system is an important factor in the planning of new facilities. In order to provide the reliability required by the dependence on continuous electric service, it is necessary that power systems be designed to be stable under any conceivable disturbance.

The ac network analyzer was used for transient stability studies to obtain the operating performance of the power network during a disturbance. The step-by-step calculations describing the operation of the machines were performed manually. The use of the digital computer to perform all computation for both the network and the machines was a natural extension of the digital load flow studies that proved so successful.

The performance of the power system during the transient period can be obtained from the network performance equations. The performance equation using the bus frame of reference in either the impedance or admittance form has been used in transient stability calculations.

In transient stability studies a load flow calculation is made first to obtain system conditions prior to the disturbance. In this calculation the network is composed of system buses, transmission lines, and transformers. The network representation for transient stability studies includes, in addition to these components, equivalent circuits for machines and static impedances or admittances to ground for loads. After the load flow calculation, therefore, the impedance or admittance matrix of the network must be modified to reflect the changes in the representation of the network.

The operating characteristics of synchronous and induction machines are described by sets of differential equations. The number of differential equations required for a machine depends on the detail needed to represent accurately the machine performance. Two first-order differential equations are required for the simplest representation of a synchronous machine.

A transient stability analysis is performed by combining a solution of the algebraic equations describing the network with a numerical solution of the differential equations. The solution of the network equations retains the identity of the system and thereby provides access to system voltages and currents during the transient period. The modified Euler and Runge-Kutta methods have been applied to the solution of the differential equations in transient stability studies.

10.2 Swing equation

In order to determine the angular displacement between the machines of a power system during transient conditions, it is necessary to solve the differential equation describing the motion of the machine rotors. The net torque acting on the rotor of a machine, from the laws of mechanics related to rotating bodies, is

$$T = \frac{WR^2}{g} \alpha \tag{10.2.1}$$

where T = algebraic sum of all torques, ft-lb
WR^2 = moment of inertia, lb-ft^2
g = acceleration due to gravity, equal to 32.2 ft/sec^2
α = mechanical angular acceleration, rad/sec^2

The electrical angle θ_e is equal to the product of the mechanical angle θ_m and the number of pairs of poles $P/2$, that is,

$$\theta_e = \frac{P}{2} \theta_m \tag{10.2.2}$$

The frequency f in cycles per second is

$$f = \frac{P}{2} \frac{\text{rpm}}{60} \tag{10.2.3}$$

Then from equations (10.2.2) and (10.2.3) the electrical angle in radians is

$$\theta_e = \frac{60f}{\text{rpm}}\,\theta_m \tag{10.2.4}$$

The electrical angular position δ, in radians, of the rotor with respect to a synchronously rotating reference axis is

$$\delta = \theta_e - \omega_0 t$$

where ω_0 = rated synchronous speed, rad/sec
t = time, sec

Then, the angular velocity or slip with respect to the reference axis is

$$\frac{d\delta}{dt} = \frac{d\theta_e}{dt} - \omega_0$$

and the angular acceleration is

$$\frac{d^2\delta}{dt^2} = \frac{d^2\theta_e}{dt^2}$$

Taking the second derivative of equation (10.2.4) and substituting,

$$\frac{d^2\delta}{dt^2} = \frac{60f}{\text{rpm}}\,\frac{d^2\theta_m}{dt^2}$$

where

$$\frac{d^2\theta_m}{dt^2} = \alpha$$

Then, substituting into equation (10.2.1), the net torque is

$$T = \frac{WR^2}{g}\,\frac{\text{rpm}}{60f}\,\frac{d^2\delta}{dt^2}$$

It is desirable to express the torque in per unit. The base torque is defined as the torque required to develop rated power at rated speed, that is,

$$\text{Base torque} = \frac{\text{base kva}\left(\dfrac{550}{0.746}\right)}{2\pi\left(\dfrac{\text{rpm}}{60}\right)}$$

where the base torque is in foot-pounds. Therefore, the torque in per unit is

$$T = \frac{\dfrac{WR^2}{g}\,\dfrac{2\pi}{f}\left(\dfrac{\text{rpm}}{60}\right)^2\dfrac{0.746}{550}}{\text{base kva}}\,\frac{d^2\delta}{dt^2} \tag{10.2.5}$$

The inertia constant H of a machine is defined as the kinetic energy at rated speed in kilowatt seconds per kilovolt-ampere. The kinetic energy in foot-pounds is

$$\text{Kinetic energy} = \frac{1}{2}\frac{WR^2}{g}\omega_0^2$$

where

$$\omega_0 = 2\pi \frac{\text{rpm}}{60}$$

and rpm is the rated speed. Therefore

$$H = \frac{\dfrac{1}{2}\dfrac{WR^2}{g}(2\pi)^2\left(\dfrac{\text{rpm}}{60}\right)^2\dfrac{0.746}{550}}{\text{base kva}}$$

Substituting in equation (10.2.5),

$$T = \frac{H}{\pi f}\frac{d^2\delta}{dt^2} \tag{10.2.6}$$

The torques acting on the rotor of a generator include the mechanical input torque from the prime mover, torques due to rotational losses (friction, windage, and core losses), electrical output torques, and damping torques due to prime mover, generator, and power system. The electrical and mechanical torques acting on the rotor of a motor are of opposite sign and are a result of the electrical input and mechanical load. Neglecting damping and rotational losses, the accelerating torque T_a is

$$T_a = T_m - T_e$$

where T_m = mechanical torque
T_e = electrical air gap torque

Thus equation (10.2.6) becomes

$$\frac{H}{\pi f}\frac{d^2\delta}{dt^2} = T_m - T_e \tag{10.2.7}$$

Since the torque and power in per unit are equal for small deviations in speed, equation (10.2.7) becomes

$$\frac{d^2\delta}{dt^2} = \frac{\pi f}{H}(P_m - P_e)$$

where P_m = mechanical power
P_e = electrical air gap power

This second-order differential equation can be written as two simultaneous first-order equations:

$$\frac{d^2\delta}{dt^2} = \frac{d\omega}{dt} = \frac{\pi f}{H}(P_m - P_e)$$

and

$$\frac{d\delta}{dt} = \frac{d\theta_e}{dt} - \omega_0 \tag{10.2.8}$$

Since the rated synchronous speed in radians per second is $2\pi f$, equation (10.2.8) becomes

$$\frac{d\delta}{dt} = \omega - 2\pi f$$

10.3 Machine equations

Synchronous machines

In transient stability studies, particularly those involving short periods of analysis in the order of a second or less, a synchronous machine can be represented by a voltage source, in back of transient reactance, that is constant in magnitude but changes its angular position. This representation neglects the effect of saliency and assumes constant flux linkages and a small change in speed. The voltage back of transient reactance is determined from

$$E' = E_t + r_a I_t + jx'_d I_t$$

where E' = voltage back of transient reactance
E_t = machine terminal voltage
I_t = machine terminal current
r_a = armature resistance
x'_d = transient reactance

The representation of the synchronous machine used for network solutions and the corresponding phasor diagram are shown in Fig. 10.1.

Saliency and changes in field flux linkages can be taken into account by representing the effects of the three-phase ac quantities of a synchronous machine by components acting along the direct and quadrature axes. The direct axis is along the centerline of the machine pole and the quadrature axis leads the direct axis by 90 electrical degrees. The position of the quadrature axis can be determined by calculating a fictitious voltage located on this axis. This is a voltage back of quadrature-axis

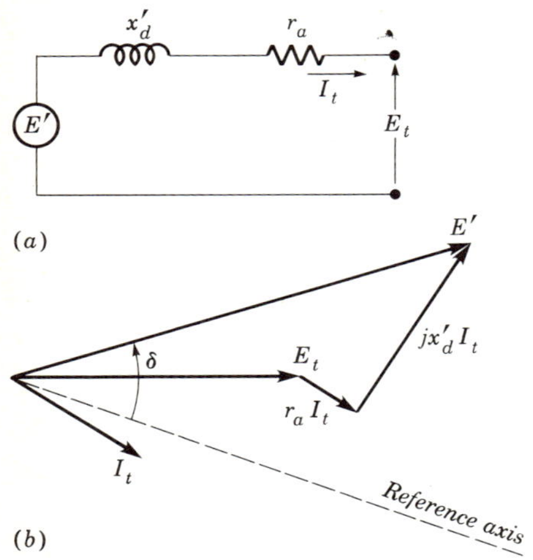

***Fig. 10.1** Simplified representation of a synchronous machine. (**a**) Equivalent circuit; (**b**) phasor diagram.*

synchronous reactance and is determined from

$$E_q = E_t + r_a I_t + jx_q I_t$$

where E_q = voltage back of quadrature-axis synchronous reactance
x_q = quadrature-axis synchronous reactance

The representation of the synchronous machine used for network solutions and the corresponding phasor diagram are shown in Fig. 10.2.

The sinusoidal flux produced by the field current acts along the direct axis. The voltage induced by field current lags this flux by 90° and, therefore, is on the quadrature axis. This voltage can be determined by adding to the terminal voltage E_t the voltage drop across the armature resistance and the voltage drops representing the demagnetizing effects along the direct and quadrature axes. Then neglecting saturation,

$$E_I = E_t + r_a I_t + jx_d I_d + jx_q I_q$$

where E_I = voltage proportional to field current
x_d = direct-axis synchronous reactance
x_q = quadrature-axis synchronous reactance
I_d = direct-axis component of machine terminal current
I_q = quadrature-axis component of machine terminal current

The phasor diagram showing E_I as well as the voltage back of transient reactance is shown in Fig. 10.3.

The quadrature component of voltage back of transient reactance from the phasor diagram is

$$E'_q = E_q - j(x_q - x'_d)I_d$$

where E'_q is the voltage proportional to the field flux linkages resulting from the combined effect of the field and armature currents. Since the field flux linkages do not change instantaneously following a disturbance, E'_q also does not change instantaneously. The rate of change of E'_q along the quadrature axis is dependent on the field voltage controlled by the regulator and exciter, the voltage proportional to the field current, and the direct-axis transient open circuit time constant and is given by

$$\frac{dE'_q}{dt} = \frac{1}{T'_{d0}}(E_{fd} - E_I)$$

where E_{fd} = term representing the field voltage acting along the quadrature axis

T'_{d0} = direct-axis transient open circuit time constant

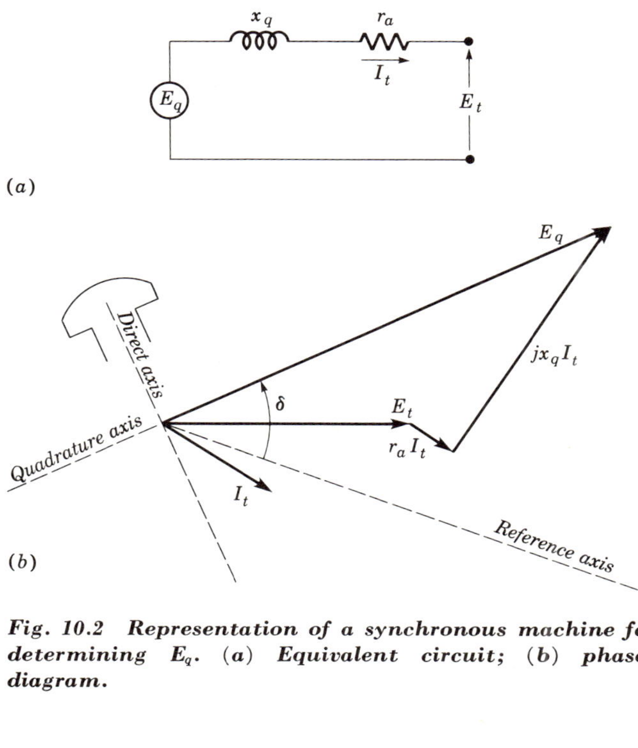

Fig. 10.2 Representation of a synchronous machine for determining E_q. (a) Equivalent circuit; (b) phasor diagram.

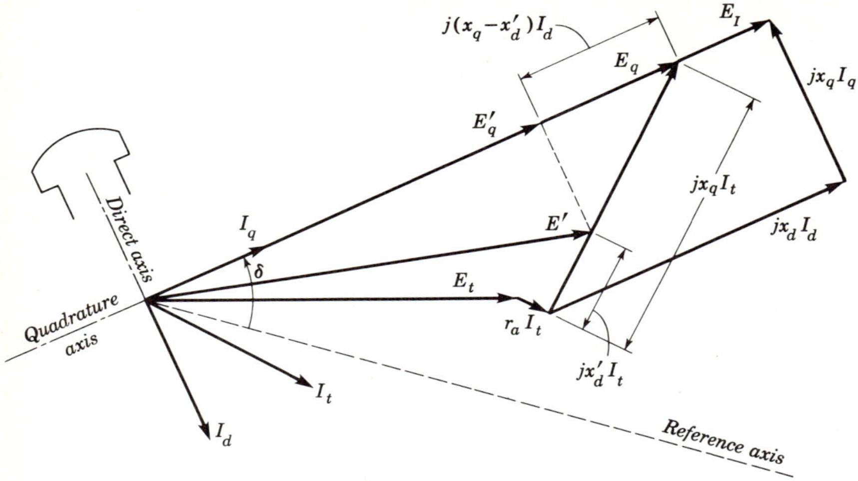

Fig. 10.3 Phasor diagram for determining the quadrature-axis component of voltage behind transient reactance.

Induction machines

In power system transient stability studies loads, including induction motors, usually can be represented adequately by shunt impedances. However, in studies involving large induction motor loads it is necessary frequently to represent the induction motors in a more detailed manner. Induction motors are used extensively in industrial processes and can have significant effects on the transient response of a power system.

A reasonable linear representation of an induction machine can be obtained by taking into account the effects of mechanical transients and rotor electrical transients. The effects of stator electrical transients on system response usually can be neglected. The equivalent circuit shown in Fig. 10.4 has been used to represent the transient behavior of an induction motor including the effects of mechanical transients and rotor electrical transients with a single time constant.

The differential equation describing the rate of change of the voltage behind transient reactance X' is

$$\frac{dE'}{dt} = -j2\pi fsE' - \frac{1}{T_0}\{E' - j(X - X')I_t\}$$

where the rotor open circuit time constant T_0 in seconds is

$$T_0 = \frac{x_r + x_m}{2\pi fr_r}$$

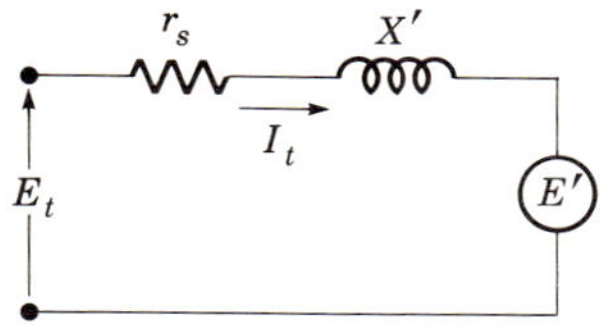

Fig. 10.4 Simplified representation of an induction machine for transient analysis.

and the terminal current is

$$I_t = (E_t - E') \frac{1}{r_s + jX'}$$

The reactances X and X' can be obtained from the conventional steady state equivalent circuit of an induction machine as shown in Fig. 10.5, where

r_s is the stator resistance in per unit
x_s is the stator reactance in per unit
r_r is the rotor resistance in per unit
x_r is the rotor reactance in per unit
x_m is the magnetizing reactance in per unit
s is the rotor slip in per unit

The resistances and reactances are all on the same kva base. The ratio of the base voltages of the stator and rotor is equal to the open circuit voltage ratio at standstill. The per unit slip is

$$s = \frac{\text{Synchronous speed} - \text{actual speed}}{\text{Synchronous speed}}$$

Since the rotor resistance r_r is small compared with the reactances, it can be neglected in the calculation of X and X'. From the steady state equivalent circuit, then, the open circuit reactance is approximately

$$X = x_s + x_m$$

The blocked rotor reactance is approximately

$$X' = x_s + \frac{x_m x_r}{x_m + x_r}$$

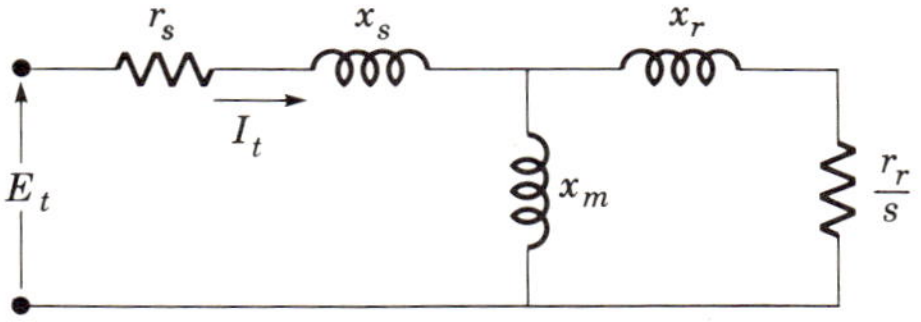

Fig. 10.5 Steady state positive sequence equivalent circuit of an induction machine.

10.4 Power system equations

Representation of loads

Power system loads, other than motors represented by equivalent circuits, can be treated in several ways during the transient period. The commonly used representations are either static impedance or admittance to ground, constant current at fixed power factor, constant real and reactive power, or a combination of these representations.

The constant power load is either equal to the scheduled real and reactive bus load or is a percentage of the specified values in the case of a combined representation. The parameters associated with static impedance and constant current representations are obtained from the scheduled bus loads and the bus voltages calculated from a load flow solution for the power system prior to a disturbance. The initial value of the current for a constant current representation is obtained from

$$I_{p0} = \frac{P_{Lp} - jQ_{Lp}}{E_p^*}$$

where P_{Lp} and Q_{Lp} are the scheduled bus loads and E_p is the calculated bus voltage. The current I_{p0} flows from bus p to ground, that is, to bus 0. The magnitude and power factor angle of I_{p0} remain constant.

The static admittance y_{p0}, used to represent the load at bus p, can be obtained from

$$(E_p - E_0)y_{p0} = I_{p0}$$

where E_p is the calculated bus voltage and E_0 is the ground voltage, equal to zero. Therefore

$$y_{p0} = \frac{I_{p0}}{E_p} \tag{10.4.1}$$

Multiplying both the dividend and divisor of equation (10.4.1) by E_p^* and separating the real and imaginary components,

$$g_{p0} = \frac{P_{Lp}}{e_p^2 + f_p^2} \qquad \text{and} \qquad b_{p0} = \frac{Q_{Lp}}{e_p^2 + f_p^2}$$

where

$$y_{p0} = g_{p0} - jb_{p0}$$

Network performance equations

The network performance equations used for load flow calculations can be applied to describe the performance of the network during a transient period (Stagg, Gabrielle, Moore, and Hohenstein, 1959). Using the bus admittance matrix with ground as reference, the voltage equation for bus p is

$$E_p = \frac{(P_p - jQ_p)L_p}{E_p^*} - \sum_{\substack{q=1 \\ q \neq p}}^{n} YL_{pq}E_q \tag{10.4.2}$$

The term $P_p - jQ_p/E_p^*$ in equation (10.4.2) represents the load current at bus p. For the constant current load representation

$$\frac{P_p - jQ_p}{(E_p^k)^*} = |I_{p0}| \underline{/\theta_p^k + \phi_p}$$

where ϕ_p is the power factor angle and θ_p^k is the angle of voltage with respect to the reference. When the constant power is used to represent the load, $(P_p - jQ_p)L_p$ will be constant but the bus voltage E_p will change every iteration. When the load at bus p is represented by a static admittance to ground, the impressed current at the bus is zero and therefore

$$\frac{(P_p - jQ_p)L_p}{E_p^*} = 0$$

In using equation (10.4.2) to describe the performance of the network for a transient analysis, the parameters must be modified to include the effects of the equivalent elements required to represent synchronous and induction machines and loads. The line parameters YL_{pq} must be modified for the new elements and an additional line parameter must be calculated for each new network element. The system shown in Fig. 10.6, which was used also to illustrate the load flow solution techniques in Sec. 8.3, has two machines and a load at each bus. Representing all loads as static admittances to ground, the voltage equation for bus 1 is

$$E_1 = -YL_{12}E_2 - YL_{13}E_3 - YL_{14}E_4 - YL_{10}E_0$$

where $YL_{12} = Y_{12}L_1$
$YL_{13} = Y_{13}L_1$
$YL_{14} = Y_{14}L_1$

The elements Y_{12}, Y_{13}, and Y_{14} from the bus admittance matrix of the network are the same as in the load flow representation. However,

$$L_1 = \frac{1}{Y_{11}}$$

where

$$Y_{11} = y_{12} + y_{13} + y_{14} + y_{10}$$

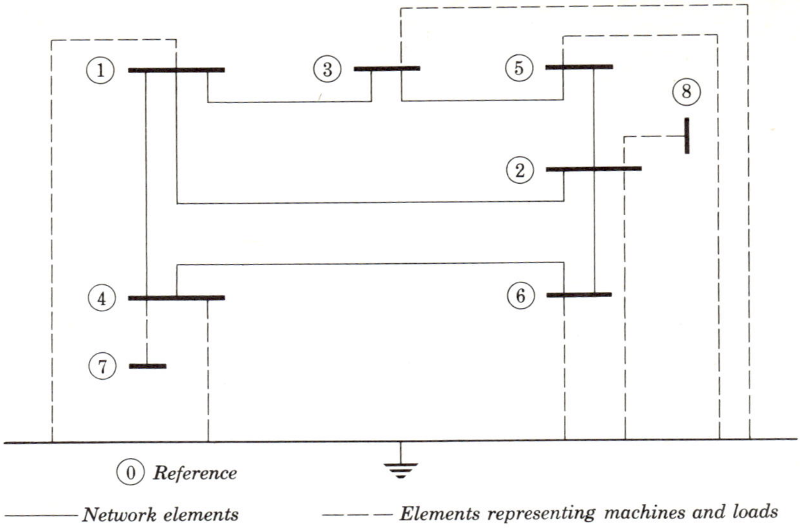

Fig. 10.6 **Single line diagram of power system for transient analysis.**

includes the static admittance representing the load. Since E_0 is zero, the line parameter YL_{10} does not have to be calculated.

The voltage equation for bus 2 is

$$E_2 = -YL_{21}E_1 - YL_{25}E_5 - YL_{26}E_6 - YL_{28}E_8$$

where bus 8 is a new bus. In this case the diagonal admittance element for bus 2 is

$$Y_{22} = y_{21} + y_{25} + y_{26} + y_{20} + y_{28}$$

where y_{20} is the static admittance representing the load and y_{28} is the machine equivalent admittance. The formulas for the Gauss-Seidel iterative solution of the network shown in Fig. 10.6 are, then,

$$\begin{aligned}
E_1^{k+1} &= -YL_{12}E_2^k - YL_{13}E_3^k - YL_{14}E_4^k \\
E_2^{k+1} &= -YL_{21}E_1^{k+1} - YL_{25}E_5^k - YL_{26}E_6^k - YL_{28}E_8 \\
E_3^{k+1} &= -YL_{31}E_1^{k+1} - YL_{35}E_5^k \\
E_4^{k+1} &= -YL_{41}E_1^{k+1} - YL_{46}E_6^k - YL_{47}E_7 \\
E_5^{k+1} &= -YL_{52}E_2^{k+1} - YL_{53}E_3^{k+1} \\
E_6^{k+1} &= -YL_{62}E_2^{k+1} - YL_{64}E_4^{k+1}
\end{aligned}$$

The initial bus voltages are obtained from the load flow solution prior to the disturbance. The initial voltages for the new buses 7 and 8 are obtained from the equivalent circuit representing the machines.

Subsequent voltages for these buses are calculated from the differential equations describing the performance of the machines.

During the iterative calculation the magnitudes and phase angles of the bus voltages behind the machine equivalent admittances are held constant. If a three-phase fault is simulated, the voltage of the faulted bus is set to zero and held constant.

If the bus impedance matrix is used for a transient stability study, ground is usually taken as reference because all network bus voltages, except at the faulted bus, change during the transient period (Brown, Happ, Person, and Young, 1965). To eliminate the need to modify the bus impedance matrix for a change in the reference bus, ground is used also as reference in the prefault load flow calculation.

When ground is used as reference for the load flow calculation and the loads are represented solely as current sources, the bus impedance matrix will include only the capacitor, reactor, and line charging elements to ground. In this case the bus impedance matrix is ill-conditioned and convergence of the solution usually is not obtained. On the other hand, if the loads are represented solely as impedances to ground to improve the convergence characteristic, then these impedances and the bus impedance matrix must be modified during the iterative solution for changes in bus voltages. To overcome this difficulty only a portion of each bus load is represented as an impedance to ground. The remaining portion of the load can be represented as a current source which varies with the bus voltage so that the total bus current satisfies the scheduled load power.

After the load flow solution is obtained, the bus impedance matrix must be modified to include the new network elements representing the machines and to account for changes in the representation of loads. These modifications can be made by using the algorithm described in Secs. 4.2 and 4.3. Each element representing a machine is a branch to a new bus, and each element representing a load change is a link to ground.

The iteration formula for the performance of the network during the transient period using ground as reference is

$$E_p^{k+1} = \sum_{q=1}^{n+m} Z_{pq} I_q \qquad \begin{array}{l} p = 1, 2, \ldots, n \\ p \neq f \end{array}$$

where n is the number of network buses, m is the number of buses behind the equivalent machine impedances, and bus f is the faulted bus. The current vector I_q is composed of load currents from either the constant current or constant power representation and the currents obtained from machine equivalent circuits.

In the application of the bus impedance matrix, only those rows and columns corresponding to machines, constant power, and constant current sources need to be retained for the network solution. All rows and columns would have to be maintained, however, if system voltages and power flows are required during the transient calculations.

The procedures described using the bus impedance and admittance matrices and representing each machine as a voltage behind the machine impedance is an application of Thevenin's theorem. An alternate method is to represent the machine as a current source between the machine terminal bus and ground and in parallel with the machine impedance. This is an application of Norton's theorem (Shipley, Sato, Coleman, and Watts, 1966). This eliminates the need to establish an additional bus behind the impedance of each machine. The machine currents are calculated by using the internal machine voltages and the machine impedances. These currents are held constant during the network iterative solution.

10.5 Solution techniques

Preliminary calculations

The first step in a transient stability study is the load flow calculation to obtain system conditions prior to the disturbance. Then the network data must be modified to correspond to the desired representation for transient analysis. In addition, the machine currents prior to the disturbance are calculated from

$$I_{ti} = \frac{P_{ti} - jQ_{ti}}{E_{ti}^*} \qquad i = 1, 2, \ldots, m$$

where m is the number of machines and P_{ti} and Q_{ti} are the scheduled or calculated machine real and reactive terminal powers. The calculated power for the machine at the slack bus and the terminal voltages are obtained from the initial load flow solution. Finally the voltages back of machine impedances must be calculated.

When the machine i is represented by a voltage source of constant magnitude back of transient reactance, the voltage is obtained from

$$E'_{i(0)} = E_{ti} + r_{ai}I_{ti} + jx'_{di}I_{ti}$$

where

$$E'_{i(0)} = e'_{i(0)} + jf'_{i(0)}$$

and $E'_{i(0)}$ is the initial value used in the solution of the differential equations. The initial internal voltage angle is

$$\delta_{i(0)} = \tan^{-1} \frac{f'_{i(0)}}{e'_{i(0)}}$$

The initial speed $\omega_{i(0)}$ in radians per second is equal to $2\pi f$ where f is the frequency in cycles per second. The initial mechanical power input $P_{mi(0)}$ is equal to the electrical air-gap power P_{ei} prior to the disturbance which can be obtained from

$$P_{ei} = P_{ti} + |I_{ti}|^2 r_{ai}$$

where $|I_{ti}|^2 r_{ai}$ represents the stator losses.

When the effects of saliency and changes in field flux linkages are taken into account a voltage back of quadrature-axis synchronous reactance is used to represent the machine. This voltage is calculated from

$$E_{qi} = E_{ti} + r_{ai} I_{ti} + j x_{qi} I_{ti}$$

where

$$E_{qi} = e_{qi} + j f_{qi}$$

The initial internal voltage angle is then

$$\delta_{i(0)} = \tan^{-1} \frac{f_{qi}}{e_{qi}}$$

As in the simplified representation, the initial speed is equal to $2\pi f$ and the initial mechanical power is equal to P_{ei}, the air-gap power.

The calculations of the voltage proportional to field current E_{Ii} and the voltage proportional to field flux linkages $E'_{qi(0)}$ are required also for this representation. These voltages are obtained from

$$E_{Ii} = E_{ti} + r_{ai} I_{ti} + j x_{di} I_{di} + j x_{qi} I_{qi}$$

and

$$E'_{qi(0)} = E_{qi} - (x_{qi} - x'_{di}) I_{di}$$

where $E'_{qi(0)}$ is the initial value used in the solution of the differential equations. Finally, the initial field voltage $E_{fdi(0)}$ is equal to E_{Ii} if saturation is neglected.

The next step is to change the system parameters to simulate a disturbance. Loss of generation, load, or transmission facilities can be effected by removing from the network the appropriate elements. A three-phase fault can be simulated by setting the voltage at the faulted bus to zero. Then, the modified network equations are solved to obtain system conditions at the instant after the disturbance occurs.

The techniques described for load flow solutions can be employed to obtain the new bus voltages for the network. In the iterative solution, however, the buses back of machine impedances are treated differently depending on the machine representation. When the machine is represented by a voltage of constant magnitude back of transient reactance, the internal machine bus voltage is held fixed during the entire iterative process. When the machine is represented by the direct and quadrature components, the internal machine bus voltage is held fixed during an iteration. However, at the end of each iteration, the voltage E_{qi} must be reevaluated to reflect the changes in the terminal voltage E_{ti}. First the new voltage for the internal bus is obtained by calculating the new machine terminal current from

$$I_{ti}^{k+1} = (E_{qi}^{k} - E_{ti}^{k+1}) \frac{1}{r_{ai} + jx_{qi}}$$

Then the new component of current along the direct axis is determined. Finally the voltage back of quadrature-axis synchronous reactance is computed from

$$E_{qi}^{k+1} = E'_{qi(0)} + (x_{qi} - x'_{di})I_{di}^{k+1}$$

where $E'_{qi(0)}$ and $\delta_{i(0)}$, the angle of E_{qi}, are held fixed.

When the network solution has been obtained, the machine terminal current becomes the initial value for the solution of the differential equations. It is used to calculate initial machine air-gap power from

$$P_{ei(0)} = \text{Re}\,(I_{ti(0)}E'^{*}_{i(0)})$$

when the magnitude of the voltage in back of transient reactance is held fixed or from

$$P_{ei(0)} = \text{Re}\,(I_{ti(0)}E^{*}_{qi(0)})$$

when the effects of saliency and changes in field flux linkages are taken into account. The initial voltage $E_{qi(0)}$ is obtained also from the network solution at the instant after the disturbance.

Modified Euler method

When a machine is represented by a voltage of constant magnitude back of transient reactance, it is necessary to solve two first-order differential equations to obtain the changes in the internal voltage angle δ_i and machine speed ω_i. Thus for an m machine problem where all machines are represented in the simplified manner, it is necessary to solve $2m$ simultaneous differential equations. These equations are

$$\begin{aligned} \frac{d\delta_i}{dt} &= \omega_{i(t)} - 2\pi f \\ \frac{d\omega_i}{dt} &= \frac{\pi f}{H_i}(P_{mi} - P_{ei(t)}) \qquad i = 1, 2, \ldots, m \end{aligned} \tag{10.5.1}$$

If no governor action is considered, P_{mi} remains constant and

$$P_{mi} = P_{mi(0)}$$

In the application of the modified Euler method the initial estimates of the internal voltage angles and machine speeds at time $t + \Delta t$ are obtained from

$$\begin{aligned} \delta_{i(t+\Delta t)}^{(0)} &= \delta_{i(t)}^{(1)} + \left.\frac{d\delta_i}{dt}\right|_{(t)} \Delta t \\ \omega_{i(t+\Delta t)}^{(0)} &= \omega_{i(t)}^{(1)} + \left.\frac{d\omega_i}{dt}\right|_{(t)} \Delta t \qquad i = 1, 2, \ldots, m \end{aligned}$$

where the derivatives are evaluated from equations (10.5.1) and $P_{ei(t)}$ are the machine powers at time t. When $t = 0$, the machine powers $P_{ei(0)}$ are obtained from the network solution at the instant after the disturbance occurs.

Second estimates are obtained by evaluating the derivatives at time $t + \Delta t$. This requires that initial estimates be determined for the machine powers at time $t + \Delta t$. These powers are obtained by calculating new components of the internal voltage from

$$\begin{aligned} e'^{(0)}_{i(t+\Delta t)} &= |E'_i| \cos \delta^{(0)}_{i(t+\Delta t)} \\ f'^{(0)}_{i(t+\Delta t)} &= |E'_i| \sin \delta^{(0)}_{i(t+\Delta t)} \end{aligned}$$

Then a network solution is obtained holding fixed the voltages at the internal machine buses. When there is a three-phase fault on bus f, the voltage E_f also is held fixed at zero. With the calculated bus voltages and the internal voltages, machine terminal currents can be calculated from

$$I^{(0)}_{ti(t+\Delta t)} = (E'^{(0)}_{i(t+\Delta t)} - E^{(0)}_{ti(t+\Delta t)}) \frac{1}{r_{ai} + jx'_{di}}$$

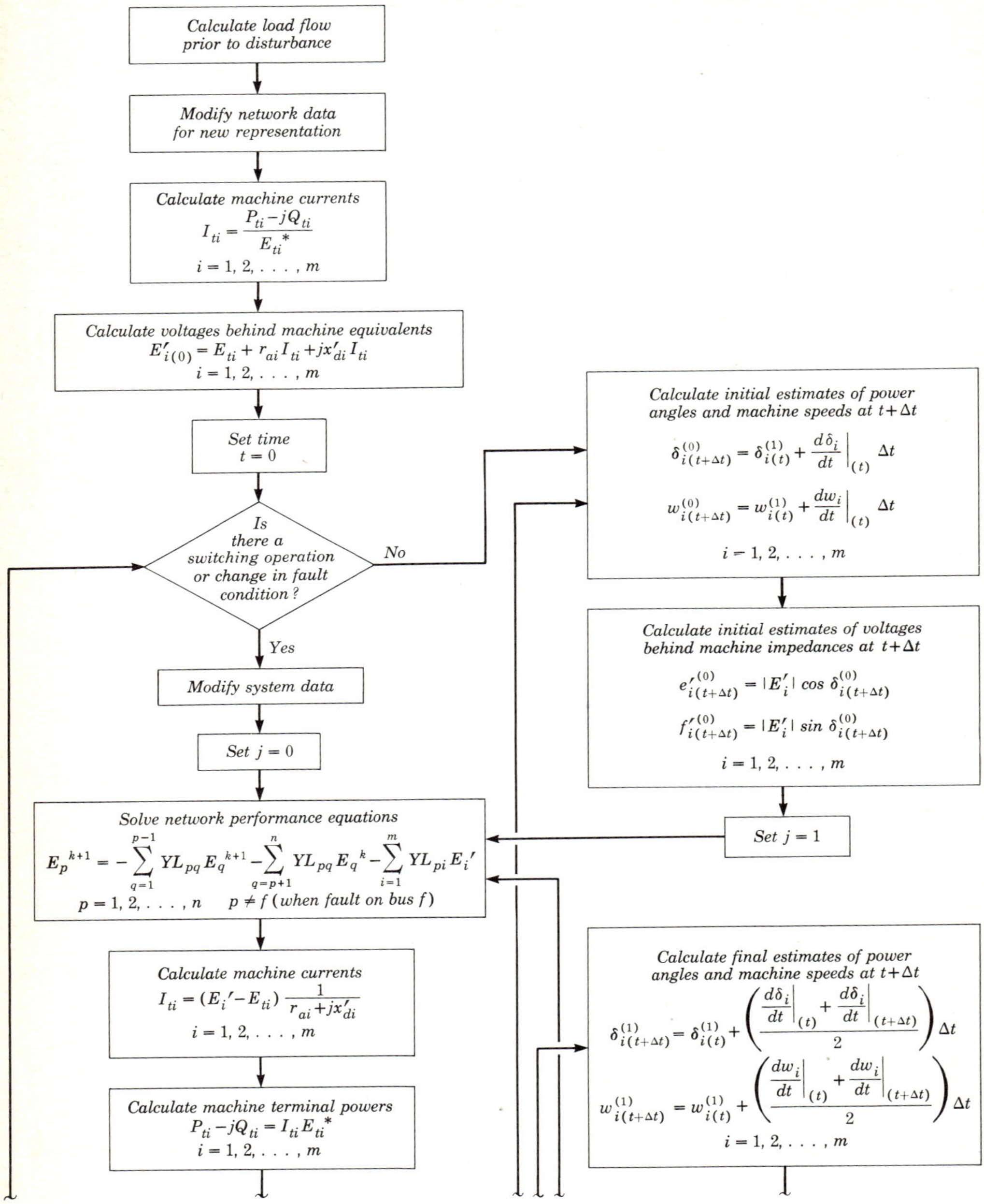

Fig. 10.7 Transient calculations using the modified Euler method.

and machine powers from

$$P^{(0)}_{ei(t+\Delta t)} = \text{Re}\,\{I^{(0)}_{ti(t+\Delta t)}(E'^{(0)}_{i(t+\Delta t)})^*\}$$

The second estimates for the internal voltage angles and machine speeds are obtained from

$$\delta^{(1)}_{i(t+\Delta t)} = \delta^{(1)}_{i(t)} + \left(\frac{\left.\dfrac{d\delta_i}{dt}\right|_{(t)} + \left.\dfrac{d\delta_i}{dt}\right|_{(t+\Delta t)}}{2}\right)\Delta t$$

$$\omega^{(1)}_{i(t+\Delta t)} = \omega^{(1)}_{i(t)} + \left(\frac{\left.\dfrac{d\omega_i}{dt}\right|_{(t)} + \left.\dfrac{d\omega_i}{dt}\right|_{(t+\Delta t)}}{2}\right)\Delta t \qquad i = 1, 2, \ldots, m$$

where

$$\left.\frac{d\delta_i}{dt}\right|_{(t+\Delta t)} = \omega^{(0)}_{i(t+\Delta t)} - 2\pi f$$

$$\left.\frac{d\omega_i}{dt}\right|_{(t+\Delta t)} = \frac{\pi f}{H_i}(P_{mi} - P^{(0)}_{ei(t+\Delta t)})$$

The final voltages at time $t + \Delta t$ for the internal machine buses are

$$e'^{(1)}_{i(t+\Delta t)} = |E'_i| \cos \delta^{(1)}_{i(t+\Delta t)}$$
$$f'^{(1)}_{i(t+\Delta t)} = |E'_i| \sin \delta^{(1)}_{i(t+\Delta t)} \qquad i = 1, 2, \ldots, m$$

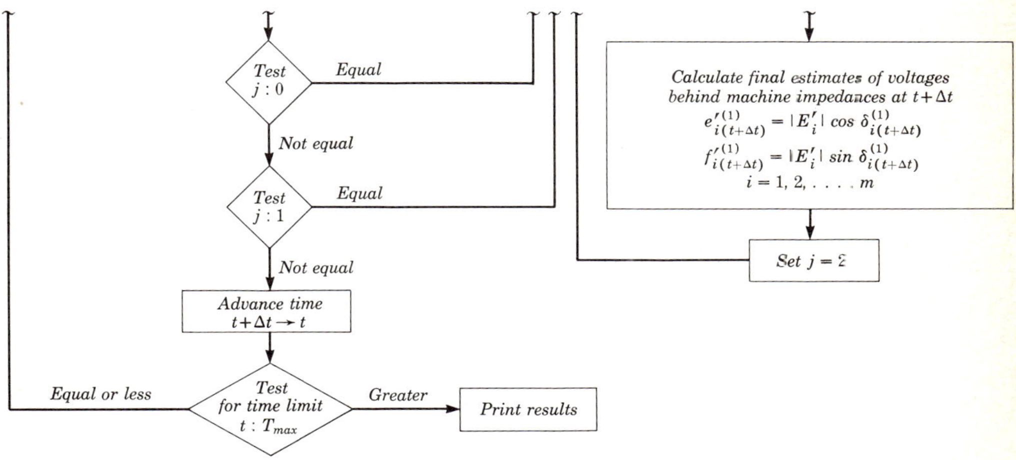

Then the network equations are solved again to obtain the final system voltages at time $t + \Delta t$. The bus voltages are used along with the internal voltages to obtain the machine currents and powers and network power flows. The time is advanced by Δt and a test is made to determine if a switching operation is to be effected or the status of the fault is to be changed. If an operation is scheduled, the appropriate changes are made in the network parameters or variables, or both. Then the network equations are solved to obtain system conditions at the instant after the change occurs. In this calculation the internal voltages are held fixed at the current values. Then estimates are obtained for the next time increment. The process is repeated until t equals the maximum time $T_{(\max)}$ specified for the study.

The sequence of steps for transient analysis by the modified Euler method and the load flow solution by the Gauss-Seidel iterative method using Y_{BUS} is shown in Fig. 10.7. Shown also are the main steps of the preliminary calculations. The procedure shown assumes that all system loads are represented as fixed impedances to ground.

When the effects of saliency and the changes in field flux linkages are to be included in the representation of the machines the following differential equations must be solved simultaneously.

$$
\begin{aligned}
\frac{d\delta_i}{dt} &= \omega_{i(t)} - 2\pi f \\
\frac{d\omega_i}{dt} &= \frac{\pi f}{H_i}(P_{mi} - P_{ei(t)}) \\
\frac{dE'_{qi}}{dt} &= \frac{1}{T'_{d0i}}(E_{fdi} - E_{Ii}) \qquad i = 1, 2, \ldots, m
\end{aligned}
\tag{10.5.2}
$$

Again, if no governor action is considered, P_{mi} remains fixed and

$$P_{mi} = P_{mi(0)}$$

If the effects of the exciter control system are not included, E_{fdi} remains constant and

$$E_{fdi} = E_{fdi(0)}$$

If each machine of the system is described by equations (10.5.2), $3m$ simultaneous equations must be solved.

Runge-Kutta method

In the application of the Runge-Kutta fourth-order approximation, the changes in the internal voltage angles and machine speeds, again for the simplified machine representation, are determined from

$$\begin{aligned}\Delta\delta_{i(t+\Delta t)} &= \tfrac{1}{6}(k_{1i} + 2k_{2i} + 2k_{3i} + k_{4i})\\ \Delta\omega_{i(t+\Delta t)} &= \tfrac{1}{6}(l_{1i} + 2l_{2i} + 2l_{3i} + l_{4i}) \qquad i = 1, 2, \ldots, m\end{aligned}$$

The k's and l's are the changes in δ_i and ω_i, respectively, obtained using derivatives evaluated at predetermined points. Then,

$$\begin{aligned}\delta_{i(t+\Delta t)} &= \delta_{i(t)} + \tfrac{1}{6}(k_{1i} + 2k_{2i} + 2k_{3i} + k_{4i})\\ \omega_{i(t+\Delta t)} &= \omega_{i(t)} + \tfrac{1}{6}(l_{1i} + 2l_{2i} + 2l_{3i} + l_{4i})\end{aligned} \tag{10.5.3}$$

The initial estimates of changes are obtained from

$$\begin{aligned}k_{1i} &= (\omega_{i(t)} - 2\pi f)\,\Delta t\\ l_{1i} &= \frac{\pi f}{H_i}(P_{mi} - P_{ei(t)})\,\Delta t \qquad i = 1, 2, \ldots, m\end{aligned}$$

where $\omega_{i(t)}$ and $P_{ei(t)}$ are the machine speeds and air-gap powers at time t. The second set of estimates of changes in δ_i and ω_i are obtained from

$$\begin{aligned}k_{2i} &= \left\{\left(\omega_{i(t)} + \frac{l_{1i}}{2}\right) - 2\pi f\right\}\Delta t\\ l_{2i} &= \frac{\pi f}{H_i}(P_{mi} - P_{ei}^{(1)})\,\Delta t \qquad i = 1, 2, \ldots, m\end{aligned}$$

where $P_{ei}^{(1)}$ are the machine powers when the internal voltage angles are $\delta_{i(t)} + (k_{1i}/2)$. Thus, before l_{2i} can be calculated, new components for the voltages for the internal machine buses must be calculated from

$$\begin{aligned}e_i'^{(1)} &= |E_i'|\cos\left(\delta_{i(t)} + \frac{k_{1i}}{2}\right)\\ f_i'^{(1)} &= |E_i'|\sin\left(\delta_{i(t)} + \frac{k_{1i}}{2}\right) \qquad i = 1, 2, \ldots, m\end{aligned}$$

Then, the network equations are solved to obtain bus voltages for the calculation of machine powers $P_{ei}^{(1)}$.

The third set of estimates are obtained from

$$\begin{aligned}k_{3i} &= \left\{\left(\omega_{i(t)} + \frac{l_{2i}}{2}\right) - 2\pi f\right\}\Delta t\\ l_{3i} &= \frac{\pi f}{H_i}(P_{mi} - P_{ei}^{(2)})\,\Delta t \qquad i = 1, 2, \ldots, m\end{aligned}$$

where $P_{ei}^{(2)}$ are obtained from a second solution of the network equations with the internal voltage angles equal to $\delta_{i(t)} + (k_{2i}/2)$ and the components of the voltages for the internal machine buses equal to

$$e_i'^{(2)} = |E_i'| \cos \left(\delta_{i(t)} + \frac{k_{2i}}{2} \right)$$

$$f_i'^{(2)} = |E_i'| \sin \left(\delta_{i(t)} + \frac{k_{2i}}{2} \right) \qquad i = 1, 2, \ldots, m$$

The fourth estimates are obtained from

$$k_{4i} = \{(\omega_{i(t)} + l_{3i}) - 2\pi f\} \, \Delta t$$

$$l_{4i} = \frac{\pi f}{H_i} (P_{mi} - P_{ei}^{(3)}) \, \Delta t \qquad i = 1, 2, \ldots, m$$

where $P_{ei}^{(3)}$ are obtained from a third solution of the network equations with internal voltage angles equal to $\delta_{i(t)} + k_{3i}$ and voltage components equal to

$$e_i'^{(3)} = |E_i'| \cos (\delta_{i(t)} + k_{3i})$$
$$f_i'^{(3)} = |E_i'| \sin (\delta_{i(t)} + k_{3i})$$

The final estimates of the internal voltage angles and machine speeds at time $t + \Delta t$ are obtained by substituting the k's and l's into equations (10.5.3). The internal voltage angles $\delta_{i(t+\Delta t)}$ are used to calculate the estimates for the components of voltages for the internal machine buses from

$$e'_{i(t+\Delta t)} = |E_i'| \cos \delta_{i(t+\Delta t)}$$
$$f'_{i(t+\Delta t)} = |E_i'| \sin \delta_{i(t+\Delta t)} \qquad i = 1, 2, \ldots, m$$

The network equations are solved then for the fourth time to obtain bus voltages for the calculation of machine currents and powers and network power flows. The time is advanced by Δt and a network solution is obtained for any scheduled switching operation and change in the fault condition. The process is repeated until t equals the maximum time $T_{\max}$.

10.6 Example of transient stability calculations

The method for determining transient stability will be illustrated with the sample power system used in Sec. 8.5 for the load flow problem. In this example the machines are represented by voltages of constant magnitudes behind direct-axis transient reactances. Loads are represented by fixed admittances to ground.

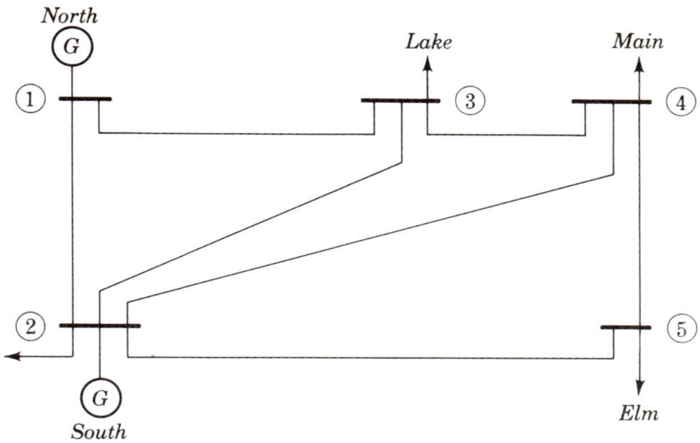

***Fig. 10.8** Sample system for transient stability calculations.*

Problem

Using the bus admittance matrix and the Gauss-Seidel iterative method for the solution of the network equations and the modified Euler method for the solution of the swing equations:

a. Determine the effects on the sample power system shown in Fig. 10.8 of a three-phase fault on bus 2 for a duration of 0.1 sec.
b. Determine the effects of the fault on bus 2 for a duration of 0.2 sec.

Solution

The results of the load flow calculation prior to the fault are given in Table 10.1. The inertia constants, direct-axis transient reactances, and equivalent admittances of the generators at buses 1 and 2 in per unit on a 100,000 kva base are given in Table 10.2.

***Table 10.1** Bus voltages, generation, and loads from load flow calculation prior to fault*

Bus code	Bus voltages	Generation		Load	
p	E_p	Megawatts	Megavars	Megawatts	Megavars
1	$1.06000 + j0.00000$	129.565	−7.480	0.0	0.0
2	$1.04621 - j0.05128$	40.0	30.0	20.0	10.0
3	$1.02032 - j0.08920$	0.0	0.0	45.0	15.0
4	$1.01917 - j0.09506$	0.0	0.0	40.0	5.0
5	$1.01209 - j0.10906$	0.0	0.0	60.0	10.0

Table 10.2 Inertia constants, direct-axis transient reactances, and equivalent admittances for generators of sample system

Bus code p–i	Inertia constant H	Direct-axis transient reactance x'_d	Equivalent admittance y_{pi}
1–6	50.0	0.25	$0.0 - j4.00000$
2–7	1.0	1.50	$0.0 - j0.66667$

a. The Gauss-Seidel iterative equations describing the performance of the network, using the bus code numbers given in Fig. 10.9, are

$$\begin{aligned}
E_1^{k+1} &= -YL_{12}E_2^k - YL_{13}E_3^k - YL_{16}E_6 \\
E_2^{k+1} &= -YL_{21}E_1^{k+1} - YL_{23}E_3^k - YL_{24}E_4^k - YL_{25}E_5^k - YL_{27}E_7 \\
E_3^{k+1} &= -YL_{31}E_1^{k+1} - YL_{32}E_2^{k+1} - YL_{34}E_4^k \\
E_4^{k+1} &= -YL_{42}E_2^{k+1} - YL_{43}E_3^{k+1} - YL_{45}E_5^k \\
E_5^{k+1} &= -YL_{52}E_2^{k+1} - YL_{54}E_4^{k+1}
\end{aligned}$$

The line parameters YL_{pq} for these equations can be obtained from the elements of the bus admittance matrix used for the load flow solution prior to the disturbance and the equivalent admittances for machines and loads.

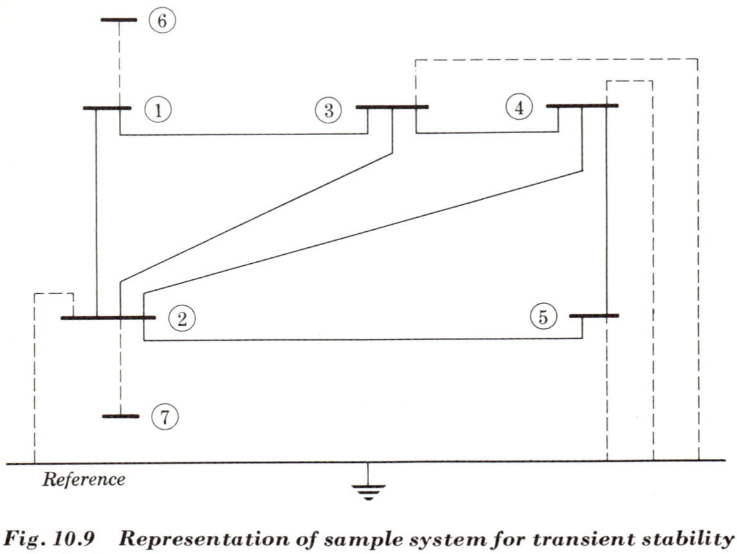

Fig. 10.9 Representation of sample system for transient stability calculations.

The bus admittance matrix is

Y_{BUS} =

	①	②	③	④	⑤
①	$6.25000 - j18.69500$	$-5.00000 + j15.00000$	$-1.25000 + j\,3.75000$		
②	$-5.00000 + j15.00000$	$10.83334 - j32.41500$	$-1.66667 + j\,5.00000$	$-1.66667 + j\,5.00000$	$-2.50000 + j\,7.50000$
③	$-1.25000 + j\,3.75000$	$-1.66667 + j\,5.00000$	$12.91667 - j38.69500$	$-10.00000 + j30.00000$	
④		$-1.66667 + j\,5.00000$	$-10.00000 + j30.00000$	$12.91667 - j38.69500$	$-1.25000 + j\,3.75000$
⑤		$-2.50000 + j\,7.50000$		$-1.25000 + j\,3.75000$	$3.75000 - j11.21000$

The line parameters are obtained from the equation

$$YL_{pq} = Y_{pq}L_p = Y_{pq}\left(\frac{1}{Y_{pp}}\right)$$

The modified line parameter for element 1–2 is

$$YL_{12} = Y_{12}\left(\frac{1}{Y_{11} + y_{16}}\right)$$

$$YL_{12} = \frac{-5.00000 + j15.00000}{6.25000 - j22.69500}$$
$$= -0.67074 - j0.03560$$

where Y_{11} and Y_{12} are elements in the bus admittance matrix and y_{16} is the equivalent admittance representing the machine at bus 1, which is given in Table 10.2. The remaining line parameters for bus 1 are obtained from the equations

$$YL_{13} = Y_{13}\left(\frac{1}{Y_{11} + y_{16}}\right)$$

$$YL_{16} = Y_{16}\left(\frac{1}{Y_{11} + y_{16}}\right)$$

where

$$Y_{16} = -y_{16}$$

The line parameter for element 2–1 is obtained from

$$YL_{21} = Y_{21}\left(\frac{1}{Y_{22} + y_{27} + y_{20}}\right)$$

where Y_{22} and Y_{21} are elements in the bus admittance matrix; y_{27} is the equivalent admittance representing the machine at bus 2 and y_{20} is the equivalent admittance to ground representing the load at bus 2. The equation for the load equivalent admittance is

$$y_{p0} = \frac{P_{Lp} - jQ_{Lp}}{e_p^2 + f_p^2}$$

and for bus 2

$$y_{20} = \frac{0.20 - j0.10}{(1.04621)^2 + (0.05128)^2}$$
$$= 0.18228 - j0.09114$$

where the bus voltage is obtained from the load flow solution and is given in Table 10.1. The line parameter YL_{21} is

$$YL_{21} = \frac{-5.00000 + j15.00000}{11.01562 - j33.17281}$$
$$= -0.45235 - j0.00052$$

The YL_{pq}'s for all elements are given in Table 10.3.

The voltages behind the equivalent admittances representing the machines are obtained from the equation

$$E'_i = E_{ti} + jx'_{di}I_{ti} \qquad i = n + 1, n + 2, \ldots, n + m$$

where

$$I_{ti} = \frac{P_{ti} - jQ_{ti}}{E^*_{ti}}$$

and n is the number of buses of the network and m is the number of machines. For the machine at bus 1

$$E_6 = 1.06 + j0.0 + j0.25\left(\frac{1.29565 + j0.07480}{1.06 - j0.0}\right)$$
$$= 1.04236 + j0.30558$$

***Table 10.3** Line parameters for transient stability representation of sample system*

Bus code p–q	YL_{pq}
1–2	$-0.67074 - j0.03560$
1–3	$-0.16769 - j0.00890$
1–6	$-0.16383 + j0.04512$
2–1	$-0.45235 - j0.00052$
2–3	$-0.15078 - j0.00017$
2–4	$-0.15078 - j0.00017$
2–5	$-0.22618 - j0.00026$
2–7	$-0.01810 + j0.00601$
3–1	$-0.09625 + j0.00089$
3–2	$-0.12833 + j0.00119$
3–4	$-0.77000 + j0.00711$
4–2	$-0.12866 + j0.00115$
4–3	$-0.77198 + j0.00687$
4–5	$-0.09650 + j0.00086$
5–2	$-0.65236 + j0.02866$
5–4	$-0.32618 + j0.01433$

where the bus voltage and generation are obtained from Table 10.1 and the machine reactance from Table 10.2. The voltage magnitude is

$$|E_6| = 1.08623$$

and the internal voltage angle is

$$\delta_6 = 16.339° \qquad \text{or} \qquad \delta_6 = 0.28517 \text{ rad}$$

The voltage behind the equivalent admittance representing the machine at bus 2 is obtained in a similar manner and is

$$E_7 = 1.50335 + j0.49981$$

The voltage magnitude is

$$|E_7| = 1.58426$$

and the internal voltage angle is

$$\delta_7 = 18.390° \qquad \text{or} \qquad \delta_7 = 0.32097 \text{ rad}$$

The fault at bus 2 is simulated by setting the voltage at this bus equal to zero. Then the network equations are solved to obtain system conditions at the instant the fault occurs. In this calculation the voltage at the faulted bus and the voltages behind the equivalent admittances representing the machines are held fixed. The calculated system voltages are given in Table 10.4.

The machine currents, with the fault, are calculated from the equation

$$I_{ti} = (E'_i - E_{ti})y_{pi}$$

Then

$$\begin{aligned} I_{61} &= \{(1.04236 + j0.30558) - (0.19234 + j0.00330)\}(0.0 - j4.0) \\ &= 1.20912 - j3.40008 \end{aligned}$$

Table 10.4 Bus voltages of sample system at the instant the fault occurs

Bus code p	*Bus voltage* E_p
1	$0.19234 + j0.00330$
2	$0.0 + j0.0$
3	$0.04707 - j0.00096$
4	$0.03758 - j0.00118$
5	$0.01226 - j0.00093$

and

$$I_{72} = \{(1.50335 + j0.49981) - (0.0 + j0.0)\}(0.0 - j0.66667)$$
$$= 0.33321 - j1.00223$$

The electrical power of the machines is calculated from

$$P_{ei} - jQ_{ei} = I_{ti}(E_i')^*$$

The real power of the machine at bus 1 is

$$P_{e6} = (1.20912)(1.04236) - (3.40008)(0.30558)$$
$$= 0.22134$$

The real power of the machine at bus 2 is zero since bus 2 is the faulted bus and its voltage is zero. Calculating the real power as a check,

$$P_{e7} = (0.33321)(1.50335) - (1.00223)(0.49981)$$
$$= 0.0000067$$

The initial estimates of the internal voltage angles and speed of the machines at $t + \Delta t$ are obtained from the differential equations. The rate of change in speed of the machines is calculated from

$$\frac{d\omega_i}{dt} = \frac{\pi f}{H_i}(P_{mi} - P_{ei(t)})$$

Then, at $t = 0$ for the machine at bus 1,

$$\left.\frac{d\omega_6}{dt}\right|_{(0)} = \frac{3.1416(60)}{50.0}(1.29565 - 0.22134)$$
$$= 4.05006$$

Similarly, for the machine at bus 2,

$$\left.\frac{d\omega_7}{dt}\right|_{(0)} = \frac{3.1416(60)}{1.0}(0.40000 - 0.0)$$
$$= 75.3984$$

Next, the initial estimates of the speed of the machines at $t + \Delta t$ are calculated from

$$\omega_{i(t+\Delta t)}^{(0)} = \omega_{i(t)}^{(1)} + \left.\frac{d\omega_i}{dt}\right|_{(t)} \Delta t$$

where $\omega_{i(t)}^{(1)}$ at $t = 0$ is the rated speed and equal to $2\pi f$ and $\Delta t = 0.02$. Then, for the machine at bus 1,

$$\omega_{6(0.02)}^{(0)} = 2(3.1416)60 + (4.05006)0.02$$
$$= 376.992 + 0.08100$$
$$= 377.07300$$

Similarly, for the machine at bus 2,

$$\begin{aligned}\omega_{7(0.02)}^{(0)} &= 2(3.1416)60 + (75.3984)0.02 \\ &= 376.992 + 1.50797 \\ &= 378.49997\end{aligned}$$

The rates of change of the internal voltage angles are calculated next from

$$\frac{d\delta_i}{dt} = \omega_{i(t)} - 2\pi f$$

Since $\omega_{i(t)}$ at $t = 0$ is equal to $2\pi f$, then for the machines

$$\left.\frac{d\delta_6}{dt}\right|_{(0)} = 0.0 \qquad \text{and} \qquad \left.\frac{d\delta_7}{dt}\right|_{(0)} = 0.0$$

The initial estimates of the internal voltage angles of the machines are calculated from

$$\delta_{i(t+\Delta t)}^{(0)} = \delta_{i(t)}^{(1)} + \left.\frac{d\delta_i}{dt}\right|_{(t)} \Delta t$$

Then, for the machines, the internal voltage angles in radians are

$$\begin{aligned}\delta_{6(0.02)}^{(0)} &= 0.28517 \\ \delta_{7(0.02)}^{(0)} &= 0.32097\end{aligned}$$

The new components of the voltages behind the equivalent admittances representing the machines are calculated from

$$e_{i(t+\Delta t)}^{\prime(0)} = |E_i'| \cos \delta_{i(t+\Delta t)}^{(0)}$$

and

$$f_{i(t+\Delta t)}^{\prime(0)} = |E_i'| \sin \delta_{i(t+\Delta t)}^{(0)}$$

These voltage components replace the previous values obtained from the load flow solution prior to the fault and again the network equations are solved. In this calculation the new voltages behind the machine equivalent admittances as well as the zero voltage at the faulted bus are held constant.

Since there is no change in the internal voltage angle for the initial estimate, the system voltages and machine currents and powers are the same as those obtained from the network solution at the instant the fault occurs. Consequently, the rates of change in the speed of the machines at $t + \Delta t = 0.02$ will be the same. Therefore,

$$\left.\frac{d\omega_6}{dt}\right|_{(0.02)} = 4.05006 \qquad \text{and} \qquad \left.\frac{d\omega_7}{dt}\right|_{(0.02)} = 75.39484$$

The final estimate for the speed of the machines at $t + \Delta t$ is calculated from

$$\omega_{i(t+\Delta t)}^{(1)} = \omega_{i(t)}^{(1)} + \left(\frac{\left.\frac{d\omega_i}{dt}\right|_{(t)} + \left.\frac{d\omega_i}{dt}\right|_{(t+\Delta t)}}{2} \right) \Delta t$$

Then, for the machine at bus 1,

$$\begin{aligned} \omega_{6(0.02)}^{(1)} &= 2(3.1416)60 + \left(\frac{4.05006 + 4.05006}{2} \right) 0.02 \\ &= 377.07300 \end{aligned}$$

Similarly, for the machine at bus 2,

$$\begin{aligned} \omega_{7(0.02)}^{(1)} &= 2(3.1416)60 + \left(\frac{75.3984 + 75.3984}{2} \right) 0.02 \\ &= 378.49997 \end{aligned}$$

The rates of change of the internal voltage angles at $t + \Delta t$ are calculated from

$$\frac{d\delta_i}{dt} = \omega_{i(t+\Delta t)}^{(1)} - 2\pi f$$

Then, for the machine at bus 1,

$$\begin{aligned} \left.\frac{d\delta_6}{dt}\right|_{(.02)} &= 377.0730 - 376.9920 \\ &= 0.08100 \end{aligned}$$

Similarly, for the machine at bus 2,

$$\begin{aligned} \left.\frac{d\delta_7}{dt}\right|_{(0.02)} &= 378.49997 - 376.9920 \\ &= 1.50797 \end{aligned}$$

The final estimates for the internal voltage angles of the machines at $t + \Delta t$ are calculated from

$$\delta_{i(t+\Delta t)}^{(1)} = \delta_{i(t)}^{(1)} + \left(\frac{\left.\frac{d\delta_i}{dt}\right|_{(t)} + \left.\frac{d\delta_i}{dt}\right|_{(t+\Delta t)}}{2} \right) \Delta t$$

Then, for the machine at bus 1,

$$\begin{aligned} \delta_{6(0.02)}^{(1)} &= 0.28517 + \left(\frac{0.0 + 0.08100}{2} \right) 0.02 \\ &= 0.28517 + 0.00081 \\ &= 0.28598 \end{aligned}$$

Similarly, for the machine at bus 2,

$$\delta_{7(0.02)}^{(1)} = 0.32097 + \left(\frac{0.0 + 1.50797}{2}\right) 0.02$$
$$= 0.32097 + 0.01508$$
$$= 0.33605$$

The internal voltage angles in degrees at $t + \Delta t = 0.02$ are

$$\delta_{6(0.02)}^{(1)} = 0.28598 \left(\frac{180}{\pi}\right) = 16.38540°$$

and

$$\delta_{7(0.02)}^{(1)} = 0.33605 \left(\frac{180}{\pi}\right) = 19.25420°$$

At $t + \Delta t = 0.02$ the final components of voltages behind the machine equivalent admittances are

$$e_6'^{(1)} = 1.08623 \cos(16.38540)$$
$$= 1.04212$$
$$f_6'^{(1)} = 1.08623 \sin(16.38540)$$
$$= 0.30641$$

and

$$e_7'^{(1)} = 1.58426 \cos(19.25420)$$
$$= 1.49564$$
$$f_7'^{(1)} = 1.58426 \sin(19.25420)$$
$$= 0.52243$$

Then the network equations are solved to obtain the final system voltages at $t + \Delta t = 0.02$. The voltages obtained from this calculation are given in Table 10.5.

Table 10.5 Bus voltages of sample system at $t + \Delta t = 0.02$

Bus code p	*Bus voltage* E_p
1	$0.19258 + j0.00353$
2	$0.0 + j0.0$
3	$0.04815 - j0.00114$
4	$0.03845 - j0.00133$
5	$0.01249 - j0.00097$

With these system voltages the machine currents and powers at $t + \Delta t = 0.02$ can be calculated. The current of the machine at bus 1 is

$$\begin{aligned} I_{61(0.02)} &= \{(1.04212 + j0.30641) - (0.19258 + j0.00353)\}(0.0 - j4.0) \\ &= 1.21152 - j3.39816 \end{aligned}$$

and the real power is

$$\begin{aligned} P_{e6(0.02)} &= (1.21152)(1.04212) - (3.39816)(0.30641) \\ &= 0.22132 \end{aligned}$$

The current of the machine at bus 2 is

$$\begin{aligned} I_{72(0.02)} &= \{(1.49564 + j0.52243) - (0.0 + j0.0)\}(0.0 - j0.66667) \\ &= 0.34829 - j0.99710 \end{aligned}$$

and the real power is zero since the fault is at bus 2.

This completes the calculations for values at $t + \Delta t = 0.02$. Then the time is set to $t = 0.02$ and the process repeated to obtain estimates at $t + \Delta t = 0.04$. When the time is advanced to $t = 0.10$, however, the network equations are solved without the fault to obtain the post fault conditions before proceeding with the normal process. In the network calculation only the voltages behind the machine equivalent admittances are held constant. The machine currents and powers obtained with the new system voltages are used to obtain new estimates at $t + \Delta t = 0.12$. The process is continued until $t = T_{\max}$.

The internal voltage angles and the ratios of actual to rated speed of the machines for the complete calculation are shown in Figs. 10.10 and 10.11, respectively. The system is stable for this disturbance.

b. The procedure for determining the transient stability of the sample system for a fault on bus 2 of duration 0.2 sec is identical except that the network solution without the fault is obtained when $t = 0.20$ instead of $t = 0.10$ as in part *a*. The internal voltage angles and the ratios of actual to rated speed of the machines for the complete calculation are shown in Figs. 10.12 and 10.13. The system is unstable for this disturbance.

10.7 Exciter and governor control systems

In the solution techniques described in Sec. 10.5 the effects of the exciter and governor control systems on power system response were neglected. In that representation the field voltage E_{fd} and the mechanical power P_m were held constant in the transient calculations. When a more detailed evaluation of system response is required or the period of analysis extends beyond one second it is important to include the effects of the exciter and governor systems.

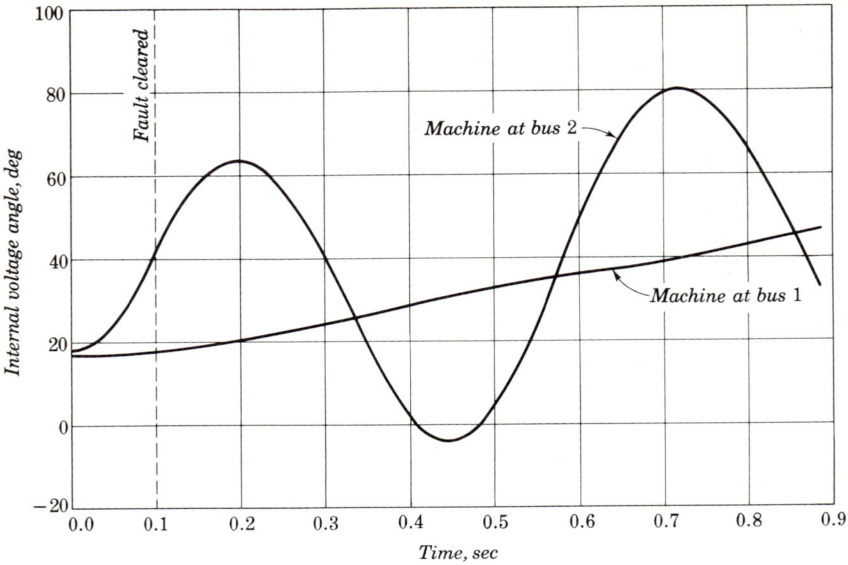

Fig. 10.10 Internal voltage angle of machine with respect to time for a fault duration of 0.1 sec.

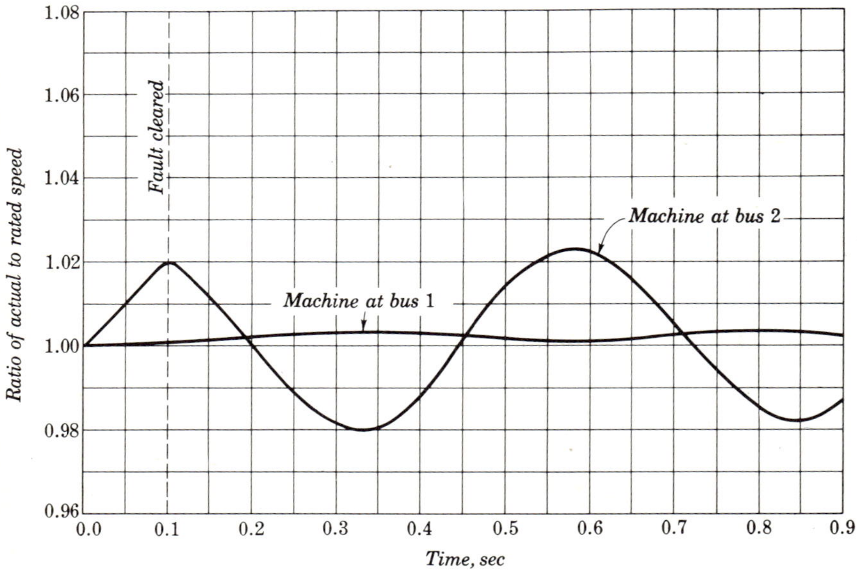

Fig. 10.11 Ratio of actual to rated speed of machine with respect to time for a fault duration of 0.1 sec.

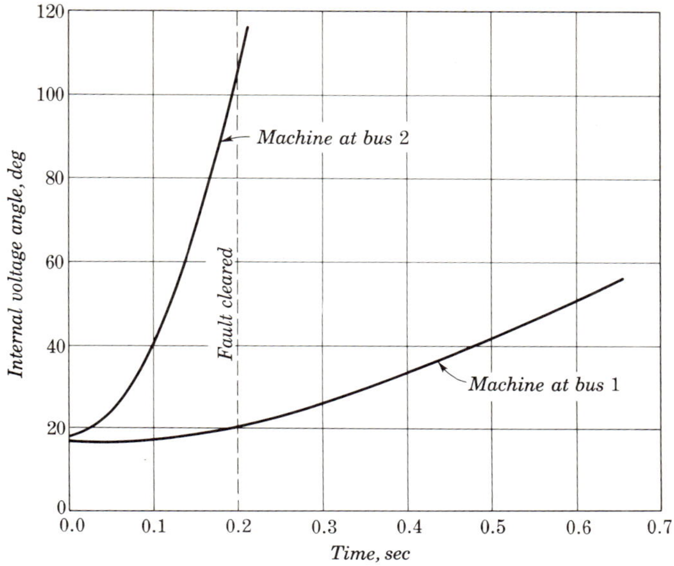

Fig. 10.12 Internal voltage angle of machine with respect to time for a fault duration of 0.2 sec.

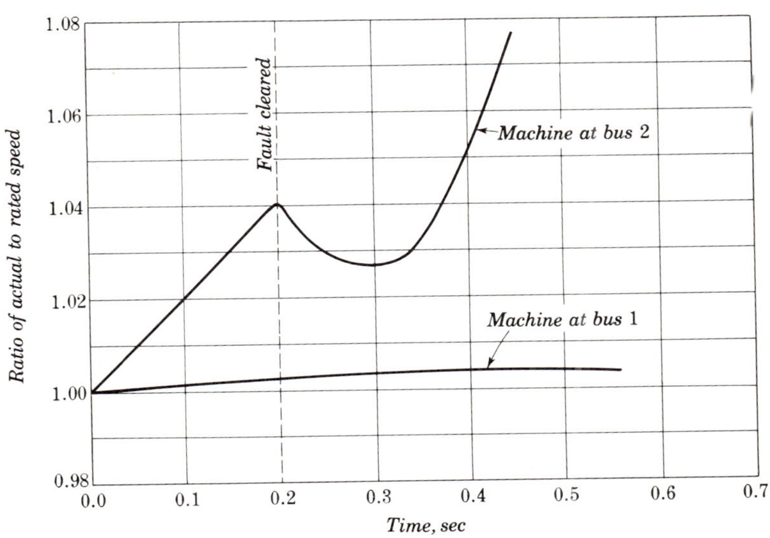

Fig. 10.13 Ratio of actual to rated speed of machine with respect to time for a fault duration of 0.2 sec.

The exciter control system provides the proper field voltage to maintain a desired system voltage, usually at the high-voltage bus of the power plant. An important characteristic of an exciter control system is its ability to respond rapidly to voltage deviations during both normal and emergency system operation. Many different types of exciter control systems are employed on power systems. The basic components of an exciter control system are the regulator, amplifier, and exciter. The regulator measures the actual regulated voltage and determines the voltage deviation. The deviation signal produced by the regulator is then amplified to provide the signal required to change the exciter field current. This in turn produces a change in the exciter output voltage which results in a new excitation level for the generator.

A convenient form of representing a control system is a block diagram that relates through transfer functions the input and output variables of the principal components of the system. A block diagram for a simplified representation of a continuously acting exciter control system is shown in Fig. 10.14. This is one of the important types of exciter control systems. This representation includes transfer functions to describe the regulator, amplifier, exciter, and stabilizing loop. The stabilizing loop modifies the response to eliminate undesired oscillations and overshoot of the regulated voltage. The differential equations relating the input and output variables of the regulator, amplifier, exciter, and stabilizing loop, respectively, are

$$
\begin{aligned}
\frac{dE^{\text{v}}}{dt} &= \frac{1}{T_R}(E_S - E_t - E^{\text{v}}) \\
\frac{dE^{\text{iii}}}{dt} &= \frac{1}{T_A}\left\{K_A\left(E^{\text{v}} + \frac{E_0^{\text{iii}}}{K_A} - E^{\text{iv}}\right) - E^{\text{iii}}\right\} \\
\frac{dE_{fd}}{dt} &= \frac{1}{T_E}(E^{\text{ii}} - K_E E_{fd}) \\
\frac{dE^{\text{iv}}}{dt} &= \frac{1}{T_F}\left\{K_F\frac{dE_{fd}}{dt} - E^{\text{iv}}\right\}
\end{aligned}
\qquad (10.7.1)
$$

where E_S = scheduled voltage in per unit

E_0^{iii} = output voltage of the amplifier in per unit prior to the disturbance

T_R = regulator time constant

K_A = amplifier gain

T_A = amplifier time constant

K_E = exciter gain

T_E = exciter time constant

K_F = stabilizing loop gain

T_F = stabilizing loop time constant

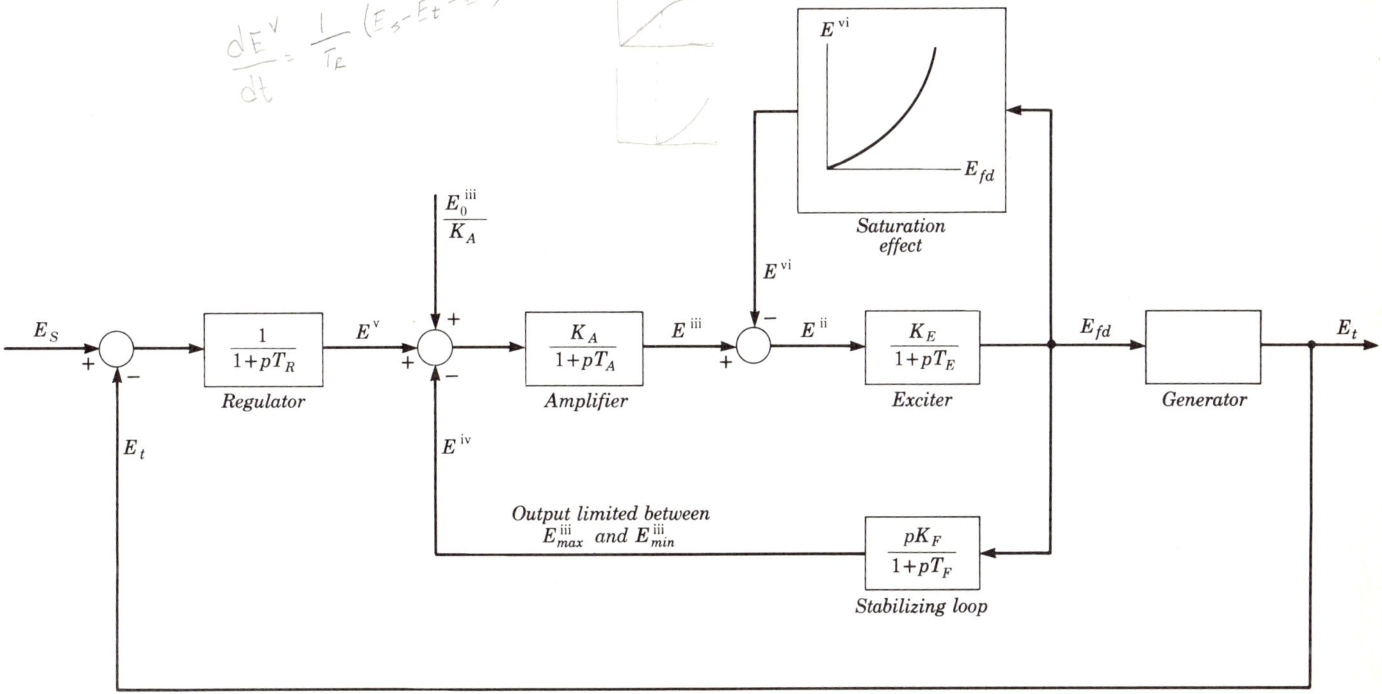

Fig. 10.14 Block diagram for a representation of an exciter control system.

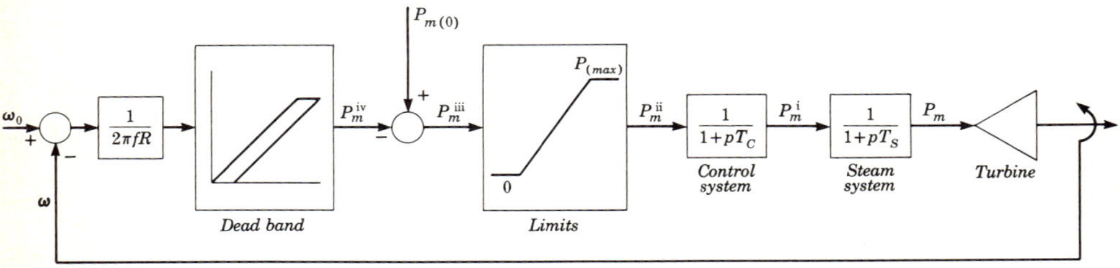

Fig. 10.15 Block diagram for a simplified representation of a speed governor control system.

and the intermediate variables are designated by E^{ii}, E^{iii}, E^{iv}, E^{v} and E^{vi}. The intermediate variable E^{ii} is

$$E^{ii} = E^{iii} - E^{vi}$$

where E^{vi} is equivalent to the demagnetizing effect due to saturation in the exciter. This is determined from

$$E^{vi} = A\epsilon^{BE_{fd}}$$

where A and B are constants depending upon the exciter saturation characteristic.

To include the effects of the exciter control system, equations (10.7.1) are solved simultaneously with the equations (10.5.2) describing the machine.

The effects of the speed governor control during transient periods can be taken into account by using the simplified representation of the governor control system shown in Fig. 10.15. This representation includes a transfer function describing the steam system with a time constant T_S and a transfer function describing the control system with a time constant T_C. The differential equations relating the input and output variables of these transfer functions, respectively, are

$$\begin{aligned} \frac{dP_m}{dt} &= \frac{1}{T_S}(P_m^{i} - P_m) \\ \frac{dP_m^{i}}{dt} &= \frac{1}{T_C}(P_m^{ii} - P_m^{i}) \end{aligned} \qquad (10.7.2)$$

where P_m is the mechanical power and the intermediate variables are designated by P_m^{i}, P_m^{ii}, P_m^{iii}, and P_m^{iv}. The variables P_m^{ii} and P_m^{iii} are related by the following:

$$
\begin{aligned}
P_m^{\text{ii}} &= 0 \qquad & P_m^{\text{iii}} &\le 0 \\
P_m^{\text{ii}} &= P_m^{\text{iii}} \qquad & 0 < P_m^{\text{iii}} &< P_{\max} \\
P_m^{\text{ii}} &= P_{\max} \qquad & P^{\text{iii}} &\ge P_{\max}
\end{aligned}
$$

where $P_{\max}$ is the maximum turbine capability. The intermediate variable P_m^{iii} is

$$P_m^{\text{iii}} = P_{m(0)} - P_m^{\text{iv}}$$

where $P_{m(0)}$ is the initial mechanical power. The intermediate variable P_m^{iv} is

$$P_m^{\text{iv}} = \frac{1}{R}\left(\frac{\omega_0 - \omega}{2\pi f} \pm DB_T\right)$$

where R is the speed regulation in per unit and DB_T is the dead band travel, that is, the change in speed required to overcome the dead band of the governor system. A typical governor characteristic is shown in Fig. 10.16.

Equations (10.7.2) are solved simultaneously with equations (10.5.2) if the effects of the governor control system are included.

10.8 Distance relays

Coordination in the planning of generation and transmission facilities and the design of an effective protective relaying system is essential for the reliable performance of a power system. The principal purpose of relays is to protect the power system from the effects of faults by initiating

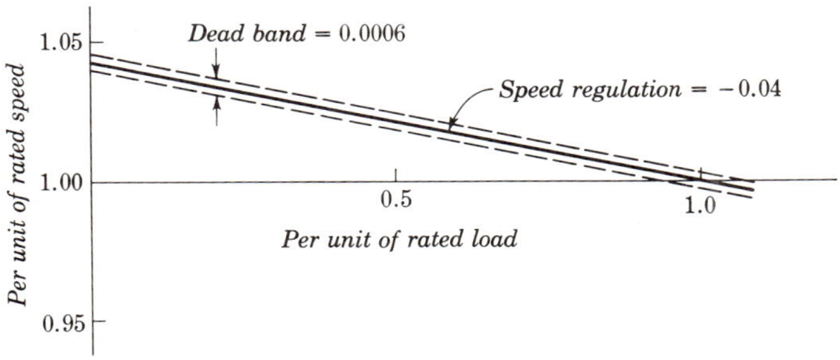

Fig. 10.16 ***Typical governor characteristic set for rated output at rated speed.***

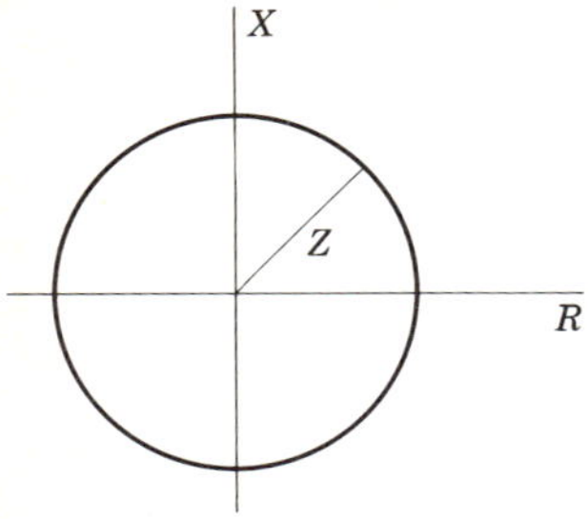

Fig. 10.17 Operating characteristic of distance relay plotted on an RX diagram.

circuit breaker operations to disconnect the faulted equipment. The design of a protective relaying system must assure proper operation so as not to disconnect additional equipment that would aggravate the effects of the disturbance and it must assure that the faulted equipment is cleared sufficiently fast to mitigate the effects of the fault. In addition, the relaying system must not limit the design capability of the generation and transmission facilities.

An important type of relay that is used for high-voltage transmission line protection is the distance relay. This relay responds to the ratio of measured voltage to measured current which can be expressed as an impedance. A convenient means of showing the operating characteristic of a distance relay is with an RX diagram on which a circle is drawn with the radius equal to the impedance setting as shown in Fig. 10.17. When the value of the impedance seen by the relay falls within the circle, the relay will operate.

To provide adequate primary and backup protection, distance relays have three units. The operating characteristic of each unit can be adjusted independently. Furthermore, the proper functioning of distance relays requires the capability to distinguish direction. This is provided by either a directional unit, as in the impedance-type distance relay, or is inherent in the operating characteristics, as in the mho-type distance relay. The operating characteristics of these two relays are shown in Fig. 10.18. The circles associated with the three units are labeled zone 1, zone 2, and zone 3.

When a fault occurs and the value of impedance seen by the relay falls within zone 1 and above the characteristic of the directional unit of the impedance type, the zone 1 contacts will close and trip the circuit breaker immediately. In this case all three units will operate because zone 1 is the smallest circle. When the impedance falls only within zones 2 and 3, or zone 3, the contacts of the associated units will close and energize a timer. At a specified time setting, the timer will close a second set of contacts associated with zone 2. If the first set of contacts associated with zone 2 is closed the circuit breaker will be tripped. If the zone 2 contacts are not closed, that is, the impedance seen by the relay is not within zone 2, then the timer, at a later specified time, will close a second set of contacts associated with zone 3. If the first set of contacts associated with zone 3 is closed then the circuit breaker will be tripped. The time delays for zones 2 and 3 can be set independently. Zones 1 and 2 provide primary protection for a transmission line section, whereas zones 2 and 3 provide backup protection in the event relays or circuit breakers of adjoining facilities fail to operate properly.

During a system disturbance and following the switching operations to clear the faulted equipment, power swings will occur on the transmission system until a new stable operating condition is established. These swings should not cause relays associated with the unfaulted equipment

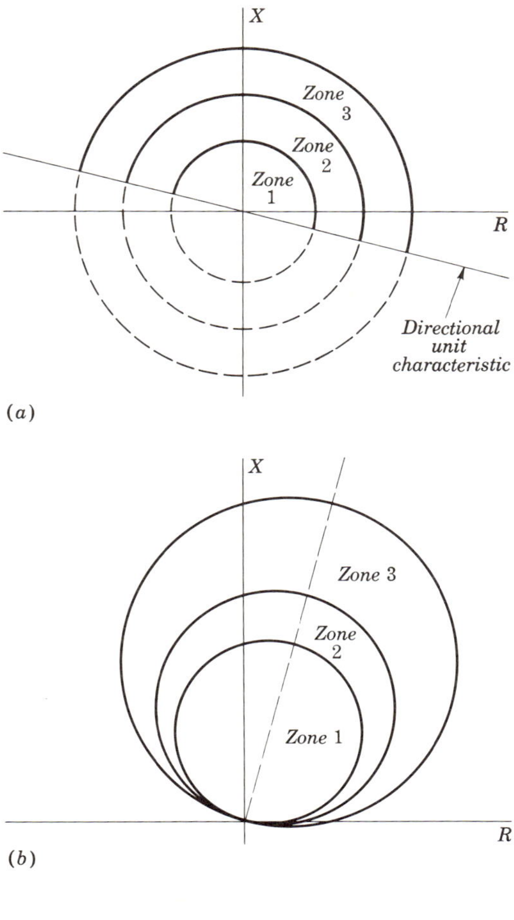

***Fig. 10.18** Operating characteristics of distance relays. (a) Impedance type; (b) mho type.*

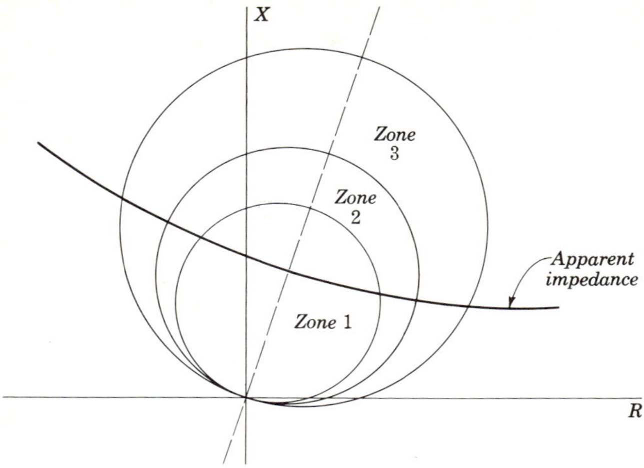

Fig. 10.19 Trajectory of apparent impedance during a power swing.

to operate. The operation of the relay system can be tested for various system disturbances by calculating during the step-by-step transient calculations the apparent impedance, that is, the impedance seen by the relay. The apparent impedance calculated at each time increment can be compared to the operating characteristics of the relay. A convenient means of making this comparison is to plot the impedance values on the RX diagram of the relay as shown in Fig. 10.19.

The apparent impedance is calculated from the final results obtained from the network solution at time $t + \Delta t$. First the current in a specified transmission line p–q is calculated from

$$I_{pq} = (E_p - E_q)y_{pq}$$

Then, the apparent impedance for bus p is

$$Z_p = \frac{E_p}{I_{pq}}$$

or in complex form,

$$R_p + jX_p = \frac{e_p + jf_p}{a_{pq} + jb_{pq}}$$

where $R_p = \dfrac{e_p a_{pq} + f_p b_{pq}}{a_{pq}^2 + b_{pq}^2}$

$$X_p = \frac{f_p a_{pq} - e_p b_{pq}}{a_{pq}^2 + b_{pq}^2}$$

The values R_p and X_p are the coordinates in per unit on the RX diagram of the apparent impedance at time $t + \Delta t$.

Normal information related to the operating characteristics of the relay includes the diameters of the circles for each zone, the angle θ with respect to the R axis of the line along which the centers of the circles lie, and the positions of the centers of the circles along this line. This information is used to determine the coordinates in per unit of the center of each circle. These coordinates are determined from

$$R_c = \left(\frac{\frac{D}{2} \times \text{base kva}}{(\text{base kv})^2 \times 10^3} \right) \cos \theta$$

$$X_c = \left(\frac{\frac{D}{2} \times \text{base kva}}{(\text{base kv})^2 \times 10^3} \right) \sin \theta$$

where D is the diameter of the circle in primary ohms. The distance d between the center of the circle C and the impedance point Z_p is

$$d = \sqrt{(\Delta R)^2 + (\Delta X)^2}$$

where

$$\Delta R = R_p - R_c \qquad \text{and} \qquad \Delta X = X_p - X_c$$

as shown in Fig. 10.20. The value of d is compared to the radius r in per unit of the circle.

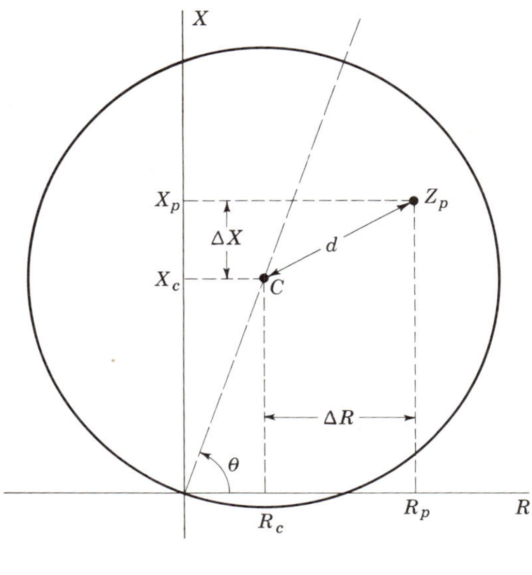

Fig. 10.20 Comparison of apparent impedance and distance relay operating characteristic.

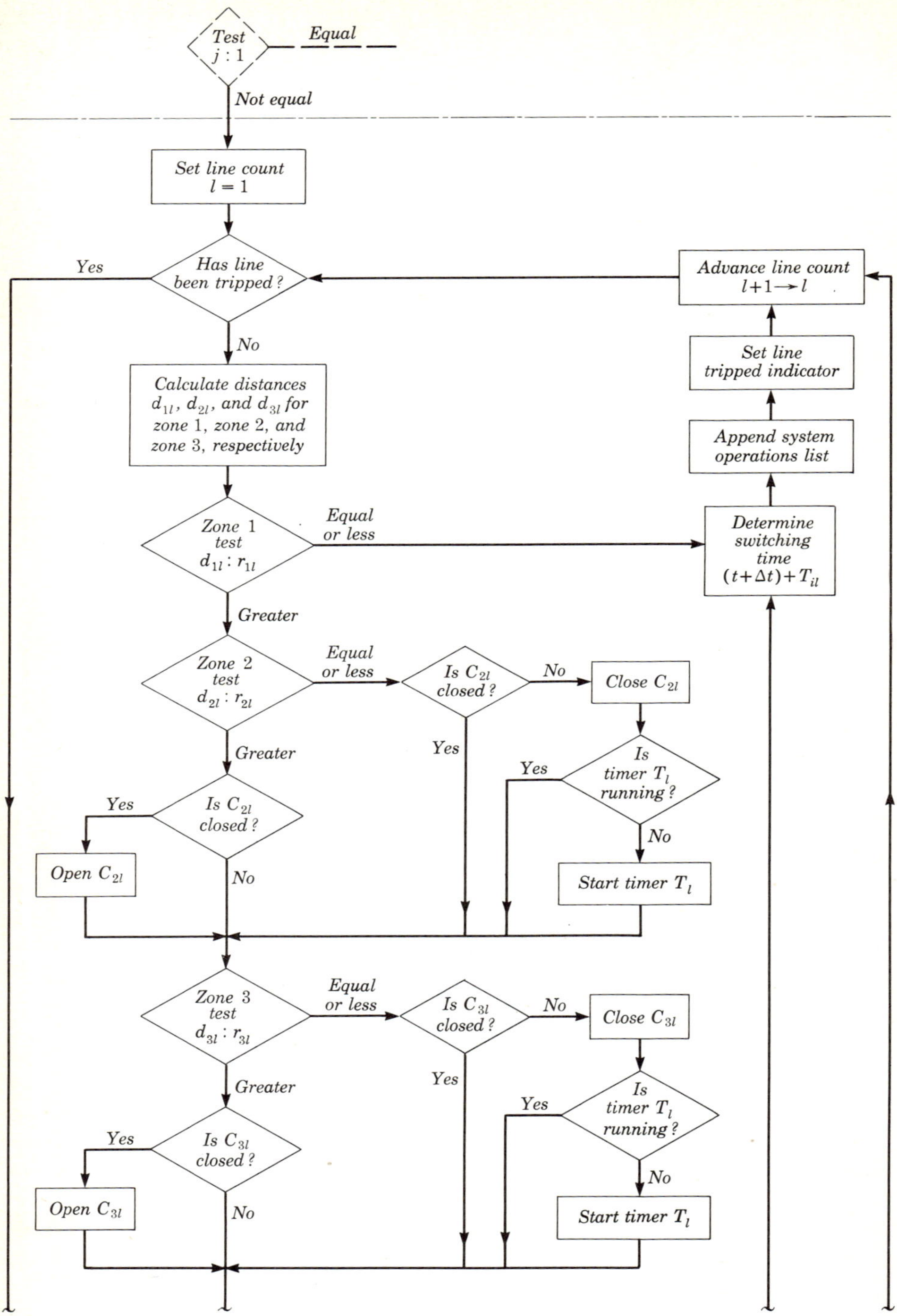

Fig. 10.21 Simulation of the operation of mho-type distance relays in a transient analysis.

The sequence of steps for simulating the operation of mho-type distance relays in a transient stability study is shown in Fig. 10.21. For a specified line l the apparent impedance calculated at $t + \Delta t$ is compared to the operating characteristic of each of the three zones. This is accomplished by calculating the distances d_{1l}, d_{2l}, and d_{3l} from the apparent impedance point to the centers of the circles for zones 1, 2, and 3, respectively. Each distance is compared to the radius of the appropriate circle, that is, d_{1l} is compared to r_{1l}; d_{2l} is compared to r_{2l}; and d_{3l} is compared to r_{3l}. If the apparent impedance is in zone 1 an immediate switching operation is initiated. If the apparent impedance is in zones 2 and 3, or zone 3, the corresponding contacts C_{2l} and C_{3l}, or C_{3l}, are closed and the timer T_l is started. When time is incremented by Δt in the transient calculations the relay timer T_l also must be advanced by Δt. When the timer reaches the time setting T_{2l} or T_{3l} for zone 2 or 3, respectively, and the corresponding contacts C_{2l} or C_{3l} are closed, a switching operation is initiated.

When an operation is initiated the switching time is determined by adding to $t + \Delta t$ the inherent relay and circuit breaker time T_{il}, that is, the time required for the relay and circuit breaker to disconnect the line.

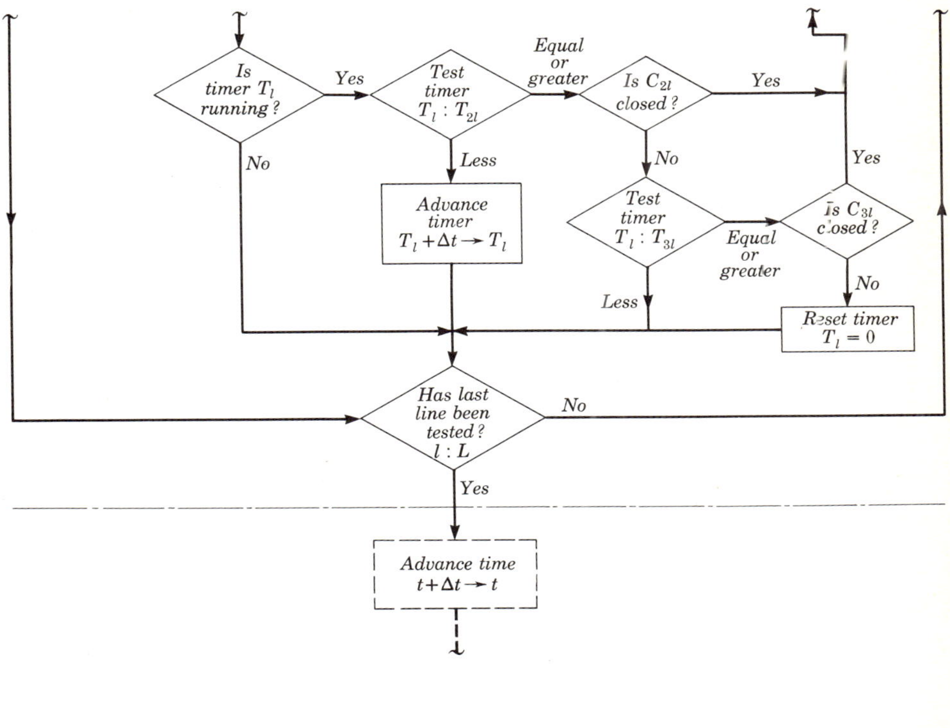

High-speed relays and circuit breakers operate in approximately 0.04 sec. The switching operation is effected in the step-by-step transient calculations at the scheduled time.

10.9 Description of transient stability program

In general, a transient stability program is developed as an extension of a load flow program. This provides the ability to obtain a load flow solution prior to the disturbance and thus the initial system values for the transient calculations. In addition, the load flow data can be used in the transient study.

The American Electric Power Transient Stability Program is composed of the following parts:

Input

The input program used for the entry of load flow data is used also to read in the additional data required for the transient stability calculations. This information includes synchronous and induction machine data, time interval, total time period, transmission lines and buses to be monitored, and sequence and time of system fault and switching operations. If the operations of impedance relays are to be simulated, the characteristics of the relays are required also.

Data assembly and modification

The data assembly and modification program links the load flow and transient stability programs. It prepares and checks all data associated with the various machine and load representations and modifies the system data by adding machine equivalent circuits, calculating internal machine voltages and currents, and converting loads to the desired representation. In addition, an initial slip for each induction motor is calculated.

Transient calculations

The transient calculations include the numerical solution of the differential equations describing machine behavior and the iterative solution of the network equations to determine the performance of the transmission system. A network solution is obtained for each estimate of machine speeds and internal voltages at the next time interval. The results of the network solution obtained with the final estimates at the next time interval are used to calculate line currents and swing impedances for preselected lines.

System operations

The system operations program modifies the system data at specified times during the transient analysis to simulate fault conditions and switching operations associated with a system disturbance. The types of operations that can be specified are:

1. Simulating a fault
2. Clearing a fault
3. Tripping a line or transformer
4. Reclosing a line or transformer
5. Switching on static loads
6. Dropping static loads
7. Tripping machines
8. Restarting motors

Any sequence and combination of these operations can be specified for a transient analysis.

Relay operations

The relay operations program compares the apparent impedances calculated for preselected lines to their distance relay characteristics. If a switching operation is effected the program automatically schedules this operation at the proper time during the transient calculation. The switching operation is performed by the system operations program.

Output

The output program uses the bus numbers assigned in the load flow calculation to identify the transient results. The first information provided includes the study title, case number, remarks, and sequence of system operations. Next, detailed results are printed for each time interval. For each machine, the type, terminal and internal voltages, current, speed, and mechanical torque are printed. For the network, the power flows and associated bus voltages as well as the swing impedances for preselected lines are printed. A sample of the output from this program is shown in Fig. 10.22.

In addition to these results, a plot of machine speeds and phase angles can be obtained for preselected machines. Letters of the alphabet are used to designate the machines and to form the curves. A sample plot of machine phase angles is shown in Fig. 10.23.

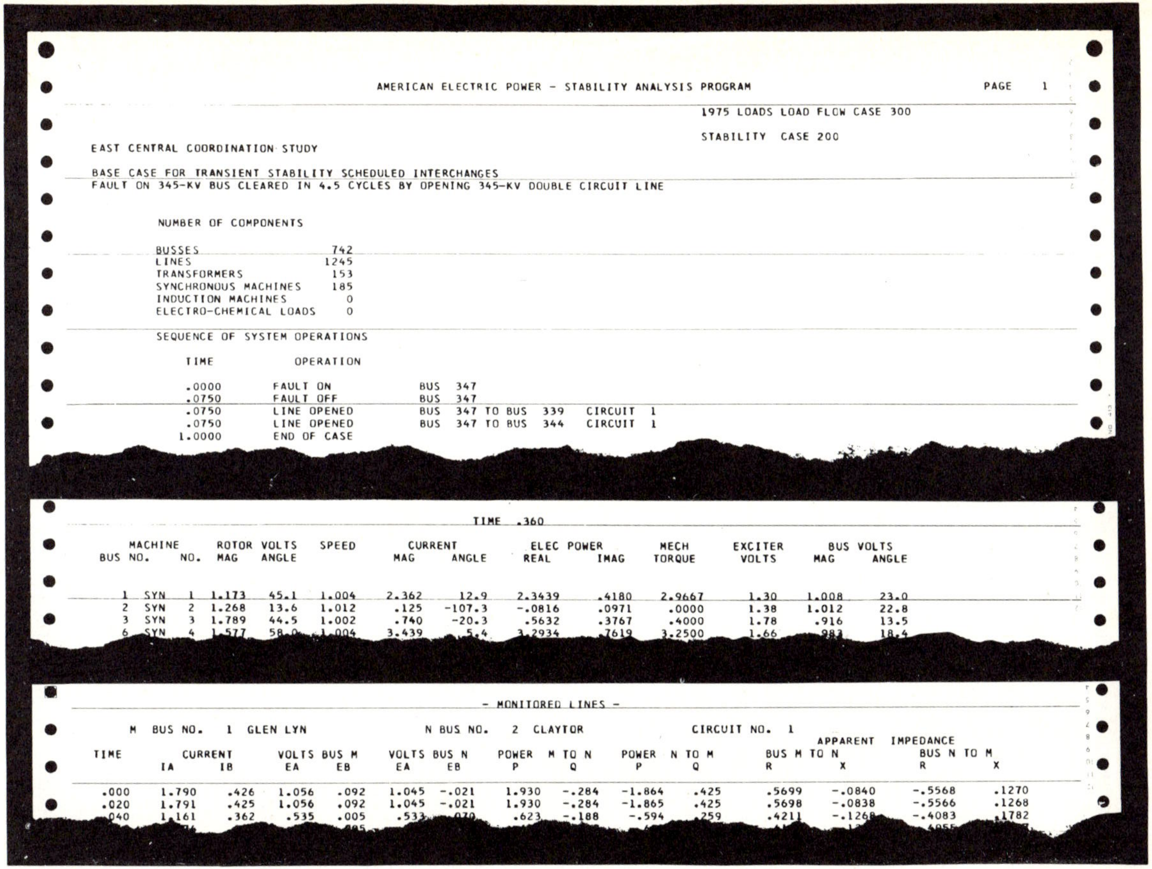

AMERICAN ELECTRIC POWER - STABILITY ANALYSIS PROGRAM PAGE 1

1975 LOADS LOAD FLOW CASE 300

STABILITY CASE 200

EAST CENTRAL COORDINATION STUDY

BASE CASE FOR TRANSIENT STABILITY SCHEDULED INTERCHANGES
FAULT ON 345-KV BUS CLEARED IN 4.5 CYCLES BY OPENING 345-KV DOUBLE CIRCUIT LINE

NUMBER OF COMPONENTS

BUSSES	742
LINES	1245
TRANSFORMERS	153
SYNCHRONOUS MACHINES	185
INDUCTION MACHINES	0
ELECTRO-CHEMICAL LOADS	0

SEQUENCE OF SYSTEM OPERATIONS

TIME	OPERATION	
.0000	FAULT ON	BUS 347
.0750	FAULT OFF	BUS 347
.0750	LINE OPENED	BUS 347 TO BUS 339 CIRCUIT 1
.0750	LINE OPENED	BUS 347 TO BUS 344 CIRCUIT 1
1.0000	END OF CASE	

TIME .360

MACHINE BUS NO.		NO.	ROTOR VOLTS MAG	ANGLE	SPEED	CURRENT MAG	ANGLE	ELEC POWER REAL	IMAG	MECH TORQUE	EXCITER VOLTS	BUS VOLTS MAG	ANGLE
1	SYN	1	1.173	45.1	1.004	2.362	12.9	2.3439	.4180	2.9667	1.30	1.008	23.0
2	SYN	2	1.268	13.6	1.012	.125	-107.3	-.0816	.0971	.0000	1.38	1.012	22.8
3	SYN	3	1.789	44.5	1.002	.740	-20.3	.5632	.3767	.4000	1.78	.916	13.5
6	SYN	4	1.577	58.0	1.004	3.439	5.4	3.2934	.7619	3.2500	1.66	.983	18.4

\- MONITORED LINES -

M BUS NO. 1 GLEN LYN N BUS NO. 2 CLAYTOR CIRCUIT NO. 1

TIME	CURRENT IA	IB	VOLTS BUS M EA	EB	VOLTS BUS N EA	EB	POWER M TO N P	Q	POWER N TO M P	Q	APPARENT IMPEDANCE BUS M TO N R	X	BUS N TO M R	X
.000	1.790	.426	1.056	.092	1.045	-.021	1.930	-.284	-1.864	.425	.5699	-.0840	-.5568	.1270
.020	1.791	.425	1.056	.092	1.045	-.021	1.930	-.284	-1.865	.425	.5698	-.0838	-.5566	.1268
.040	1.161	.362	.535	.005	.533	[illegible]	.623	-.188	-.594	.259	.4211	-.1268	-.4083	.1782

Fig. 10.22 ***Sample output of the American Electric Power Transient Stability Program showing the sequence of operations and machine and power system conditions.***

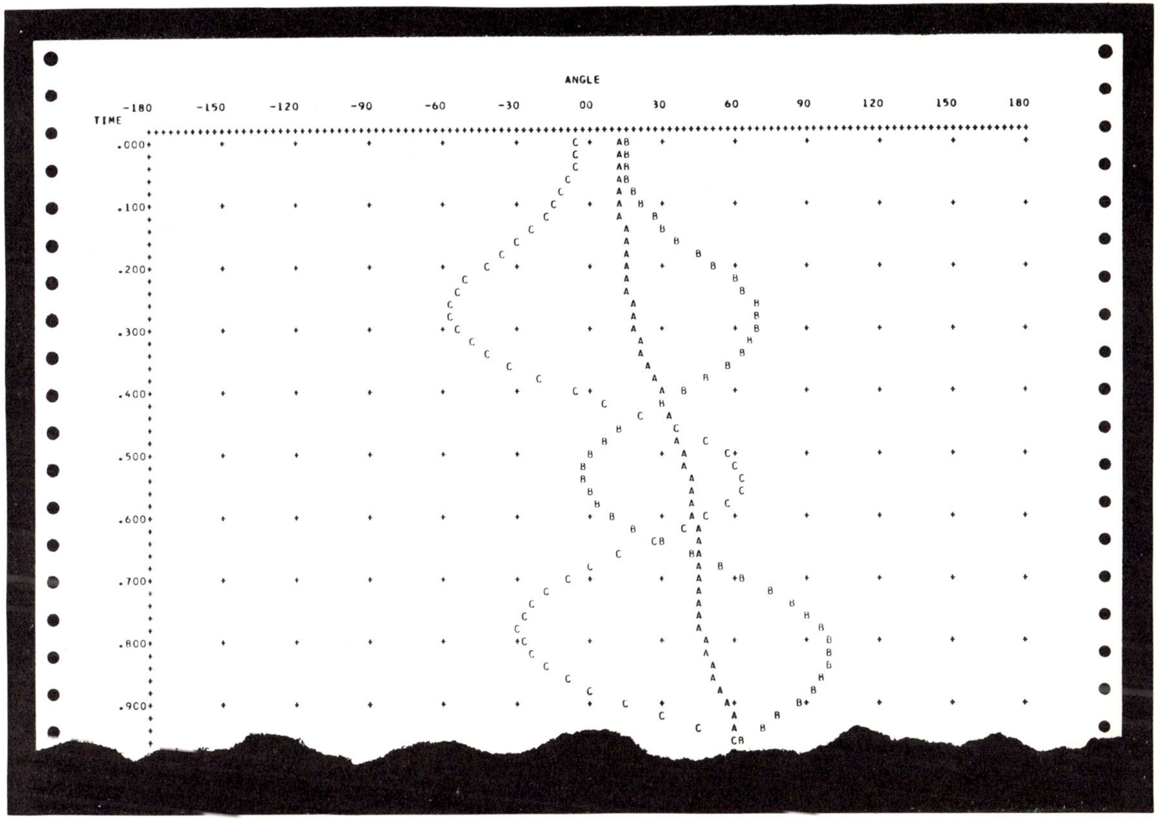

Fig. 10.23 Sample output of the American Electric Power Transient Stability Program showing a plot of machine phase angles during the disturbance.

Problems

10.1 The synchronous machine shown in Fig. 10.24 is generating 100 megawatts and 50 megavars. The voltage of the infinite bus q is $1.0 + j0$ and the line reactance is 0.05 per unit on a 100,000 kva base. The machine transient reactance is 0.20 and the inertia constant is 3.5 per unit on a 100,000 kva base. Calculate the changes in phase angle and speed of the generator for a three-phase fault at bus p for a duration of 0.1 sec. Use the modified Euler method with time increments of 0.02 sec and maximum time of 0.2 sec.

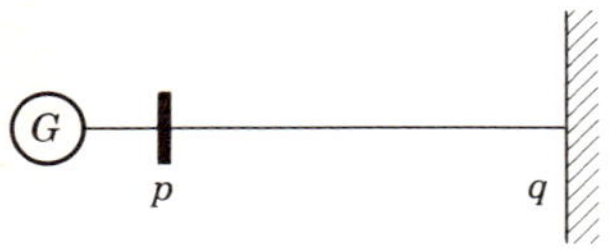

Fig. 10.24 Sample system for Prob. 10.1.

10.2 The load flow results for the sample system shown in Fig. 10.25 are given in Table 10.6. The machine data in per unit on a 100,000 kva base is given in Table 10.7. The effects of saliency and changes in field flux linkages of the machines are to be taken into account in the transient calculations. The loads are to be represented as fixed impedances to ground. Calculate the values of all parameters and variables needed to convert the system representation to that required for transient analysis.

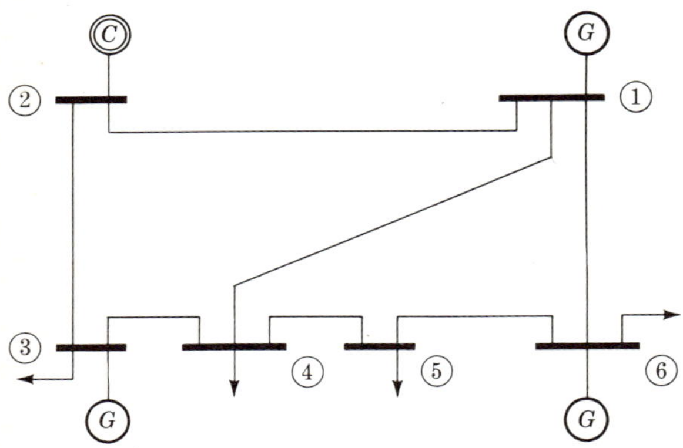

Fig. 10.25 Sample system for Prob. 10.2.

Table 10.6 *Bus voltages, generation, and loads from load flow calculation for sample system for Prob. 10.2*

Bus code	Bus voltages		Generation		Load	
p	Magnitude	Angle	Megawatts	Megavars	Megawatts	Megavars
1	1.040	0	296.7	49.4	0	0
2	1.030	−1.5	0	11.1	0	0
3	0.985	−5.9	40.0	41.6	100.0	30.0
4	0.961	−6.5	0	0	160.0	90.0
5	0.961	−6.7	0	0	90.0	60.0
6	1.015	−4.5	325.0	71.0	300.0	30.0

Table 10.7 *Synchronous machine data for sample system for Prob. 10.2*

Bus code p	Inertia constant H	Direct-axis transient open circuit time constant T'_{d0}	Direct-axis transient reactance x'_d	Direct-axis synchronous reactance x_d	Quadrature-axis synchronous reactance x_q
1	35.00	3.5	0.06	0.22	0.19
2	0.25	7.0	0.80	3.20	2.50
3	4.50	4.0	0.50	1.60	1.50
6	22.00	3.5	0.10	0.34	0.30

Table 10.8 *Voltages from load flow calculation at the instant following the fault for sample system for Prob. 10.3*

Bus code p	Terminal voltage E_t	Voltage back of quadrature-axis synchronous reactance E_q	Field voltage E_{fd}
1	$0.618 + j0.094$	$1.795 + j0.860$	1.30
2	$0.482 + j0.063$	$2.469 - j0.065$	1.38
3	$0 + j0$	$3.460 + j0.901$	1.79
4	$0.320 + j0.005$		
5	$0.445 + j0.025$		
6	$0.662 + j0.063$	$1.651 + j1.097$	1.66

10.3 The results of the load flow calculation for the sample power system used in Prob. 10.2 at the instant following a three-phase fault on bus 3 are given in Table 10.8. Using a time increment of 0.02 sec, calculate the first estimates of machine speeds, phase angles, and voltages proportional to the field flux linkages at $t = 0.02$.

10.4 The load flow data in per unit on a 100,000 kva base for the sample power system shown in Fig. 10.26 is given in Tables 10.9 and 10.10. The slack is bus 1. The machine data in per unit on a 100,000 kva base is given in Table 10.11. The loads are to be represented as fixed impedances to ground. For a time increment of 0.02 sec and a maximum time of 1.0 sec calculate the changes in phase angles and speeds of the generators for a three-phase fault at the East bus for a duration of 0.1 sec. Use the following methods and compare the results:

a. Euler's method
b. The modified Euler method
c. Runge-Kutta method

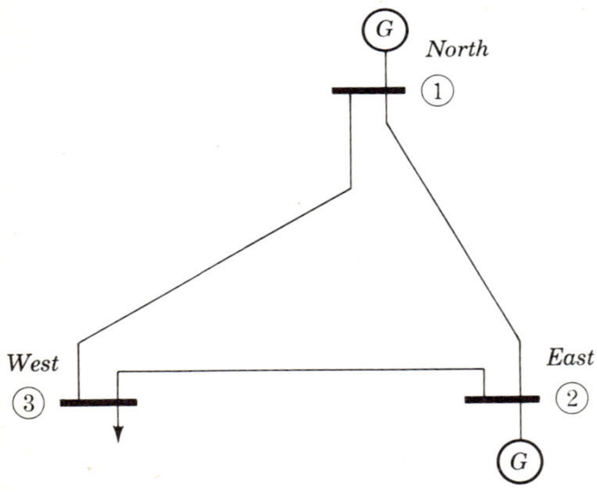

Fig. 10.26 Sample system for Prob. 10.4.

10.5 The terminal voltage E_t of a synchronous generator connected to a power system is 1.0 per unit and the terminal output power is zero. The machine reactances are

$$x_d = 1.2$$
$$x_q = 0.8$$

Taking into account the effects of saliency and assuming the magnitude of the voltage proportional to the field current E_I remains fixed, calculate the per unit terminal output power when the quadrature axis is advanced 30° with respect to the terminal voltage.

Table 10.9 Impedances for sample system for Prob. 10.4

Bus code p–q	Impedance z_{pq}	Line charging $y'_{pq}/2$
1–2	$0.04 + j0.16$	0.15
1–3	$0.02 + j0.08$	0.07
2–3	$0.03 + j0.10$	0.04

Table 10.10 Scheduled generation and loads and assumed bus voltages for sample system for Prob. 10.4

Bus code	Assumed	Generation		Load	
p	bus voltage	Megawatts	Megavars	Megawatts	Megavars
1	$1.04 + j0$	0	0	0	0
2	$1.00 + j0$	100	70	50	20
3	$1.00 + j0$	0	0	250	150

Table 10.11 Synchronous machine data for sample system for Prob. 10.4

Bus code p	Inertia constant H	Direct-axis transient reactance x'_d
1	160.0	0.1
2	3.0	0.3

10.6 A power system with two generating units is supplying a total load of 200 megawatts at rated frequency. The data for the generating units is given in Table 10.12. Assuming there is no load frequency control system to change the governor setting and that there is no change in load with frequency, calculate the following:

a. The new system frequency when the load increases 30 megawatts

b. The new megawatt loading of the generating units

10.7 A synchronous generator, rotating at synchronous speed, has an inertia constant of 3.0 and speed regulation of -0.05. A load of constant power equal to 25 percent of generator rating is connected

Table 10.12 Generator data for Prob. 10.6

Unit number	Megawatt rating	Megawatt loading	Per unit speed regulation R
1	100	50	−0.05
2	200	150	−0.04

suddenly to the machine terminals. Neglecting the time constants of the governor control system and the steam system, calculate the speed of the generator as a function of time.

10.8 The initial operating conditions of a synchronous generator are:

Megawatt output = 0.50 per unit
Terminal voltage = $1.0 + j0$
Power factor = 1.0

The generator data is

$x'_d = 0.2$
$x_d = 1.2$
$x_q = 1.1$

A load of 0.1 per unit and 0.8 power factor is connected suddenly to the generator terminals. Assuming that there is no change in speed and that the generator field voltage remains constant, calculate the generator terminal voltage as a function of time. Represent the load as a fixed impedance to ground.

Bibliography

Brereton, D. S., D. G. Lewis, and C. C. Young: Representation of Induction-motor Loads During Power-system Stability Studies, *Trans. AIEE*, vol. 76, pt. III, pp. 451–461, 1957.

Brown, H. E., H. H. Happ, C. E. Person, and C. C. Young: Transient Stability Solution by an Impedance Matrix Method, *Trans. IEEE on Power Apparatus and Systems*, vol. 84, pp. 1204–1214, 1965.

Clarke, Edith: "Circuit Analysis of A-c Power Systems," vol. 2, John Wiley & Sons, Inc., New York, 1950.

Crary, Selden B.: "Power System Stability," vol. 2, John Wiley & Sons, Inc., New York, 1947.

Dyrkacz, M. S., and D. G. Lewis: A New Digital Transient Stability Program, *Trans. AIEE*, vol. 78, pt. III-A, pp. 913–919, 1959.

Dyrkacz, M. S., C. C. Young, and F. J. Maginniss: A Digital Transient Stability Program Including the Effects of Regulator, Exciter, and Governor Response, *Trans. AIEE*, vol. 79, pt. III, pp. 1245–1257, 1960.

Gabbard, J. L., Jr., and J. E. Rowe: Digital Computation of Induction-motor Transient Stability, *Trans. AIEE*, vol. 76, pt. III, pp. 970–977, 1957.

Johnson, D. L., and J. B. Ward: The Solution of Power System Stability Problems by Means of Digital Computers, *Trans. AIEE*, vol. 75, pt. III, pp. 1321–1329, 1956.

Kimbark, Edward Wilson: "Power System Stability," vol. 1, John Wiley & Sons, Inc., New York, 1948.

Lane, C. M., R. W. Long, and J. N. Powers: Transient Stability Studies II—Automatic Digital Computation, *Trans. AIEE*, vol. 77, pt. III, pp. 1291–1295, 1958.

Lokay, H. E., and R. L. Bolger: Effect of Turbine-Generator Representation in System Stability Studies, *Trans. IEEE on Power Apparatus and Systems*, vol. 84, pp. 933–942, 1965.

Lokay, H. E., and J. W. Skooglund: Power System Stability—Digital Analysis Showing Effect of Generator Representation, Types of Voltage Regulators, and Speed Governor Systems, *IEEE Power Ind. Computer Appl. Conf. Proc., Clearwater, Fla.*, pp. 230–241, 1965.

Mason, C. Russell: "The Art and Science of Protective Relaying," John Wiley & Sons, Inc., New York, 1956.

Olive, D. W.: New Techniques for the Calculation of Dynamic Stability, *Trans. IEEE on Power Apparatus and Systems*, vol. 85, pp. 767–777, 1966.

Rindt, L. J., R. W. Long, and R. T. Byerly: Transient Stability Studies Part III, Improved Computational Techniques, *Trans. AIEE*, vol. 78, pt. III-B, pp. 1673–1677, 1959.

Savant, C. J., Jr.: "Basic Feedback Control System Design," McGraw-Hill Book Company, New York, 1958.

Shankle, D. F., C. M. Murphy, R. W. Long, and E. L. Harder: Transient-stability Studies—I, Synchronous and Induction Machines, *Trans. AIEE*, vol. 73, pt. III-B, pp. 1563–1580, 1954.

Shipley, R. B., N. Sato, D. W. Coleman, and C. F. Watts: Direct Calculation of Power System Stability Using the Impedance Matrix, *Trans. IEEE on Power Apparatus and Systems*, vol. 85, pp. 777–782, 1966.

Stagg, G. W., A. F. Gabrielle, D. R. Moore, and J. F. Hohenstein: Calculation of Transient Stability Problems Using a High-speed Digital Computer, *Trans. AIEE*, vol. 78, pt. III-A, pp. 566–574, 1959.

Stanley, H. C.: An Analysis of the Induction Machine, *Trans. AIEE*, vol. 57, pp. 751–757, 1938.

Index

Index